Research Approaches to Sustainable Biomass Systems

Research Approaches to Sustainable Biomass Systems

Seishu Tojo and Tadashi Hirasawa
Tokyo University of Agriculture and Technology, Tokyo, Japan

AMSTERDAM • BOSTON • HEIDELBERG • LONDON
NEW YORK • OXFORD • PARIS • SAN DIEGO
SAN FRANCISCO • SINGAPORE • SYDNEY • TOKYO

Academic Press is an Imprint of Elsevier

ELSEVIER

Academic Press is an imprint of Elsevier
The Boulevard, Langford Lane, Kidlington, Oxford, OX5 1GB, UK
225 Wyman Street, Waltham, MA 02451, USA

First published 2014

Notice
No responsibility is assumed by the publisher for any injury and/or damage to persons or
property as a matter of products liability, negligence or otherwise, or from any use or
operation of any methods, products, instructions or ideas contained in the material
herein. Because of rapid advances in the medical sciences, in particular, independent
verification of diagnoses and drug dosages should be made

British Library Cataloguing in Publication Data
A catalogue record for this book is available from the British Library

Library of Congress Cataloguing in Publication Data
A catalog record for this book is available from the Library of Congress

ISBN: 978-0-12-404609-2

For information on all Academic Press publications
visit our website at www.store.elsevier.com

Printed and bound in the United States

13 14 15 16 17   10 9 8 7 6 5 4 3 2 1

Working together
to grow libraries in
developing countries

www.elsevier.com • www.bookaid.org

Contents

4.   **Production Technology for Bioenergy Crops and Trees**
*Tadashi Hirasawa, Taiichiro Ookawa, Shinya Kawai,
Ryo Funada and Shinya Kajita*

5. Soil Fertility and Soil Microorganisms
Haruo Tanaka, Akane Katsuta, Koki Toyota and Kozue Sawada

6. Machinery and Information Technology for Biomass Production
Tadashi Chosa, Takeshi Matsumoto and Masahiro Iwaoka

7. Pretreatment and Saccharification of Lignocellulosic Biomass
Eika W. Qian

8. Energy-Saving Biomass Processing with Polar Ionic Liquids
Yukinobu Fukaya and Hiroyuki Ohno

9. Enzymes for Cellulosic Biomass Conversion
Takashi Tonozuka, Makoto Yoshida and Michio Takeuchi

10. Ethanol Production from Biomass
Haruki Ishizaki and Keiji Hasumi

11. Co-Generation by Ethanol Fuel
Hideo Kameyama

14. Evaluation of Biomass Production and Utilization Systems

Chihiro Kayo, Seishu Tojo, Masahiro Iwaoka and Takeshi Matsumoto

15. Local Activity of Biomass Use in Japan

Hiroshi Yoshida, Toshio Nomiyama, Nobuhide Aihara, Ryoichi Yamazaki, Sachiho Arai and Hiroyuki Enomoto

Our present life depends largely on burning fossil fuels for energy, but reserves of fossil fuels may be exhausted in the near future if the current trend of fossil fuel consumption continues. Additionally, burning fossil fuels is actually returning the CO_2 that had been fixed by ancient organisms eons ago to the atmosphere; the Intergovernmental Panel on Climate Change (IPCC) has warned that human consumption of fossil fuels is increasing the atmospheric CO_2 concentration so that the resulting greenhouse effect may seriously impact global climate and human life in the near future. Hence, new energy resources are being explored as substitutes for fossil fuels.

Use of atomic energy had been expected to be a safe and reliable energy source because it emits much less CO_2. However, since the nuclear disaster in the aftermath of the 2011 earthquake and tsunami in Japan, the safety and reliability of atomic energy has been questioned, and the expansion of atomic energy is being reconsidered in many countries.

Under these circumstances, many countries are planning to increase the use of renewable energy in the near future. Biomass energy has some advantages as compared with other renewable energy sources such as solar and wind power. Many forms of energy, e.g. heat, liquid and gas fuels, and electricity can be easily derived from various types of available biomass, including forestry by-products, plant residues from cropped fields, plant by-products created by food-processing industries, and animal and food waste, as well as plants grown and harvested mainly for energy.

Biomass energy is carbon neutral. Although using grain for energy may compete with food, such competition can be avoided if liquid fuels are derived from lignocelluloses that are abundant in various biomass materials from forests and fields as either raw biomass or reused woody products. Growing and using biomass resources provide opportunities to maintain and strengthen the functions of agriculture and forestry while achieving conservation of land and environments.

Biomass is unique in that it can be used not only for renewable energy generation but also for other non-energy purposes. Creation of technological and social systems is important for increasing both the potential for plants to fix solar energy and the use of biomass. Developing such systems for biomass use is one of the great challenges of this century and may facilitate the establishment of material-recycling and sustainable human societies.

Our main concern in this book is to address the ways in which industries generating energy from sustainable biomass may be established and become commercially viable. Some hold the view that only large-scale enterprises are likely to be successful. The high volume of biomass required by such industries is indeed available in some countries, for example corn in the USA and sugar cane in Brazil. However, this is not always the case in other countries. Many Asian communities (including those in Japan) are local to some degree because they are separated from one another by major landform features such as mountain ranges. Transporting biomass from such regions to a centralized processing hub would be economically prohibitive. Thus, there is a need to establish techniques suitable for small-scale application for using the energy locked into biomass, and improving the efficiencies of biomass production, harvesting, energy conversion, and energy use are equally important. Further, the improvement must be sensitive to geographical restraints and the social traditions of the communities involved in order to warrant the maximum overall benefit from biomass production and use. Any successful approach must blend the talents of those working in various disciplines of natural and social sciences; this would allow sustainable biomass production to be integrated with biomass use in community settings.

This book is intended principally for graduate students, young scientists, and government officials who are interested in exploring the potential of future biomass application for renewable fuels. We emphasize the need for sustainable biomass production and use in individual communities and thus put much value on the accomplishments of both natural and social scientists in this area. Biomass production and use are outlined from the standpoint of the social systems and ecosystems in Chapters 1–3. Chapters 4–13 deal with efficient and sustainable biomass production, energy production from biomass, and other uses of biomass. The methods that may be used to evaluate the success of biomass production and use, and the acceptance of such systems by the societies they serve, are outlined in Chapters 14 and 15 respectively.

We seek to not only describe the latest developments in basic science and technology, but also include relevant case studies with exciting recent research findings highlighted in "boxes". We hope that these "real-world" examples will make our contentions easier to understand, and that readers will learn the basic science and technology of biomass creation and use so that they will become committed to the principle of sustainability.

Seishu Tojo
Tadashi Hirasawa

Acknowledgments

Our ongoing research program, entitled "Green Biomass Research for Improvement of Local Energy Self-sufficiency" commenced in mid-2000. At that time, several faculty members of Tokyo University of Agriculture and Technology decided to integrate their separate studies into a comprehensive research framework for the construction of model systems with energy self-sufficiency and effective material recycling at the farm or local community scale. We thank the Ministry of Education, Culture, Sports, Science, and Technology of Japan for funding the program for 5 years, commencing in 2008. We also acknowledge Tokyo University of Agriculture and Technology for supporting our research. This book is but one outcome of our program.

We thank Kattie Washington and Tiffany Gasbarrini at Elsevier for their professional advice and support. We also thank Hiroshi Nishiyama for assistance given when the book was in the early stages of preparation. We thank all of the contributors for sharing their thoughts and expertise with us and want to extend a special thanks to Dr. Atsushi Chitose for his editorial efforts. Finally, we would like to acknowledge Dr. Allen C. Chao, who reviewed all manuscripts with his extensive knowledge and language skill.

Nobuhide Aihara, Waseda University Organization for Japan-US Studies

Sachiho Arai, Graduate School of Agriculture, Tokyo University of Agriculture and Technology

Takuya Ban, Graduate School of Agriculture, Tokyo University of Agriculture and Technology

Atsushi Chitose, Graduate School of Agriculture, Tokyo University of Agriculture and Technology

Tadashi Chosa, Graduate School of Agriculture, Tokyo University of Agriculture and Technology

Ryo Funada, Graduate School of Agriculture, Tokyo University of Agriculture and Technology

Keiji Hasumi, Graduate School of Agriculture, Tokyo University of Agriculture and Technology

Tadashi Hirasawa, Graduate School of Agriculture, Tokyo University of Agriculture and Technology

Haruki Ishizaki, Graduate School of Agriculture, Tokyo University of Agriculture and Technology

Masahiro Iwaoka, Graduate School of Agriculture, Tokyo University of Agriculture and Technology

Shinya Kajita, Graduate School of Bio-Applications and Systems Engineering, Tokyo University of Agriculture and Technology

Hideo Kameyama, Graduate School of Engineering, Tokyo University of Agriculture and Technology

Shuhei Kanda, Graduate School of Agriculture, Tokyo University of Agriculture and Technology

Hitoshi Kato, National Agriculture and Food Research Organization

Shinya Kawai, Graduate School of Agriculture, Tokyo University of Agriculture and Technology

Yuichi Kobayashi, National Agriculture and Food Research Organization

Takeshi Matsumoto, Graduate School of Agriculture, Tokyo University of Agriculture and Technology

Shoji Matsumura, Graduate School of Agriculture, Tokyo University of Agriculture and Technology

Takashi Motobayashi, Graduate School of Agriculture, Tokyo University of Agriculture and Technology

Toshio Nomiyama, Graduate School of Agriculture, Tokyo University of Agriculture and Technology

Hiroyuki Ohno, Graduate School of Engineering, Tokyo University of Agriculture and Technology

Taiichiro Ookawa, Graduate School of Agriculture, Tokyo University of Agriculture and Technology

Eika W. Qian, Graduate School of Bio-Applications and Systems Engineering, Tokyo University of Agriculture and Technology

Jun Shimada, Graduate School of Agriculture, Tokyo University of Agriculture and Technology

Michio Takeuchi, Graduate School of Agriculture, Tokyo University of Agriculture and Technology

Haruo Tanaka, Graduate School of Agriculture, Tokyo University of Agriculture and Technology

Hiroto Toda, Graduate School of Agriculture, Tokyo University of Agriculture and Technology

Seishu Tojo, Graduate School of Agriculture, Tokyo University of Agriculture and Technology

Takashi Tonozuka, Graduate School of Agriculture, Tokyo University of Agriculture and Technology

Koki Toyota, Graduate School of Bio-Applications and Systems Engineering, Tokyo University of Agriculture and Technology

Kenichi Yakushido, National Agriculture and Food Research Organization

Masaaki Yamada, Graduate School of Agriculture, Tokyo University of Agriculture and Technology

Ryoichi Yamazaki, Graduate School of Agriculture, Tokyo University of Agriculture and Technology

Tadashi Yokoyama, Graduate School of Agriculture, Tokyo University of Agriculture and Technology

Hiroshi Yoshida, Graduate School of Agriculture, Tokyo University of Agriculture and Technology

Makoto Yoshida, Graduate School of Agriculture, Tokyo University of Agriculture and Technology

Chapter 1

Background of Bioenergy Utilization Development in Japan

Tadashi Yokoyama and Hiroshi Yoshida

1.1. ENVIRONMENTAL CHANGE WITH HUMAN ACTIVITIES

Tadashi Yokoyama

In recent times, the global temperature and carbon dioxide level in the atmosphere have increased rapidly. Although over at least the last 650,000 years, atmospheric temperatures and carbon dioxide levels have periodically increased and decreased in a cyclical pattern (Jouzel et al., 2007), many people feel that today's climate change is different from previous climate changes based on several observations such as the remarks issued by the US Environment Protection Agency (EPA) on current climate change. First, none of the natural causes of climate change, including variations in the Sun's energy and the Earth's orbit, can fully explain the observed climate changes. Second, burning huge quantities of fossil fuels, including coal, oil, and natural gas, to satisfy domestic and industrial energy needs is overloading the atmosphere with carbon dioxide that contributes to the greenhouse effect. Human activities are also adding other heat-trapping greenhouse gases (GHGs), such as methane and nitrous oxide, to the atmosphere. For hundreds of thousands of years, the concentration of carbon dioxide in the atmosphere stayed between 200 and 300 parts per million; however, today it has increased to nearly 400 parts per million, and is still rising. Along with other GHGs, this extra atmospheric

Research Approaches to Sustainable Biomass Systems. http://dx.doi.org/10.1016/B978-0-12-404609-2.00001-5

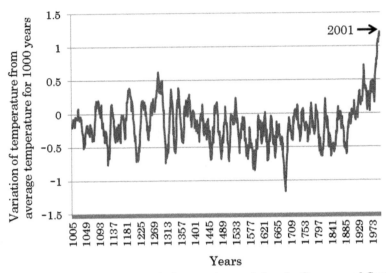

FIGURE 1.1 **A thousand-year record of temperature variations for Germany and Central Europe.** *(Source: Glaser and Riemann, 2009.)*

carbon dioxide is trapping solar heat to cause climate change (http://www.epa.gov/climatechange/science/indicators/weather-climate/).

Figure 1.1 illustrates a thousand-year record of temperature variations for Germany and Central Europe based on documentary data obtained from Glaser and Reimann (2009); it suggests that after around 1900, temperature has certainly increased from the baseline that represents the average temperature for the past 1000 years in Central Europe. Figure 1.2 shows the average carbon dioxide concentration in December in Manua Los, Hawaii (Keeling et al., 2009). In 1958, the atmospheric carbon dioxide averaged 314.67 ppmv; it increased 22.4% to 385.02 ppmv in 2008 over a period of 50 years.

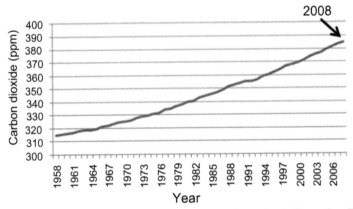

FIGURE 1.2 **Average carbon dioxide concentration every December in Manua Los, Hawaii.**

The conference held in Villach, Austria, 1985 was regarded as an important milestone in the international climate change debate. A major scientific assessment, "The role of carbon dioxide and of other greenhouse gases (GHGs) in climate variations and associated impacts", presented at the conference was based on the investigation made by a small group of environmental scientists and research managers associated with nongovernmental organizations from 1983 to 1985. A general conclusion of the conference is the anticipation of an unprecedented rise of global mean temperature during the first half of the twenty-first century.

With the scientific evidence accumulated, climate change appeared to be one of the important challenges to future economic prosperity in the international community during the late 1980s. In 1988, the Intergovernmental Panel on Climate Change (IPCC) was established jointly by the World Meteorological Organization and United Nations Environment Program. The IPCC emphasized that the emissions resulting from human activities may be changing the Earth's climate in its first assessment report published in 1990 (Houghton et al., 1990). In the wake of increased concerns about environmental issues, including global warning, the United Nations Conference on Environment and Development (UNCED) was held in Rio de Janeiro, Brazil in June 1992. This conference, also known as the Earth Summit, led to the United Nations Framework Convention on Climate Change (UNFCCC), together with the Rio Declaration on Environment and Development and other documents also known as "Agenda 21". The ultimate objective of the UNFCCC is to stabilize the atmospheric GHG concentrations at a level that would prevent dangerous anthropogenic interference with Earth's climate system. Subsequently, the Kyoto Protocol, which set a target to reduce GHG emissions by 5.2% below the 1990 levels by 2012, was agreed upon as a protocol of the UNGCCC in December 1997. However, it was not until February 2005 that the Protocol eventually entered into force after long negotiations among the parties involved (United Nations, 1998). Industrialized countries are required to ratify and implement the Protocol in order to reduce GHG emissions below the quantified targets assigned to each country during the first commitment period (2008–2012).

1.2. JAPANESE BIOMASS UTILIZATION POLICY

Hiroshi Yoshida

According to the Kyoto Protocol, Japan needs to achieve a 6% reduction of GHG emissions to prompt the government of Japan to reconsider its energy policy (Government of Japan, 2006). Based on various responses and inputs from the general public during the 1970s, the government decided to promote the utilization of biomass fuel because it is a carbon-neutral energy source. The first oil crisis in 1973 also aroused public interest in the use of biomass energy; however, the widespread use of biomass has not been realized

TABLE 1.1 Fundamental Principles and Governmental Policies of the Basic Law on the Promotion of the Biomass Use, Japan

Fundamental principles on biomass use	Governmental policies
1. Promoting policies comprehensively, collectively, and effectively (article 3)	1. Establishing national and local plans for expanding biomass use (articles 20, 21)
2. Promoting policies for preventing climate change (article 4)	2. Preparing proper infrastructure for biomass use (article 22)
3. Promoting policies for organizing a recycling-based society (article 5)	3. Supporting biomass suppliers (article 23)
4. Contributing to the development and improvement of international competitiveness (article 6)	4. Promoting and spreading technological developments (article 24)
5. Contributing to activating agriculture, forestry, fishing, and rural areas (article 7)	5. Securing sufficient human resources (article 25)
6. Using biomass as much as possible according to its properties (article 8)	6. Encouraging the use of biomass products (article 26)
7. Diversifying the energy sources (article 9)	7. Promoting voluntary action by private businesses and organizations (article 27)
8. Promoting voluntary action by participants and coordination between them (articles 10, 18)	8. Promoting action of local municipalities (article 28)
9. Developing social understanding for the biomass use (article 11)	9. Promoting international cooperation (article 29)
10. Preventing adverse effects on the stable supply of food (article 12)	10. Collecting and providing information about biomass, domestically and internationally (article 30)
11. Considering environmental protection (article 13)	11. Promoting public understanding about biomass (article 31)

because the government prioritized the development of nuclear power or natural gas in the national energy policy. The promotion of biomass utilization in Japan since the early 2000s has been motivated not only by GHG emissions reduction but also by other issues. First, international crude oil prices substantially increased during the early 2000s, which motivated the development of renewable energy sources as alternatives for foreign oil. Second, recycling organic wastes such as animal waste, sewage sludge, and other biomass disposals is urgently needed in accordance with the national move to construct a recycle-oriented society. Third, the effective utilization of abundant biomass resources available in rural areas is expected to revitalize rural economies that have been adversely affected during the long recession by generating new industries and jobs in rural regions. With increasing public concerns over these issues, the government of Japan established the Biomass Nippon Strategy (BNS) in 2002 (MAFF, 2009).

The BNS has four ultimate goals: (i) the prevention of global warning; (ii) the creation of a recycling-oriented society; (iii) the promotion of new competitive strategic industries for bioproducts, biofuel and biopower; and (iv) the revitalization of the agricultural, forestry, and fisheries industries as well as local communities. In line with these goals, initial specific targets were laid out, with 2002 as the base year and 2010 as the target year. These targets include increasing (i) biofuel use for the transportation sector from virtually zero to 500,000 KL (crude oil basis); (ii) biomass energy use capacity from 68,000 KL to 3.08 million KL; and (iii) the recovery rate of unused biomass materials from 20% to more than 25%.

In 2009, the government enacted the Basic Law on the Promotion of Biomass Utilization by addressing fundamental principles, defining responsibilities of the concerned institutions, and setting the fundamental conditions of the policy for comprehensive biomass utilization. Table 1.1 summarizes the fundamental principles on biomass utilization and governmental policies of this law. Because the Basic Law addresses only the conceptual framework of policies, detailed policies are defined in the "Basic Plan for the Promotion of Biomass Utilization" established in 2010. The Basic Plan has provided basic strategies and various action plans to improve the performance of biomass-related projects with three specific objectives: (i) the revitalization of 600 local communities by establishing local biomass plans; (ii) the strengthening of international competency by achieving 500 billion yens of output in biomass-related industries; and (iii) the prevention of climate change and creation of a recycling-oriented society by achieving 26 million tons of carbon equivalents of emission reductions.

REFERENCES

Glaser, R., & Riemann, D. (2009). A thousand year record of climate variation for Central Europe at a monthly resolution. *Journal of Quaternary Science, 24*, 437–449.

Government of Japan. (2006). *Baiomasu nippon sogo senryaku (Biomass nippon comprehensive strategy)*. <http://www.maff.go.jp/j/biomass/pdf/h18senryaku.pdf> Accessed 29.06.2012 [in Japanese].

Houghton, J. T., Jenkins, G. J., & Ephraums, J. J. (Eds.). (1990). *Climate change: The IPCC scientific assessment. Report prepared for Intergovernmental Panel on Climate Change by Working Group I*. Cambridge: Cambridge University Press. <http://www.ipcc.ch/ipccreports/far/wg_I/ipcc_far_wg_I_full_report.pdf> Accessed 11.06.2012.

Jouzel, J., Masson-Delmotte, V., Cattani, O., Dreyfus, G., Falourd, S., Hoffmann, G., Minster, B., Nouet, J., Barnola, J. M., Chappellaz, J., Fischer, H., Gallet, J. C., Johnsen, S., Leuenberger, M., Loulergue, L., Luethi, D., Oerter, H., Parrenin, F., Raisbeck, G., Raynaud, D., Schilt, A., Schwander, J., Selmo, E., Souchez, R., Spahni, R., Stauffer, B., Steffensen, J. P., Stenni, B., Stocker, T. F., Tison, J. L., Werner, M., & Wolff, E. W. (2007). Orbital and Millennial Antarctic Climate Variability over the Past 800,000 Years. *Science, 317*(5839), 793–796.

Keeling, R. F., Piper, S. C., Bollenbacher, A. F., & Walker, J. S. (2009). *Atmospheric Carbon Dioxide Record from Mauna Loa.* <http://cdiac.ornl.gov/trends/co2/sio-mlo.html>. doi: 10.3334/CDIAC/atg.035, 2009.

Ministry of Agriculture. (2009). *Forestry, and Fisheries (MAFF).* Japan: Japanese biomass policy. <http://www.biomass-asia-workshop.jp/biomassws/06workshop/presentation/01Saigou.pdf> [in Japanese].

United Nations. (1998). *Kyoto Protocol to the United Nations Framework on Climate Change.* <http://unfccc.int/resource/docs/convkp/kpeng.pdf> Accessed 20.06.2012.

United Nations Framework Convention on Climate Change (UNFCC). <http://unfccc.int/essential_background/items/6031.php> Accessed 20.06.2012.

Chapter 2

Biomass as Local Resource

Toshio Nomiyama, Nobuhide Aihara, Atsushi Chitose,
Masaaki Yamada and Seishu Tojo

2.1. DEFINITION AND CATEGORY OF BIOMASS

Toshio Nomiyama and Nobuhide Aihara

2.1.1. Definition of Biomass

Biomass is a collective term applied to the abiotic organic matter and the living plants that are partially integrated into the ecosystem. Plants grow by performing photosynthesis in the presence of sunlight to convert atmospheric carbon dioxide and water in the soil into carbohydrate. Other living organisms use the energy stored in plants to live by degrading carbohydrate into carbon dioxide and water. Plants are important sources of food for human beings and other herbivores in the ecosystem.

Research Approaches to Sustainable Biomass Systems. http://dx.doi.org/10.1016/B978-0-12-404609-2.00002-7

In Japan, the roles of biomass as an essential resource in human life including its energy use are emphasized in legal documents. The word "biomass" as a legal term was first mentioned in the "Enforcement Ordinance on the Law Concerning Special Measures to Promote the Use of New Energy", which was revised in 2002. In the revision, Article 1, Number 1 stipulates that the use of biomass is "to create fuel produced from organic matter derived from animals and plants which can be utilized as energy sources (not including crude oil, petroleum gas, inflammable natural gas, coal, and products manufactured from these resources)." In addition, Numbers 2 and 6, which address the organic matter as "biomass", stipulate that biomass be used "to gain heat" and "to use it to generate power" as new energy. This interpretation of biomass use is shared by the Basic Law on the Promotion of the Biomass Utilization, which was enacted in 2002. Defined in Article 2, Section 1, biomass is "organic resources derived from animals or plants" excluding fossil resources such as crude oil, petroleum gas, inflammable natural gas and coal. The Basic Law stipulates in Section 2 that biomass be used as raw materials for products (not including indirect uses of biomass as raw materials of products or the use of agricultural and fishery products for originally intended purposes) or as energy sources."

2.1.2. Characteristics and Categories of Biomass Resources with Emphasis on Energy Use

There are three categories of biomass resources based on their source: (1) waste biomass including organic waste such as livestock excrement and food residues discharged from households and food-related industries; (2) unused biomass composed of inedible parts of agricultural products such as logging residues, rice straw, and rice husk; and (3) resource crops (dedicated energy crops) that can be used to produce bioethanol and biodiesel fuel, such as rapeseed, corn, and rice. The biomass is produced in agriculture as well as forestry and fishery industries as either main crops or waste and unused materials.

Biomass resources are converted into the following six types of end-products: (1) fertilizers and feed; (2) raw materials of chemical products such as amino acids and useful chemical substances; (3) materials such as plastics and resins; (4) fuel such as ethanol, diesel, wood pellets, biogas, and solid fuel; (5) thermal energy; and (6) electric energy. These products are generally used in the following five categories: (1) recycled use; (2) use of products; (3) power generation; (4) use of heat; and (5) use of fuel.

Biomass used for energy is expected to contribute to establishing a lower-carbon recycling society to mitigate global warming. Specifically, biofuel is viewed as a promising renewable energy that contributes to carbon savings as compared with liquid fossil fuel of higher carbon intensity. This is because theoretically biofuel is carbon neutral based on the fact that all biomass carbon has recently been taken from atmospheric carbon in photosynthesis during the plant growth period.

Among renewable energy sources, biomass occupies a unique position due to the following characteristics. First, both historically and geographically, biomass is used almost all over the world; it is available almost anywhere on earth. Second, biomass energy is diverse in terms of both feedstocks and end-use products. A variety of biomass can be converted into different products in the form of solid, liquid, gas, and power. Third, among the many renewable energy sources, only biomass is transportable and storable so that its production can be adjusted to accommodate changes in output demand. Fourth, its use is not limited to energy; fuel is only one option among the many other potential uses. Over human history, biomass has been used for multiple purposes: food, animal feed, fertilizer, fiber, fuel, medicines, and materials for various crafts and buildings. Fifth, waste biomass that has been viewed as inherently detrimental to society can be used as feedstock for energy use.

Although biomass energy has various advantages, it also has disadvantages. First, biomass energy is in general of lower energy intensity than fossil fuel because of its lower bulk density. Second, biomass is less efficient in the conversion into liquid fuel than fossil fuel. The production of liquid biofuel entails greater energy loss in the conversion process. Third, biomass resources are usually distributed in small amounts across wide areas except for high-yielding resource crops. This causes higher feedstock procurement costs (including collection and transportation costs). Fourth, due to the first three characteristics, biofuel production is unlikely to be economically viable using the current level of technology. Fifth, because biomass possesses a feature of possible multiple uses, enhancing biomass energy production would likely lead to unintended repercussions regarding its other uses (e.g. food and animal feeds) that may adversely impact the food supply or biodiversity, causing social and environmental problems.

2.2. POTENTIAL AMOUNT OF BIOMASS AND USE OF BIOENERGY

2.2.1. Potential Availability of Biomass in the World

There are abundant amounts of biomass growing on land and in the ocean. The total amount of biomass on land is about 1.8 trillion tons based on dry weight and about 4 billion tons in the ocean. Also, the soil contains biomass of almost equal amount to terrestrial plants. Because the biomass includes food for human consumption and animal feed, estimating only the amount of convertible biomass for energy sources is meaningful and important.

The available amount of biomass for energy and other uses is subject to the net balance between biomass production and self-consumption by living organisms on a flow basis. The flow biomass is usually called primary biomass produced from the stock biomass. Net primary production (NPP) is a widely used indicator to show the net balance; it is defined as the difference between the rate at which the plants in an ecosystem produce biomass and the rate at which they consume some of the resulting biomass for respiration. Cao and Woodward (1998) estimated that global NPP was 57 Pg-C y^{-1} with 640 Pg-C of carbon stocks in vegetation and 1358 Pg-C of carbon stocks in soil. According to Janzen (2004), atmospheric CO_2 enters terrestrial biomass via photosynthesis at a rate of about 120 Pg-C y^{-1} (gross primary productivity). However, about half of it is immediately released as CO_2 by plant respiration, so that the NPP is about 60 Pg-C y^{-1}. Undoubtedly, sustainable biomass utilization is subject to a lower rate than the NPP. In addition to the excessive use of biomass, inadequate use of biomass should be avoided because inappropriate management in biomass use may lead to adverse effects on a local ecosystem such as the degradation of biodiversity and the diminished capacity to provide other environmental services.

Regarding the potential amount of biomass for bioenergy, according to an estimate by the Japan Institute of Energy (2009), the standing stock of waste-based biomass per year in the world is 51 EJ in agriculture, about 47 EJ in livestock farming, and about 40 EJ in forestry, making the total amount about 138 EJ. The waste biomass from cattle manure is about 23 EJ, accounting for the largest proportion in the breakdown, followed by logging waste that is about 22 EJ. By regions, Asia has large potential, accounting for 58% of availability in crop biomass and 47% in livestock biomass.

2.2.2. Potential Amount of Biomass and the Amount Available for Use in Japan

The available quantity of biomass and the target rate of biomass use in 2020 based on its type in Japan estimated by the Ministry of Agriculture, Forestry and Fishery (MAFF), Japan are shown in Table 2.1. This estimated amount of biomass expressed in carbon equivalents is based on the information available in the Basic Plan for the Promotion of Biomass Utilization. Most types of biomass are used for resources and raw materials. Meanwhile, the data in the table indicate slow progress in the use of non-edible biomass such as food waste, inedible parts of agricultural crops, and residual materials from forestry. Table 2.2 summarizes the potential annual bioenergy production in 2020 in Japan. The use of biomass energy in 2020 was predicted by MAFF for converting the potential biomass into energy.

2.3. BIOMASS SYSTEMS

Atsushi Chitose, Masaaki Yamada, and Seishu Tojo

TABLE 2.1 Potential Amount of Biomass and Target Rate of its Use in 2020 by Biomass Type in Japan

Type	Potential amount (10^3t)	Potential amount (10^3t-C)	2020 Target rate of use	Maximum amount available for use (10^3t-C)* (Ratio used as energy)		Direction of application
				2020	Use ratio 100%	
Livestock excrement	88,000	5250	90% → 90%	1310 (25%)	1840 (35%)	Energy from methane fermentation, compost
Sewage sludge	78,000	900	77% → 85%	190 (21%)	320 (36%)	Energy from bio-gasification, construction materials
Black liquor	14,000	4660	100% → 100%	4660 (100%)	4660 (100%)	Energy at lumber mills
Paper	27,000	10,340	80% → 85%	520 (5%)	2070 (20%)	Energy, return paper into ethanol, biogas, recycled paper
Food waste	19,000	800	27% → 40%	210 (26%)	690 (86%)	Energy from methane fermentation, fertilizers, feed
Residual material at lumber mills	3400	1700	95% → 95%	1020 (60%)	1110 (65%)	Energy, raw materials for paper, boards

(Continued)

TABLE 2.1 Potential Amount of Biomass and Target Rate of its Use in 2020 by Biomass Type in Japan—cont'd

Type	Potential amount (10³t)	Potential amount (10³t-C)	2020 Target rate of use	Maximum amount available for use (10³t-C)* (Ratio used as energy)		Direction of application
				2020	Use ratio 100%	
Lumbers produced in construction sites	4100	1810	90% → 95%	850 (47%)	940 (52%)	Energy, recycling of wood pulp, boards
Crops as emergency food	14,000	4980	30% → 45% (excluding plow down)	750 (15%)	1250 (25%)	Energy, fertilizer, feed
Residual materials in forests	8000	4000	Almost no use → more than about 30%	1200 (30%)	4000 (100%)	Raw materials of paper, board
Crops residue		400†	Nearly zero → 400,000 t-C	400 (100%)	400 (100%)	Biofuel from crops and microalgae
Total	34,840			11,110	17,280	

*Maximum amounts available for use are calculated based on an assumption that (1) target use ratio in 2020 and (2) 100% use ratio are all achieved by increasing the uses as energy.
†Crops are target amount of production in 2020.
Source: MAFF (2012). Current status and problems of biomass. Reference in the 9th Meeting of the Biomass Business Strategy Exploration Team of the MAFF.

TABLE 2.2 Target of Bioenergy Production in 2020 in Japan

	2020	Use ratio 100%
Gross calorific value	460 PJ y^{-1}	720 PJ y^{-1}
1. Energy* (Electricity to households)[†]	13 billion kWh 2,800,000 households	22 billion kWh 4,600,000 households
2. Fuel (crude oil equivalent) (Gasoline for automobiles)[†]	11,800,000 kL 13,200,000 cars	18,500,000 kL 20,800,000 cars
3. GHG to be reduced[‡] (GHG emissions in Japan)[§]	40,700,000 t-CO_2 3.2%	63,400,000 t-CO_2 5.0%

*1 PJ = 25,800 kL (equivalent of crude oil) = 278 million kWh. Available amount to be used as electricity is calculated while taking into account of power generation loss and efficiency of gasification, etc.
[†]Calculations are based on conditions that electricity consumption per household is 4734 kWh y^{-1}, and the amount of gasoline consumption per automobile is 1000 L y^{-1}.
[‡]Converted maximum amount available for use (t-C) based on the ratio of atomic/molecular weights of C and CO_2.
[§]The value in the quick report in FY 2010 was 1.256 billion tons.
Source: MAFF (2012). Current status and problems of biomass. Reference in the 9th Meeting of the Biomass Business Strategy Exploration Team of the MAFF.

2.3.1. Components of a Biomass System

When biomass is used as a resource for human activities, the concept of a biomass system should be established (Kitani, 2004). Effective utilization of biomass requires a comprehensive framework that integrates all stages of the supply chain of biomass material flows – i.e. production, collection, transport, conversion, distribution, and consumption – into a well-organized and structured system. The concept of this supply chain is schematically illustrated in Figure 2.1.

In the production stage, identifying the various types of biomass suitable for local geographical and meteorological conditions is required. Biomass can be produced mainly in artificial forest, farm land, lake, and coastal areas. These spaces and fields are located in a continuous area or regions that are linked organically. In the collection and transport stage, biomass will be collected to centralize conversion facilities. The collector and transporter of biomass along with the disposal contractor and the distributor of products in the distribution stage play leading roles as linkers of respective space and facility.

In the conversion stage, the conversion facilities of biomass to biofuel and other end-use products, which are important components of the biomass system, play central roles in supplying energy and industrial raw materials to

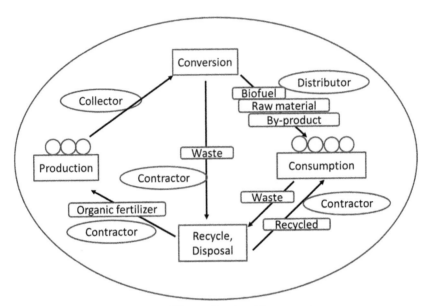

FIGURE 2.1 Biomass flow in a biomass system.

consumers. The waste generated in the conversion process should be disposed of by returning it to the original fields in a manner in harmony with the local ecosystem. Biomass supply will be maintained by keeping the fields adequately in the ecosystem because the production and use of biomass fundamentally rely on the function of nature everywhere.

In the consumption stage, the consumer of biofuel and other end-products performs the anchor role in the biomass system. The biomass system would be managed under the governing regulations and rules, and a well-managed system coordinates various fields, spaces, facilities, stakeholders and agents for enhancing local industries without imparting additional environmental loads by cooperating with various local organizations through the establishment of an effective social network.

2.3.2. Prerequisites for a Biomass System

The construction of an optimal biomass system requires thorough consideration from a holistic point of view with attention to the following location-specific characteristics in terms of natural endowments as well as social and economic conditions.

First, the biomass system is subject to the biomass feedstock available in both type and volume. Thus, the optimal biomass system would depend on the natural resources in a particular region. The optimal combination of potential biomass resources in use is determined by climate, geography, water, land resources, and other natural conditions. Second, the biomass system is conditional on the stage of economic development. The promotion of biomass utilization is expected to generate social and economic benefits in addition to its contribution to the reduction of GHG emissions. However, these expected benefits will vary according to local social and economic conditions with different system components. Also, the stage of economic development governs the conditions surrounding the biomass supply chain. Generally social infrastructure such as roads, water supply, and electric distribution is better provided in higher-income than in lower-income regions. Third, the biomass system may be subject to social, cultural, and institutional rules including formal laws, acts, other legal regulations and codes, as well as informal constraints such as local customs, norms and cultural beliefs. The adoption and diffusion of new biomass use to a larger extent may require institutional innovation that will induce a society to establish an appropriate social system to facilitate an effective use of local biomass resources.

Given these location-specific conditions, the optimal technology applied to the biomass system can be determined. Technology is associated with all stages of the biomass supply chain, spanning production, collection, transport, conversion, and distribution. Specifically, the biomass system requires the development of an optimal set of bio-related technologies that adapts local conditions in the production and conversion stages. Furthermore, because the enhancement of biomass utilization may bring about adverse environmental and social impacts and consequences, the biomass system should suffice the environmental assessment and other evaluations in the planning stage.

It is clear that the construction of a biomass system requires a thorough assessment in all stages of the biomass supply chain from various points of view. In principle, the sustainable operation of the biomass system should satisfy the conditions such that it is economically viable, environment friendly, and socially acceptable. More specifically, the prerequisites of the biomass system are: (a) not to compete with food production; (b) to produce no additional environmental load; (c) to contribute to the construction of a lower-carbon society; (d) to be adaptable to a local society; (e) to be managed throughout the biomass system; and (f) to be economically viable from a societal point of view.

With these prerequisites, the priority of using biomass resources may differ among regions; however, it seems legitimate that, in practice, the priority should be assigned in the following order: first on food, second on the material use for higher-value products that will contribute more to increases in rural income, third on animal feed or fertilizer, and perhaps last on fuel. The main reason that fuel production would be given the lowest priority in biomass utilization is its low profitability in a commercial operation. With the current state of technology, biofuel cannot compete with fossil fuel in profitability with only a few exceptions such as the case of sugar cane-based ethanol in Brazil. This implies that economic viability is an important prerequisite for the development of biofuel production.

Although in practice there are various challenges for the adoption and development of biomass utilization in society, the Japanese experience suggests that, currently, limited profitability and inadequate legal provision are recognized as the major bottleneck. The generation of new industries associated with biomass use has been impeded by the current legislative framework that is likely to discourage the new entry of agricultural and forestry industries into the energy business. Moreover, the lack of evidence of economic viability for ongoing biomass projects casts doubts on the sustainability of the biomass system without government subsidies. Because lower profitability is more likely to be viewed as a serious problem, the essential factors underlying lower profitability with focus on the case of biofuel production will be briefly discussed in the following paragraphs.

Needless to say, profit is positive if the unit price of the produced biofuel is higher than its unit production cost. On the demand side, the biofuel price is influenced heavily by the prices of other substitutable fuels such as oil as well as by government support. However, the dependence of government support in the form of subsidies and tax exemption will not last long; thus, an optimal biomass system that functions without government support in the long run is needed.

On the supply side, the production cost roughly consists of the feedstock procurement cost and the conversion cost. The procurement cost involves the stages of procurement of feedstock for producing, collecting, and transporting it to the processing sites. The procurement cost relies on production environments of feedstock such as climate and geographic conditions, production technology (e.g. genetics and farming practices), infrastructure (e.g. irrigation facilities and roads), as well as social institutions (e.g. land tenure institutions). The conversion cost that is composed primarily of the cost of conversion from feedstock to biofuel is subject to technology (e.g. fermentation and bioengineering) and infrastructure (e.g. water supply and electricity). Because both types of costs involve the transportation of raw materials or final products, these costs are influenced by the accessibility to markets as long as biofuel is produced on a commercial basis.

The two costs are more likely to have a tradeoff relationship; a higher feedstock procurement cost coincides with a lower conversion cost, and vice versa. The first-generation biofuel feedstocks such as corn and rapeseed are viewed as an example of the case of a higher procurement cost and a lower conversion cost. Conversely, the use of cellulose such as rice straw or other second-generation biofuel feedstocks is characterized by a lower procurement cost and a higher conversion cost, given the present state of technology. From an economic point of view, the production of first-generation biofuel is subject primarily to types of feedstocks and their production environments. In fact, given current technology, the cost of feedstock constitutes more than half of biofuel production cost (World Bank, 2007). In contrast, the production of the second-generation biofuel relies mostly on the levels of bioengineering technology that plays a critical role in the conversion process. This implies that the key to attaining economic viability of the first-generation biofuel production is a rational choice of feedstocks coupled with the development of higher biomass cultivars under favorable production environments, whereas the second-generation biofuel production relies on technological innovation in the conversion process for practical use.

Because technological innovation in biology and bioengineering will lower either the procurement or the conversion cost, it is the key to the success of the biomass system. The subsequent chapters of this book discuss the current advancements in biomass-related science and technology that will contribute to the construction of an ideal biomass system to assist in ultimately developing a lower-carbon society. However, it should also be kept in mind that identifying the optimal combination of technologies associated with all stages of the biomass supply chain, subject to location-specific characteristics in terms of both natural endowments and socio-economic conditions, is of equal importance to the success of the biomass system.

REFERENCES

Cao, M., & Woodward, F. I. (1998). Net primary and ecosystem production and carbon stocks of terrestrial ecosystems and their responses to climate change. *Global Change Biology, 4,* 185–198.

Janzen, H. H. (2004). Carbon cycling in earth systems; a soil science perspective. *Agriculture, Ecosystems and Environment, 104,* 399–417.

Japan Institute of Energy. (2009). *Biomass handbook* (2nd ed.). Tokyo: Ohmusha [in Japanese].

Kitani, O. (2004). *Biomass – Bioresources and environment.* Tokyo: Coronasha.

Ministry of Agriculture, Forestry, and Fisheries (MAFF), Japan (2012). Current status and problems of biomass. Reference material distributed in the 9th Meeting of the Biomass Business Strategy Exploration Team of MAFF [in Japanese].

World Bank. (2007). *World Development Report 2008: Agriculture for development* (Chapter 2). Washington, DC: World Bank.

Chapter 3

Ecosystems and Biomass Systems

Hiroto Toda, Tomoe Shimizu, Jun Shimada and Takashi Motobayashi

3.1. NATURAL ECOSYSTEMS AND SATOYAMA UTILIZATION

Hiroto Toda

3.1.1. Forest Ecosystems

a. Forest Productivity and Self-Fertilization

Organic materials such as fallen leaves and branches are continuously circulated in forests to add nutrients to the soil that will be taken up by growing plants.

Research Approaches to Sustainable Biomass Systems. http://dx.doi.org/10.1016/B978-0-12-404609-2.00003-9

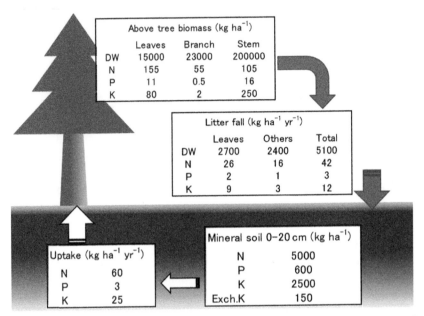

FIGURE 3.1 Nutrient circulation of *Sugi* (Japanese cedar; *Cryptomeria japonica*) and *Hinoki* (Japanese cypress; *Chamaecyparis obtusa*) artificial forests. *Tokyo University of Agriculture and Technology, Field Science Center, Field Museum Ohya-san, the east of Gunma Pref., Japan. Data from Toda et al. (1991).*

This is called self-fertilization of forests and maintains the productivity and soil fertility of forest land (Figure 3.1). Natural and artificial forests in deep mountains maintain nutrient cycling, or self-fertilization, unless they suffer huge natural disasters such as volcanic eruptions and landslides caused by earthquakes and/or heavy rain.

Cutting down trees for timber production disturbs nutrient cycling in artificial forests in addition to jeopardizing the functions of water conservation and biological diversity. Additionally, the tree cutting speeds up the decomposition of surface organic matter whereby a portion of carbon accumulated in forest soil is released into the atmosphere. However, these influences may be short-lived if the biological diversity as well as nutrient cycling and other public functions are restored to their former state by conducting proper forest regeneration and scheduling the cutting cycle at sufficient intervals (Figure 3.2).

b. Forest Biomass Utilization in Deep Mountains

In Japan, the demand for timber increased rapidly from 1945 to 1965 because of the postwar reconstruction work after World War II. This paralleled the "fuel revolution" to replace fuel wood and charcoal with gas and petroleum. Hence,

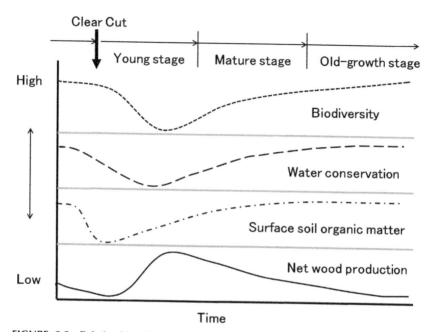

FIGURE 3.2 Relationships between the growth stage and public functions of forest ecosystems.

expansive forestation to plant *Sugi* (Japanese cedar; *Cryptomeria japonica*) and *Hinoki* (Japanese cypress; *Chamaecyparis obtusa*) artificially spread to deep mountains where deciduous broad-leaved trees had been cut down to make charcoal. As a result, Japan has 10 million hectares of *Sugi* and *Hinoki* forest resources in deep mountains for timber production.

Deep mountains have steep slopes with no modern forest road networks; high costs to transport wood from these deep mountains make logging unprofitable. However, logging for timber production along with utilizing the waste materials such as the slash produced during the sawing process is cost-effective and beneficial. However, the established *Sugi* and *Hinoki* forests accumulate a great amount of nitrogen and phosphorus with quantities equivalent to 4–7 years of the annual uptake. Hence, the leaves and branches after timber harvest should be left on the forest ground in order to maintain soil productivity (Toda, 2004).

c. Huge Disturbance and Fertilizer Trees

When the disappearance of a large amount of forest vegetation occurs due to huge disturbances caused by natural disasters such as volcano eruptions, the restoration of forest ecosystems and ecosystem services is the first priority. On Miyake island, one of the Izu islands, forest ecosystem disturbances caused by

Mt. Oyama's eruptions have been replicated at an interval of about 20 years for many decades.

Ohbayashabusi (*Alnus sieboldiana*) is a popular pioneer tree in the Izu islands. It has numerous root nodules consisting of symbiotic bacteria with the capability of fixing nitrogen from the atmosphere. This species is capable of growing well in soil with poor nutrients (Toda et al., 2007, 2008). *Ohbayashabusi* originally grew in the Izu islands; the characteristics of the tree have been developed to replicate the forest in regions that have undergone a massive disturbance (Kamijo et al., 2002).

Ohbayashabusi is often used as a fertilizer tree for forests in Honshu because its leaves have a rich nitrogen concentration so that the leaf litter is capable of rapidly replenishing the soil nitrogen. The growth of this fertilizer tree declines after about 20 years, and the area is then replaced by subsequent stages of tree species. Therefore, fertilizer trees are useful to restore both forest vegetation and soil fertility. If fertilizer trees are properly managed, sustainable forest biomass resources and soil fertility can be maintained.

3.1.2. Satoyama Systems

a. What is Satoyama?

Satoyama is a region with secondary forest and artificial forest surrounding a village. Compared with natural deep mountains, it is the secondary natural environment created and maintained by various human activities. Forests in Satoyama regions have long been the source of biomass for raw materials and energy such as fuel wood and charcoal; the fallen leaves are made into compost as fertilizer.

In the forests of Satoyama, harvesting wood and coppicing are done at a short rotation of 15–20 years because small trees with diameters of 5–20 cm are used to make fuel wood and charcoal (Figure 3.3). Also, the fallen leaves are collected and used as fertilizer for farmland so that the forest's self-fertilization system is exploited. After decades of such management, the fertility of soil in Satoyama has been depleted (Figure 3.4). Therefore, trees that are capable of growing in soil are planted so that the forests in Satoyama now have strong regeneration power.

Trees growing in the secondary natural environment of Japan's Satoyama include deciduous broad-leaved oak, i.e. *Konara* (*Quercus serrata*) and *Kunugi* (*Quercus acutissima*), and confiers, i.e. *Akamatu* (red pine: *Pinus densiflora*). When the practice of weeding and gathering fallen leaves is carried out regularly, the forest floor is incapable of supporting the growth of various plants inhabited by small animals. With the primeval natural forest in the deep mountains left untouched, Satoyama that was created by human activities has played an important role in maintaining the biological diversity of Japan.

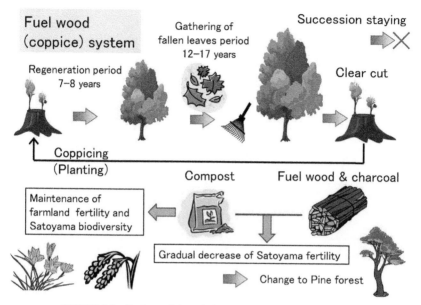

FIGURE 3.3   Fuel wood (coppice) system of Satoyama pre-1960s.

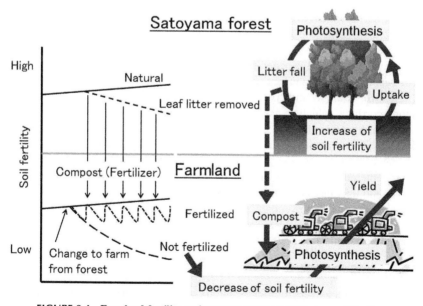

FIGURE 3.4   Farmland fertility maintenance system by Satoyama utilization.

Fertile soil site　　　　　　　　Infertile soil site
Few ectotrophic mycorrhizae　　Many ectotrophic mycorrhizae
　　　　　　　　　　　　　　　We can see many Hartig nets.

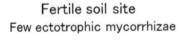

FIGURE 3.5　Ectomycorrhiza's Hartig net of *Konara* root. The concentration of ectomycorrhiza in fertile soil sites is small. On the other hand, at infertile soil sites, the concentration is greatly increased.

b. Strategy of Nutrient Acquisition in Satoyama Forests

Removing fallen leaves and cutting trees for short rotation periods have made *Konara* and *Kunugi* the major species growing in Satoyama because the short-term practice prevents the progressive growth of succession species. Repeated excessive cutting will degenerate the succession tree species; hence, *Akamatu* will become the predominant species (Figure 3.3).

　　Konara, *Kunugi*, and *Akamatu* can grow on such infertile land because ectomycorrhiza, which is a type of mycorrhizal fungi and has hypha tissue called Hartig nets (Hiroki, 2002) (Figure 3.5), have a symbiotic relationship with tree roots. Mycorrhiza in the host's roots receives photosynthetic products generated by the host to grow, and supplies the host plant with nutrients and water gathered by its hyphae that extend under the ground (Box 3.1). The hypha is long and thin and has a larger surface area than the fine root of the host tree. In particular, mycorrhiza contributes much in promoting the uptake of phosphorus and building up tolerance of the host tree to resist drought and heavy metals (Smith and Read, 1997).

c. Satoyama and Farmland Fertility

Gathering fallen leaves from Satoyama for providing nutrients to farmland has maintained the productivity of the farmland (Inui, 1996) (Figure 3.4). The ash

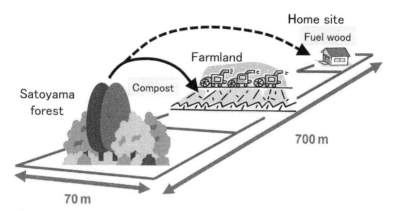

FIGURE 3.6   Land-use pattern in San-tomi area, west of Saitama prefecture, Japan.

after burning fuel wood and charcoal was also applied to farmland as soil conditioner or fertilizer.

The San-tomi area located to the west of Saitama prefecture was reclaimed as an agricultural field in the Edo era. The land use pattern for home site, farmland, and Satoyama forest that adjoin one another on the rectangular 70 m × 700 m land (Figure 3.6) remains unchanged. The fallen leaves have been gathered and utilized to make compost on germination beds since around 1800, when sweet potatoes were extensively cultivated there. Results of the authors' investigation reveal that the forest is a typical coppice of Musashino with *Konara* and a few *Kunugi*. About 10–30% of the annual nutrient uptake is estimated to be reduced by the gathering of fallen leaves in this managed Satoyama site (Figure 3.7). This site shows an obvious lower amount of soil nutrients than an adjoining control site where fallen leaves have not been gathered for several decades. However, a decline of *Konara* is not observed, although the gathering of fallen leaves has continued in the Satoyama forest. Hence, maintaining the growth of *Konara* should be suggested to enrich the soil with nutrients generated by mycorrhiza.

3.1.3. Sustainable Use and Management of Satoyama

a. Crisis of Satoyama

As mentioned before, intensive use of Satoyama with enhanced growth of fungi and production of edible wild plants helps the survival of spring ephemerals that has played an important role in restoring the original food culture and biodiversity in the Satoyama village. However, the use of Satoyama, such as leaf litter compost, fuel wood and charcoal, was decreased by the development of chemical fertilizer and the fuel revolution (Duraiappah et al., 2012) (Figure 3.8).

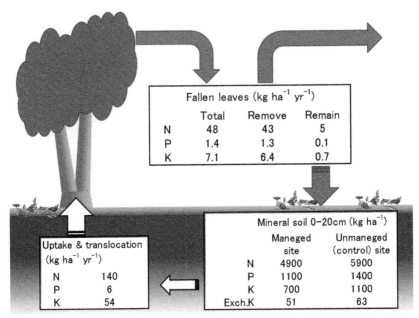

Fallen leaves (kg ha^{-1} yr^{-1})			
	Total	Remove	Remain
N	48	43	5
P	1.4	1.3	0.1
K	7.1	6.4	0.7

Uptake & translocation (kg ha^{-1} yr^{-1})	
N	140
P	6
K	54

Mineral soil 0–20cm (kg ha^{-1})		
	Maneged site	Unmaneged (control) site
N	4900	5900
P	1100	1400
K	700	1100
Exch.K	51	63

FIGURE 3.7   Amounts of nutrients of fallen leaves and surface soil in Konara secondary forests in San-tomi area, west of Saitama prefecture, Japan.

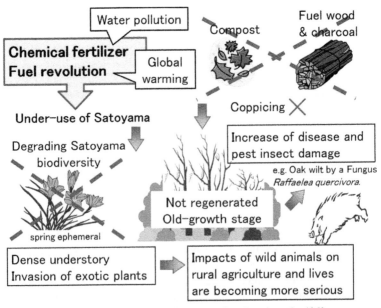

Water pollution

Compost

Fuel wood & charcoal

Chemical fertilizer
Fuel revolution

Global warming

Coppicing ✕

Under-use of Satoyama

Increase of disease and pest insect damage

e.g. Oak wilt by a Fungus *Raffaelea quercivora.*

Degrading Satoyama biodiversity

Not regenerated Old-growth stage

spring ephemeral

Dense understory Invasion of exotic plants

Impacts of wild animals on rural agriculture and lives are becoming more serious

FIGURE 3.8   Underuse of Satoyama situation since the 1960s.

Managed site	Unmanaged site
Sparse understory and good visibility	Dense understory and bad visibility

FIGURE 3.9   Comparison of understory managed site and unmanaged site in Motegi-mach's Satoyama, east of Tochigi prefecture, Japan. Wild animals, such as deer, boars and monkeys, enter the village due to the bad visibility in Satoyama.

Satoyama forest became underused and the understory vegetation was altered by the advances of invading exotic bamboo species, as well as dense thickets and climbing plants. Because this alteration leads to changes in the forest floor environment from light-filled to dark space, the growth of spring ephemerals has declined (Figure 3.9). Wild animals of various sizes including deer, boars, and monkeys are also adversely affected due to the poor visibility in Satoyama. Human lifestyle is also affected because the buffer zone between wild animal habitats and human living space as well as many other functions of Satoyama have vanished. The unmanaged and ungenerating Satoyama forests have changed to a region with old growth of trees raged by diseases. Oak wilt, which is caused by a fungus, is spreading from Satoyama into deep mountain areas (Kuroda, 2001).

Therefore, underuse of Satoyama forests has brought about many problems such as biodiversity degradation, a threat to rural lives caused by wild animals, spread of tree diseases, decrease of water conservation, and poor soil erosion control.

b. Reconstruction of Satoyama Systems

The above observations illustrate the importance of protecting both the primeval nature and secondary natural environments such as the Satoyama area. As mentioned before, Satoyama biomass had been used and managed to provide compost and energy for a long period; it is used less nowadays because socio-economic conditions have become efficiency oriented. Most Satoyama forests have become unmanaged secondary forests, or have been replaced with artificial *Sugi* and *Hinoki* forests. The Satoyama Initiative to promote sustainable use and management of natural resources in human-influenced natural environments was implemented in 2010.

The woody materials of Satoyama forests are expected to be used as new economic resources for the development of biomass technologies in the coming

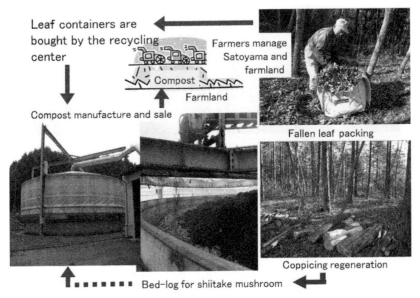

FIGURE 3.10 Reconstruction of the Satoyama system with an organic recycling center in Motegi-mach, east of Tochigi prefecture, Japan.

years. For example, woody biomass has been used as pellets and other forms of fuel by local communities. In some areas, a system to recycle organic matter has been developed by compost manufacturing companies (Figure 3.10). Fallen leaves are gathered and packed in containers and shipped to recycling centers to be composted with other organic wastes, and then sold at a reasonable price to the farmers of the area.

The capability of forest to absorb CO_2 will be enhanced by practicing proper forest management, e.g. thinning *Sugi* and *Hinoki* forests and regenerating the *Konara* forest. These management practices produce woody biomass while contributing to the reduction of CO_2 emissions. The national government and some regional governments have established systems for verifying the amount of CO_2 absorption and emission reduction as credits, known as the "Offsetting Scheme (J-VER Scheme)".

Reconstruction of the Satoyama system is a noteworthy model of local production for local consumption. Therefore, this practice should be greatly promoted to extend the experience on sustainable use and management of Satoyama to other forests.

3.2. BIOMASS PRODUCTION IN CULTIVATED FIELDS AND CONSERVATION OF ECOSYSTEMS

Tomoe Shimizu and Jun Shimada

3.2.1. Continuous Use and Ecosystems of Cultivated Fields

Cultivated fields in Japan are divided roughly into paddy fields and upland fields; both types of fields have efficient production of biomasses, including crops. Therefore, crops of the same line and the same variety are often grown in cultivated fields on a large scale by removing many other species that used to be components of the original ecosystem in order to protect the crop. In cultivated fields, because the intention is to keep the biota simple, and to keep succession at an early stage in cultivated fields, the cultivated field is essentially an unstable agricultural ecosystem. In addition, because the stem, leaf, root, and seed or fruit of crops is harvested and removed from the ecosystem, the material cycle is cut off. Hence, the current practice is to implement strong management based on recent advances in technology to maintain a stable ecosystem and sustainable cultivation of the field.

In order to utilize farmland sustainably, it is desirable to keep the soil environment the same before and after harvesting the crop. Compensating the loss of soil components absorbed by plants removed from the field as crops, and rapid decomposition of the organic matter (e.g. root secretions and crop residues) left in the field are essential. If these tasks are efficiently performed, "injury by continuous cropping" is alleviated. However, the technology to restore the soil conditions immediately after harvesting has not been developed at present. Hence, the problem of continuous cropping injury is avoided by the introduction of a crop rotation system by planting a different type of crop after some crops have been harvested. Monoculture systems in which only a single type of crop is planted may lead to outbreaks of particular insect pests or pathogens depending on the crop. The same is true for soil ecosystems; continuously cropping some plants causes the soil pest density to increase gradually and jeopardize the crop. As a result of injury by continuous cropping of a single type of crop, imbalance of soil nutrients is caused by the lack of certain nutrient components that have been depleted excessively but unnecessarily. The accumulation in cultivated soil of particular substances that inhibit plant growth, such as root secretions, have been considered. The injury caused by practicing continuous cropping does not happen in paddy rice fields even if the rice plant is grown every year because pathogenic microbes and soil nematodes cannot survive in the paddy soil that is submerged in irrigated water. A practice known as "paddy–upland rotation" can be implemented to use a cultivated field as a paddy rice field and an upland field alternatively. Because the species that can survive in the presence of irrigation water are different, paddy–upland rotation is effective for the control of weeds and pests.

Decreasing the energy input while maintaining or improving productivity is important to Japan's agriculture. For cultivation of crops, much energy and advanced technology are needed to remove harmful components and maintain the condition of nutrient balance in the soil. The organism-preserving ability of the farming ground is a key to achieving these goals. However, because

BOX 3.1 Concentration of Ectomycorrhiza in Symbiosis with Roots of *Konara*

The relationship between the concentration of ectomycorrhiza in symbiosis with roots and the soil fertility has been studied by observing the growth of *Konara* seedlings. The pots for growing *Konara* seedlings were prepared by adding various quantities of fertilizer to soil to result in different combinations of nitrogen and phosphorus concentrations. In the following set of samples, symbols N and P refer to nitrogen and phosphorus respectively, whereas the numbers refer to total mg of N or P in 100 mg soil. For example, "N200" indicates 200 mg nitrogen in 100 g soil whereas "N200 + P80" refers to 200 mg nitrogen and 80 mg phosphorus in 100 mg soil. Ten samples each with ten replications were prepared using urea as the nitrogen fertilizer and phosphatic manure as the phosphorus source; these samples are N200 + P40, N200 + P80, N200 + P120, N500 + P40, N500 + P80, N500 + P120, N1000 + P40, N1000 + P80, and N1000 + P120. Concentrations of nitrogen and phosphorus were 110 mg-N per 100 g soil and 0.03 mg-P per 100 g soil respectively in the non-fertilized (NF) pot containing untreated natural soil (control pot). One *Konara* seedling was planted in each pot to grow for about 6 months. The numbers of fine roots with Hartig nets were then counted, and the percentages of these numbers to the total fine roots were calculated (Momiyama et al., unpublished data).

The percentages of fine roots with Hartig nets are significantly lower in N500 + P80, N500 + P120, N1000 + P40, N1000 + P80, and N1000 + P120 pots than NF pots (Figure 3.11). Roughly speaking, the percentages decrease with higher nitrogen and phosphorus concentrations in the soil. The results suggest that the concentration of ectomycorrhiza in symbiosis with *Konara* roots is dependent on soil fertility.

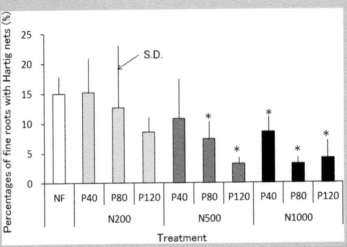

FIGURE 3.11   Percentages of fine roots with Hartig nets to total number of fine roots of *Konara* seedlings that were fertilized differently in soil. NF: not fertilized (control) soil. N200, N500, and N1000: 200 mg N, 500 mg N, and 1000 mg N per 100 g fertilized soil respectively. P40, P80, and P120: 40 mg P, 80 mg P, and 120 mg P per 100 g fertilized soil respectively. *Significantly different from NF at $P < 0.05$ in t-test.

efficient production is pursued, the agricultural field is managed to exclude beneficial organisms for the sake of keeping only cultivated crop in the field without considering the benefit and damage to crops.

Soil is the site where various nutrients and water are supplied to support plant growth, and plant residues are resolved on-site to return nutrients to the field. In most ecosystems, the resources of the energy and nutrients originate from remains of flora and fauna that are resolved or decomposed by microorganisms, nematodes, annelids, and micro-arthropods living in the soil. Finally, the excrement and remains of these organisms are resolved by microorganisms to inorganic or organic nutrients that are available for plants to use. In the soil, there are numerous species of animals feeding on the fungi, bacteria, and the flora and fauna remains; these soil-dwelling animals play an important role in the resolution process in the ecosystem. Soil nematodes maintain the microbial population for a long time to speed up the resolution of an organism by microorganisms (Trofymow and Coleman, 1982); hence, there is a close relationship between the cluster structure of a nematode and soil fertility (Bongers, 1990).

The proliferation of microorganisms in soil is relatively slow. It is thought that small arthropods such as mites and collembolans activate microbes, expanding the surface area by fragmentation of litter to disperse microorganisms (Schowalter, 2000). Ants and earthworms mix the soil's organic and inorganic components by feeding on the soil to increase the soil pore space (Crossley, 1977; Seastedt and Crossley, 1984; Whitford, 2000). Through those processes, soil animals and fauna have a significant direct and/or indirect impact on the capability of land to produce primary production.

In cropland, soil organisms play an important role in crop protection. Carabid beetles are not only the litter transformer in pedogenesis but also a generalist predator in the upland field. The generalist predators included in macrofauna suppress the density of pests in the crop field. The density of specialist predators and parasitoids that have high ability of finding and killing target pest is nearly dependent on the density of pest because of their feeding habit. Consequently, predators and parasitoids usually become populated in the cropland when the pest is abundant, and leave when the pest density is diminished. On the other hand, generalist predators can use various types of prey that survive in the vicinity of the field. Hence, they are regarded as the most important natural enemy to pests during the first period when the latter invade the field and propagate. Results of some research show that fungivorous nematodes, collembolans, and mites are effective in controlling plant diseases caused by fungi (Ishibashi and Choi, 1991; Nakamura et al., 1991; Matsuzaki and Itakura, 1992). Therefore, soil fauna may be an important factor for producing food safely and efficiently.

Since 1999, our research team started an investigation to clarify the role of soil fauna in upland ecosystems in Japan. The study is conducted in a long-term experimental upland field (1.5 ha area) established in 1992 for growing dent corn and soybeans alternately in summer and for growing grass in winter. It is located in Field Museum Fuchu, Field Science Center, Faculty of Agriculture, Tokyo University of Agriculture and Technology. The experimental field is divided into 20 plots to conform with the two-way ANOVA analysis of the results. Each plot is managed using different tillage methods, i.e. conventional tillage (T) or reduced tillage (RT), combined with various fertilizer applications, i.e. chemical fertilizer (F) or manure application (M). As a result of our research in the long-term experimental field, we will discuss the effects of soil management on the cropland food web based on our long-term field experimental results.

3.2.2. Detritus Food Web in Upland Fields

In terrestrial ecosystems, the main resources necessary for organisms to live in the soil are detritus, fragments of dead plant organic matter such as plant leaves, and roots' peeling cells. Around the detritus, microorganisms and microphagous species form a complex degradation loop known as the "detritus cycle", which is joined by higher predators or large soil animals that eat litter directly to form a soil detritus food. The detritus food web leads to the grazing food chain that plays an important role in supporting the food web of terrestrial ecosystems.

In the upland field ecosystem, any approach to till the soil for crop production disturbs the soil food web. Tillage will not only disrupt the habitat of soil organisms but also help microbial degradation by mixing the soil with crop residue. Introducing compost is a good way to invest organic resources into the detritus food web.

No tillage or reduced tillage management increases under-decomposed organic matter in the field and accumulates crop residues on the soil surface. Through the food web, tillage methods or manure application have a great impact on soil fauna.

In 1992, when the experimental field was first developed, there was no difference in the number of individual organisms in the tilled and untilled fields. In 1997 and 1999, reduced tillage plots had significantly higher populations of mites and collembola than fully tillage plots. The influence of fertilization is reflected by the number of individual organisms in 1999, when the populations of mites and collembola were significantly higher in manured plots than chemically fertilized plots. In 1999, research on the population of arthropods living on the ground and in crops was carried out. Although there is no statistical significance, numbers of both generalist predators (natural enemy) captured using pitfall traps and herbivores (insect pest) captured using Roth-type traps show some correlation with soil management. When the same tillage method was used, manured plots had larger populations of the predator and smaller populations of the herbivores than chemically fertilized plots. Also, an antagonistic relationship is observed between the populations of prey and predator (Figure 3.12).

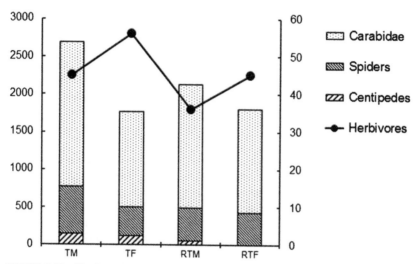

FIGURE 3.12   Number of individuals of prey and predator in each plot. F: chemical fertilizer; M: manure; RT: reduced tillage; T: tillage. Bar graph presents the number of predators captured by pitfall trap and the solid line with filled circles the number of herbivores captured by Roth-type trap in the long-term experimental field during soybean cultivation period in 1999.

3.2.3. Effects of Manure Application on Biota in Upland Fields

a. Microorganisms and Protozoa

Sato and Seto (2000) investigated the relationship between microbial biomass and method of tillage or fertilizer application. Their results show that continuous use of manure increases the fungi biomass although no difference is observed in the bacterial biomass. The order of fungi biomass is RTM > TM > RTF > TF. The continuous practice of manure application and reduced tillage is thus assumed to have a synergistic effect on biomass fungi.

Protozoa of heterotrophic and autotrophic species account for 70% of the total soil respiration rate by the soil fauna, 14–66% of the carbon mineralization, and 20–40% of the nitrogen mineralization (Griffiths, 1994; Ekelund et al., 2001). This fact allows us to assume that protozoa play an important role in crop production in upland ecosystems. A single ciliate can consume 300–400 individual bacteria whereas an amoeba can consume 2000–4000 bacteria (Ekelund and Ronn, 1994).

Griffiths (1994) reported that protozoa use approximately 40% of the nutrients absorbed from prey microorganisms for cell regeneration, and discharge the remaining 60% either directly or indirectly through respiration as inorganic components. The inorganic components released by protozoa will be used by microorganisms, plants, and animals. Concerning the role of protozoa in the rhizosphere, Clarholm (1994) considered that bacteria fixed on the surface of soil organic matter are in a steady state because they do not grow due to limited carbon source. Polysaccharide emitted from the tip of the roots and detached root cell during the growth of plant roots is easily biodegradable, and is used by the bacteria as a carbon source.

In addition, bacteria mineralize and uptake soil organic matter into cells as bacterial nitrogen; protozoa that are attracted to the CO_2 released by bacteria feed on the bacteria to excrete inorganic nitrogen. Hence, bacteria-derived inorganic nitrogen becomes plant fertilizer near the roots, and in the presence of protozoa plant roots can intake the nitrogen efficiently due to the presence of bacteria.

Other organisms such as fungivorous nematodes, mites and collembolans, and protozoa contribute to the control of plant diseases because of their feeding habits. Foissner (1987) demonstrated that the inoculation of protozoa suppresses the potato soft-rot disease infected by *Bacillus aroideae*, and a genus of ciliate, *Colpoda* sp., inoculated to a cotton field decreases infection of wilt disease caused by *Verticillium dahliae* by 7–10% to increase the cotton yield by 30% (Foissner, 1987).

Wardle (1995) pointed out that the pressure exerted by protozoa on the bacterial biomass is higher than the pressure exerted by the fungivorous species on the fungus biomass in the detritus food chain. However, as described above, although the fungus biomass is significantly increased by the continuous use of manure, results of the long-term upland field study (Sato and Seto, 2000) do not show increasing bacterial biomass. We have investigated the number of individual protozoa living on these microorganisms in order to re-examine the effect of the difference in crop management on the number of bacteria and actinomycetes. The colony-forming units (CFUs) of aerobic bacteria and actinomycetes contained in 1 g dried soil were counted using a dilution plate technique using albumin agar medium. The protozoan population was enumerated using the most probable number (MPN) method designed by Ronn et al. (1995).

The proliferation of several typical ciliate and flagellate species among the species isolated from the upland field soil subject to long-term studies using the dilution plate technique were purified and subcultured in the albumin agar to enumerate their MPNs. Figures 3.13 and 3.14 show the CFUs of bacteria and actinomycetes respectively during the periods before and during soybean cultivation. Both manure and chemical fertilizer were applied on June 15; the soil was tilled on the same day, and seeded on June 18. CFUs of bacteria remain the same in the manured and fertilized plots. This observation is consistent with the results reported by Sato and Seto (2000). The influence of fertilization on the CFUs of bacteria appeared 10 days later and disappeared after 20 days. The transition density of the flagellate is shown in Figure 3.15. Ciliate density is about one hundredth of the flagellate and the same trend remains. Flagellate density rises from around 10 days after manure application to reach its highest level at 20 days. Despite an increase in bacterial population observed in the plots with chemical fertilizer application, the flagellate density is observed not to increase but decrease.

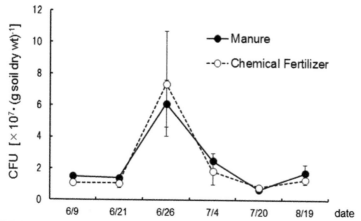

FIGURE 3.13   Number of colony-forming units of soil bacteria for albumin age medium during the soybean cultivation period in 2004.

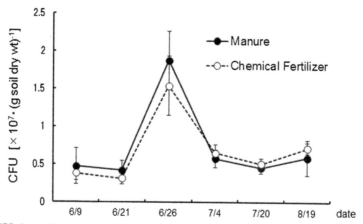

FIGURE 3.14   Number of colony-forming units of soil actinomycetes for albumin age medium during the soybean cultivation period in 2004.

In general, introduction of the substrate through fertilization increases the initial population of microorganisms temporarily. However, because of some limiting factors such as substrate reduction and living environmental degradation, the microbial biomass has a tendency to decrease. Although there is a possibility that this observation only applies to our study, other factors such as weather conditions and tillage methods may also promote potential activities of bacteria and actinomycetes. Protozoa are sensitive to changes in the environment; they are transformed into a dormant cyst under unfavorable conditions. Typical culture methods cannot differentiate between cysts and active cells, whereas direct microscopy is incapable of observing small protozoa.

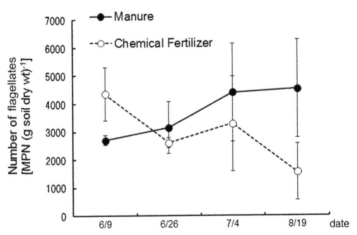

FIGURE 3.15 Number of soil flagellates counted using the MPN technique during the soybean cultivation period in 2004.

To recognize the role of protozoa in upland fields, the development of molecular biology techniques and other ex-situ experimental model systems is needed.

b. Soil Nematodes

In order to study soil nematodes, soil samples were collected from each plot, i.e. TM, TF, RTM, and RTF, of the long-term upland field where dent corn was planted. Two samples were collected at depths of 0–5 and 20–25 cm respectively from the surface. Nematodes were extracted from each sample using the Baermann funnel technique and classified into groups of free-living or plant parasites; their numbers were then counted. The nematode density was calculated and expressed as number of nematodes per gram of soil dried at 80°C for 48 h. The density of free-living nematodes in the soil layer at 20–25 cm depth is higher in the manure plot than the chemical fertilizer plot, and higher in the reduced tillage plot than the tillage plot (Table 3.1).

The observations demonstrate that manure application and reduced tillage management lead to a favorable soil environment for nematodes. On the contrary, the density of plant-feeding nematodes is higher in the chemical fertilizer plot than the manure plot, suggesting that the combination of conventional tillage and chemical fertilizer has a synergistic effect on the number of nematodes. Also, a significantly negative correlation exists between the density of total nematodes and the proportion of plant-feeding nematodes in each sample (Figure 3.16). These results suggest that an environment that has an increased number of free-living nematodes becomes unfavorable to plant-feeding nematodes.

TABLE 3.1 Number of Nematode Individuals in Each Treatment Plot (mean ± standard deviation per gram of soil dry weight)

	Date							
	2001/7/5		8/2		8/30		10/4	
Depth (cm)	0–5	20–25	0–5	20–25	0–5	20–25	0–5	
Total*								
TM	24.7 ± 3.65^b	38.6 ± 4.5^a	88.5 ± 13.6^{ab}	74.9 ± 14.5^a	83.1 ± 10.8^a	26.9 ± 3.1^a	99.6 ± 7.5^b	27
TF	22.5 ± 2.71^b	29.6 ± 2.0^a	30.8 ± 4.0^b	25.2 ± 3.7^b	28.6 ± 1.6^b	20.3 ± 3.7^a	40.4 ± 3.0^c	27
RTM	54.8 ± 8.46^a	15.2 ± 2.5^b	100.2 ± 12.6^a	15.0 ± 2.8^b	104.6 ± 17.8^a	17.4 ± 3.0^{ab}	159.0 ± 6.6^a	19
RTF	37.5 ± 3.81^{ab}	7.6 ± 1.1^b	36.2 ± 3.8^b	9.8 ± 2.1^b	52.7 ± 8.3^b	7.7 ± 1.4^b	75.8 ± 15.5^{bc}	16
Plant parasites†								
TM	0.4 ± 0.1^a	1.3 ± 0.2^b	0.2 ± 0.1^a	0.3 ± 0.1^b	0.6 ± 0.2^a	0.3 ± 0.2^a	0.9 ± 0.2^b	1
TF	0.6 ± 0.1^a	6.5 ± 0.7^a	0.2 ± 0.1^a	3.6 ± 1.1^a	0.8 ± 0.3^a	1.9 ± 0.6^a	3.4 ± 0.6^a	6
RTM	0.3 ± 0.1^a	0.6 ± 0.2^b	0.3 ± 0.1^a	0.4 ± 0.1^b	0.7 ± 0.1^a	0.4 ± 0.2^{ab}	0.7 ± 0.1^b	0
RTF	0.6 ± 0.2^a	$0.6 \pm 0.2b$	$0.5 \pm 0.2a$	0.7 ± 0.4^b	0.8 ± 0.2^a	0.8 ± 0.3^{ab}	4.2 ± 0.7^a	1

F: chemical fertilizer; M: manure; RT: reduced tillage; T: tillage.
There are significant differences between the different characters of the same sampling date ($P < 0.05$).
*The combined number of free-living and plant parasites.
†The number of plant parasites only.

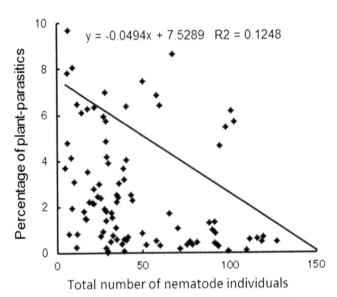

FIGURE 3.16 Relationship between the total number of nematode individuals and the percentage of plant parasite species in each soil sample during soybean cultivation period in 2001 (per gram of soil dry weight).

c. Soil Mesofauna

The effects of long-term manure application and reduced tillage in cultivated fields on the population dynamics of soil mesofauna were examined. Soil samples were collected from the four types of cultivated field, i.e. TM, RTM, TF and RTF, managed with different combinations of manure or chemical fertilizer application and conventional or reduced tillage. Soil mites and collembolans were extracted using Tullugren apparatus; they were classified into several groups mainly on the basis of family in order to enumerate the density for each group.

Fungivorous species of both mites and collembolans are dominant in every field. Densities of both mites and collembolans are higher in the manure plots than the chemical fertilizer plot. In particular, the plot with combined manure application and reduced tillage has remarkably higher densities of fungivorous mites and collembolans than plots with other combinations of nutrient application and tillage.

These results correlate positively with the fungal biomass in each treatment plot, and the density of predatory mites also correlates positively with the density of fungivorous species. Hence, manure application or reduced tillage will maintain the population of predatory macrofauna at a higher trophic level by increasing mites and collembolans that serve as the food for macrofauna.

d. Soil Macrofauna and Insect Pests

Soil macro-arthropods and flying invasive insects were caught by using the pitfall trap or funnel-shaped trap in the above four types of field for enumeration, in addition to carrying out the aforementioned examination of soil mesofauna. The trapped organisms were classified into several functional groups based on their food habits before their numbers were enumerated.

The total number of arthropods tends to be greater in the field with reduced tillage than the conventionally tilled field, and greater in the field with manure application than the field with chemical fertilizer application. This tendency depends on the number of crickets (Gryllidae) or "other insects" that are neither herbivores nor predators. The number of phytophagous arthropods of two orders, i.e. Lepidoptera and Hemiptera, in the field with chemical fertilizer application is higher than in the field with manure application. Additionally, spiders, which have predatory food habits, are more numerous in the manured field than in the field with chemical fertilizer application. Few centipedes and earwigs (*Labidura riparia japonica*) were caught in the reduced tillage field, probably because of physical differences in the soil.

The number of phytophagous insect pests is reduced as a result of long-term manure application and reduced tillage. However, there is no correlation between the number of herbivores and that of predators. In nature, organisms may excrete volatile substances that function as allelochemicals to fend off natural enemies. In several cases, herbivorous insects were observed to perceive and avoid the volatile substances released from the general predaceous insect. Hence, such a relationship between organisms and their natural enemy, e.g. insect pests, may prevent the outbreak of specific insect pests that are harmful to crops. The manure application and the reduced tillage management are effective in maintaining the population of arthropods, which are prey of the general predaceous insects, so that sufficient allelochemicals will be produced by arthropods to prevent the density of harmful insects from becoming too high in the field.

3.3. BIOMASS PRODUCTION AND BIODIVERSITY IN PADDY FIELD ECOSYSTEMS

Takashi Motobayashi

Since paddy fields are constantly flooded with irrigation water, many characteristics (soil environment, biota, nutrient cycle, etc.) associated with paddy field ecosystems are different from those associated with upland field ecosystems. Paddy fields are typically puddled; the flooded soil is plowed and harrowed to destroy soil aggregates. Puddling eliminates water transmission pores, reduces the percolation rate, and drastically reduces gaseous exchange between the soil and the atmosphere. The puddling also results in the formation of a soil hardpan that restricts downward water flow to prevent the loss of nutrients by leaching.

This practice also assists in transplanting rice seedlings and controlling weeds (Sharma and De Datta, 1985; Buresh and De Datta, 1990). Generally, rice plants are cultivated in paddy fields during late spring or early summer in Japan, and the fields are most likely irrigated with river water through channels during the growing season; during the fallow season, the irrigation water is drained from the paddy fields so that the paddy soil becomes dried. Therefore, the environment of wetland (such as shallow marsh) and upland appears periodically in paddy fields. The unique biological community established in the paddy fields has adapted to this rapidly changing environment. These communities in paddy field systems during rice plant growth and after rice harvest play an important role in building up and maintaining paddy field ecosystems.

3.3.1. Rice Production in Paddy Fields Characteristic of Paddy Field Ecosystems

a. Nutrient Supply by Irrigation Water

Japanese rivers, which tend to be shorter and swifter than continental rivers, contain relatively smaller amounts of various dissolved mineral nutrients as compared with the latter. However, silica is relatively abundant in Japanese rivers; many other types of dissolved mineral nutrients for rice plant growth are moderate (see Tables 3.2 and 3.3). These mineral nutrients accumulate in paddy soil and are absorbed by rice plants.

b. Nutrient Supply and Cycling by Microorganisms (Photoplankton, Bacteria)

Soil fertility mainly depends on the availability of nutrient elements such as nitrogen, phosphate, and potassium contained in the soil. Among the major elements, nitrogen is of primary significance to affect crop production. The rice plant obtains a great portion of its nitrogen requirement from the organic nitrogen pool in the soil (Broadbent, 1979).

Nitrogen fixation is an important process to supply nitrogen for rice plants. Cyanobacteria and other bacteria (*Azospirillum, Azotobacter, Clostridium,*

TABLE 3.2 Comparison of Water Quality Between Japanese and World River (Takayama, 1986)

	Ca	Mg	Na	K	CO_3	SO_4	Cl	$NH_4 + NO_3$	SiO_2	Fe_2O_3	PO_4
Japanese river	8.8	1.9	6.7	1.2	15.2	10.6	5.8	0.31	19.0	0.34	0.02
Average of the world	15.0	4.1	6.3	2.3	28.3	11.2	7.8	1.0	13.1	0.96	—

TABLE 3.3 Balance of Some Mineral Nutrition in Paddy Fields (Sekiya, 1992)

Factor	N	PO_4	K	$CaCO_3$	Mg	SiO_2
Input						
Irrigation water	0.48	0.03	2.79	22.5	4.5	30.5
Fertilizer	10.54	10.93	9.71	45.0	4.5	52.5
Rice straw	3.51	1.5	11.6	3.0	1.7	70.0
Precipitation	1.34	0.27	0.34	0.91	0.08	–
Nitrogen fixation	2.0	–	–	–	–	–
Total input	17.9	12.7	24.6	71.4	10.8	141.3
Output						
Paddy water	0.53	0.06	0.44	1.52	0.29	–
Percolation	1.5	0.15	3.75	75.0	9.0	30.0
Rice plant	9.8	4.64	13.4	3.57	2.64	100.0
Denitrification	4	–	–	–	–	–
Total output	15.8	4.85	17.6	80.1	11.9	130.0
Balance	2.1	7.8	7	−8.7	−1.1	11.0

Beijerinckia, etc.) present in the paddy water or the paddy soil are capable of fixing nitrogen from the air contained in the paddy water (Firoza et al., 2002; Choudhury and Kennedy, 2004). The total nitrogen fixation capability of paddy fields has been estimated to be $1–6.8 \text{ g m}^{-2}$ during the cropping season without artificial nutrient application (Hirano, 1958; Nishigaki and Shioiri, 1959).

Photoplankton takes in nitrogen and phosphate in paddy water and produces various organic substances. Although there is no exact information on the synthesis of photoplankton biomass in experimental plots, approximately $5–6 \text{ kg DW ha}^{-1}$ is estimated to exist in the paddy field during the early stage of rice cultivation (Ichimura, 1954; Kimura et al., 2004).

Actually most organic matter applied artificially to the paddy soil annually is residue of weeds and rice stubble. The amount of weed and rice stubble biomass that was incorporated by spring plowing ranges from 1300 to 2300 kg DW ha^{-1} and 930 to 1500 kgDW ha^{-1} respectively (Kimura et al., 2004).

As mentioned above, many types of nutrients for rice plants are stored in the paddy soil as organic matter; it is then decomposed by various microorganisms, mainly bacteria, in the paddy soil. Because of a shortage of oxygen in the paddy soil caused by flooding water, the decomposition rate of the organic matter in the paddy soil during the flooding period is slower than that in the upland fields. As a result of accumulation of organic matter in the paddy soil, the surplus nutrients are recycled back to the paddy water while maintaining the fertility of the paddy soil so that outflow of the surplus nutrients from the paddy field to rivers and lakes is prevented.

c. Nitrogen Metabolism During Flooding Season

During most flooding periods, paddy fields develop a "reduced soil layer" as a result of active microbial metabolism. Surface paddy water contains oxygen dissolved from the atmosphere or produced by the photosynthetic activities of hydrophytes. This oxygen is supplied to surface soil through diffusion or with the downward movement of water; an "oxidized soil layer" thus develops in the uppermost part of the paddy soil (Takai and Kamura, 1966). The thickness is generally a few millimeters for the oxidized soil layer and 10–15 cm for the underlying reduced soil layer. The rice plant obtains a great part of its nitrogen requirement from the organic nitrogen pool of the soil (Broadbent, 1979). The organic nitrogen contained in the soil is mineralized to ammonium (NH_4^+) by microorganisms; the ammonium is stable under reduced soil conditions and adheres to the soil particles. Rice plants take up ammonium from the soil by absorption through their roots. However, if the ammonium exists in the oxidized soil layer, it is first converted to nitrite (NO_2^-), primarily by ammonia-oxidizing bacteria, and then oxidized into nitrate (NO_2^-) by nitrite-oxidizing bacteria. Nitrate is more soluble in water than nitrite; it moves from the oxidized soil layer to the reduced soil layer with the downward movement of the water. In the reduced soil layer, nitrate is denitrified by denitrifying bacteria into nitrogen gas (N_2), which eventually escapes to the atmosphere in a process known as denitrification (Araragi, 1978). The rice paddy soil is known to have strong denitrifying activities (Nishimura et al., 2004); hence, about 20–25% of nutritional nitrogen for rice plant growth is lost to the atmosphere (Broadbent and Tusneem, 1971; Koyama et al., 1973) (see Table 3.3). However, application of nitrogen fertilizer to the reduced soil layer can suppress the microbial denitrification and improve the efficiency of fertilizer. Therefore, the uniform application of fertilizer to topsoil has been developed and implemented extensively, with the excess nitrogen being removed from the irrigation water by denitrification. Thus, some research on using paddy fields as a system to treat nutrient-polluted river water has been conducted (Zhou and Hosomi, 2008).

d. Injury by Continuous Cropping

The injury caused by continuous cropping is a serious problem for upland field crops. It is caused by the deficiency of particular nutrients in the soil, accumulation of toxic substances in the soil, and prolific growth of soil-borne pest organisms (pathogenic microorganisms, nematodes) in the fields stimulated by continuous cropping. In particular, the increase of soil-borne pest organisms is the most severe problem caused by continuous cropping (Sekiya, 1992).

The soil in upland fields is continuously aerobic. Thus, aerobic microorganisms including plant pathogenic fungi or nematodes dominate in the community of microorganisms inhabiting the soil. As a result, crops cultivated continuously in the same field may be vulnerable to infestation by soil-borne pathogens.

As mentioned above, the reduced soil layer develops in the paddy soil during the flooding period, whereas the oxidized soil layer develops during the non-flooding period. Therefore, during the flooding period, aerobic microorganisms including plant pathogenic microorganisms become extinct or inactive, and anaerobic microorganisms dominate in the soil microorganism community in the paddy field. During the non-flooding period, anaerobic microorganisms are replaced by aerobic microorganisms in the soil microorganism community.

Thus, a particular soil-borne pathogenic microbial population may not develop continuously in paddy fields. Moreover, the accumulation of toxic substances in the soil may be suppressed by flooding water. Therefore, rice plants may be cultivated in the same paddy field continuously without causing injury by continuous cropping. The Japanese paddy field system is thus considered an effective and sustainable production system because nutrients are derived from irrigation water and converted into appropriate forms by various microorganisms. Therefore, the paddy field achieves effective recycling of surplus nutrients in paddy water and suppression of injury by continuous cropping. However, the paddy fields may operate more effectively if rotation cropping of rice and other field crops (such as soybean, wheat, vegetables) is implemented (Sekiya, 1992).

3.3.2. Biodiversity of Paddy Fields in Japan

For a long time in Japan, the paddy field has played important roles as agricultural land to produce rice and also to provide a habitat for various animals and plants. Example species and numbers of animals and plants detected in paddy fields are discussed in the following section.

Kobayashi et al. (1973) reported that 450 invertebrate species were detected in paddy fields in the Tokushima prefecture; 11 species of frogs were reported to use paddy fields as a breeding site (Maeda and Matsui, 1989). Additionally, 24 species of fish (Saito et al., 1988), 101 species of bird (Hidaka, 1998), and 186 species of plants (Kasahara, 1951) live in paddy fields either temporary or continuously. The paddy field ecosystem has more abundant species richness or species diversity than any other ecosystem. However, these data may have methodological or taxonomical problems (Hidaka, 1998).

Moriyama (1997) hypothesized that Japanese paddy fields were mostly developed in riverside wetlands; however, Japanese rice cultural systems had maintained the environment of the original wetlands before paddy fields were developed. Therefore, the paddy field has played a role of alternative habitats for organisms originally living in shallow wetlands located along the riverside.

Hidaka (1998) indicated that the Japanese paddy fields are temporal and spatial stable wetlands, and have higher sustainability than the original wetlands. He also suggested that there are great varieties in the time to flood the paddy fields, water content of paddy soils before flooding among various paddy fields, and various species may have been adapted to take advantage of these diverse paddy environments.

From the viewpoint of the life cycle of species using paddy fields, some species complete their life cycle in the paddy field; however, most species need the surrounding environments such as irrigation ponds and forests or groves for completing their life cycle. For example, *Sympetrum*, a dragonfly, uses paddy fields to lay eggs in autumn; the eggs hatch at puddling time next spring, and the larvae grow in paddy water during spring and early summer. Adult dragonflies then move from the paddy to adjacent or highland forests to live through summer. At the beginning of autumn, they return to paddy fields for breeding (Taguchi and Watanabe, 1985).

Hibi et al. (1998) and Saijo (2001) indicated that many species of predatory aquatic insects such as water stick, which is a Japanese water scavenger beetle, live in irrigation ponds adjacent to paddy fields. In other words, the adults of these insects migrate from ponds to puddled paddy fields for breeding and oviposition. The larvae grow in paddy water, where they prey upon abundant spices such as small invertebrates, tadpoles, and small fishes during summer. Adult dragonflies of the next generation return to irrigation ponds to hibernate at the beginning of autumn.

The landscape of "Satochi-Satoyama" consists of heterogeneous ecosystems including paddy fields, ridges between paddy fields, rivers, channels, forests, and upland fields. This complicated and diverse environment is one of the primary factors supporting various species in the paddy fields in Japan (Hidaka, 1998; Kiritani, 2000).

The total area of paddy fields in Japan is 2,600,000 ha, which accounts for 7% of the nation's rural area and 20% of total flat land. Thus, the conservation of biodiversity in paddy fields is very important to conserve the biodiversity in Japan.

However, the serious problem caused by the disappearance of various species, especially aquatic species, inhabiting the paddy fields in Japan has been recognized recently. Causes of this disappearance include the drier paddy soils during the non-flooding period, disappearance of irrigation ponds due to adjustment of paddy fields, pollution of irrigation water by excessive use of pesticides and/or chemical fertilizers, acceleration of cultivation period of rice, and increasing non-cultivated or abandoned paddy fields. Hidaka (1998) has classified the rice cultivation practices in relation to the temporal and spatial stability of paddy fields. He suggested that the present paddy field environments are becoming unstable temporally and spatially, and therefore some permanent residents such as fishes and shellfishes that are vulnerable to unstable environment cannot survive in those unstable paddy field environments. Moreover, he hypothesized that migrants including Japanese crested ibis, stork, Japanese water beetle and giant water bug are more resistant to the unstable environment but they do not inhabit the paddy field. This is because their living cannot be supported by the decreasing population of permanent resident species in paddy fields. However, there has been little research conducted on the relation between the stability of paddy field environments and the population of individual species populations. More studies are needed on this in the future.

3.3.3. Coexistence of Biomass Production and Maintenance of Biodiversity in Paddy Fields

a. Increasing Non-Cultivation and Abandoned Paddy Fields

Rice consumption has been declining in Japan, and rice production has been restricted to a balanced rice supply and demand since 1970. The rice production adjustment policy has been continued by the National Government. In 2011, the area of the paddy fields without rice cropping reached about 900,000 ha (MAFF, 2012a) that is about 35% of total paddy fields in Japan.

The declining and aging farmer population in recent years causes a substantial number of paddy fields, especially those located in mountainous and hilly areas, to be laid fallow (MAFF, 2010). On the other hand, diversion of agricultural lands to residential or business sites and the abandoned paddy fields are increasing in suburban areas (see Figures 3.17 and 3.18).

As mentioned earlier, the Japanese paddy field system is an effective and sustainable food and biomass production system, and increasing fallow paddy fields implies that a large portion of the agricultural production potential in Japan is lost.

Moreover, most of the non-cultivation or abandoned paddy fields are not periodically flooded and the aquatic organisms living in paddy fields disappear.

FIGURE 3.17 Non-cultivation or abandoned paddy fields located in mountainous and hilly areas.

FIGURE 3.18   Non-cultivation or abandoned paddy fields located in residential or business areas.

Thus, increasing non-cultivation or abandoned paddy fields is becoming one of the serious problems causing declining biodiversity in Japanese paddy eco-systems to adversely affect the national land conservation (flood control, soil conservation) and maintenance of the agricultural landscape in Japan.

b. Production of Whole Crop Silage (WCS), High-Yielding Rice in Non-Cultivation Rice Paddy

As mentioned in the above section, the area of paddy fields without rice cropping has reached about 1,000,000 ha. On the other hand, the percentage of self-sufficient food supply is only 39% on a calorie basis in Japan (MAFF, 2012b). Furthermore, 74% of the feed for livestock is imported from overseas (MAFF, 2012c). This striking shortage of self-sufficient supply against demand of feed causes serious problems for the livestock industry, similar to other problems including damage to the environment by animal waste and the risk of livestock diseases through imported feed. The enhancement of self-sufficient feed supply and the establishment of a recycling system between the live-stock industry and agriculture are strongly recommended in order to solve these problems. The Ministry of Agriculture, Forestry and Fisheries (Japan) started a research project in 1999 to utilize rice as whole crop silage (WCS). As a result of this project, some new rice cultivars that have high productivity and are suitable for growing low-cost rice grain for WCS uses have been developed (Sakai et al., 2003). The area of paddy field cropping for WCS reached about 23,000 ha in 2011 (MAFF, 2012d). Furthermore, the production of rice grain for livestock feed has been increased. The area of paddy field cropping for this type of rice reached about 34,000 ha in 2011 (MAFF, 2012c).

Incidentally, bioethanol is widely used in Brazil and the USA as a biofuel additive for gasoline. The bioethanol can be made from very common crops such as corn and sugar cane. However, there have been concerns about producing ethanol fuel from food crops because the competition between bioethanol and food production may cause food shortages, especially for poor nations, in the future.

In Japan, because rice production has been in excess, using the surplus rice grain as raw material for bioethanol seems to be logical for the time being but may lead to future shortage problems. Hence, developing cellulosic ethanol production technique using rice straw for bioethanol production will be a preferred alternative method to produce bioethanol.

Thus, the rice cropping for WCS or bioethanol in non-cultivation or abandoned paddy fields may sustain the productive potential of Japan's paddy fields in addition to conserving habitats for organisms inhabiting the paddy fields. Additionally, this practice may also be an effective means for maintaining the national land conservation and agricultural landscape in Japan.

c. Subjects of WCS and High-Yielding Rice Production

At present, the production cost of WCS is 80 yen kg^{-1}, twice as much as the cost of importing hay from overseas. The rice grain production cost is 133 yen kg^{-1}, which is about three times the cost of importing corn from overseas. Thus, the breeding of new cultivars that have higher-yielding grain or whole biomass will be one of the primary technological innovations for achieving this objective (Horie, 2009). Nemoto (2010) indicated that, in addition to high yielding, the cultivars also need to be lodging resistant, disease resistant, insect and pest resistant, and suitable for direct sawing.

Moreover, with regard to rice cultivation practices for low-cost production, direct sawing that reduces the use of chemicals such as fertilizers and pesticides needs to be developed by conducting future research. The construction of a nutrient recycling system including paddy fields and livestock is also needed.

REFERENCES

Araragi, M. (1978). Denitrification in paddy field. In K. Kawaguchi (Ed.), *Soil science of paddy field* (pp. 256–263). Tokyo: Kobunnsha [in Japanese].

Bongers, T. (1990). Comparison of soil surface arthropod populations in conventional tillage and old field systems. *Agro-Ecosystems, 8,* 247–253.

Broadbent, F. E. (1979). Mineralization of organic nitrogen in paddy soils. In *Nitrogen and rice* (pp. 105–118). Los Banos, Philippines: IRRI.

Broadbent, F. E., & Tusneem, M. E. (1971). Losses of nitrogen from some flooded soils in tracer experiments. *Soil Science Society of America Proceeding, 35,* 922–926.

Buresh, R. J., & De Datta, S. K. (1990). Denitrification losses from puddled rice soils in the tropics. *Biology and Fertility of Soils, 9,* 1–13.

Choudhury, A. T. M. A., & Kennedy, I. R. (2004). Prospects and potentials for systems of biological nitrogen fixation in sustainable rice production. *Biol. Fertil. Soils, 39,* 219–227.

Clarholm, M. (1994). The microbial loop in the soil. In K. Ritz, J. Dighton, & K. E. Giller (Eds.), *Beyond the biomass compositional and functional analysis of soil microbial communities* (pp. 221–230). Chichester: John Wiley.

Crossley, D. A., Jr. (1977). The roles of terrestrial saprophagous arthropods in forest soils. In W. J. Mattoson (Ed.), *The role of arthropods in forest ecosystems* (pp. 49–56). New York: Springer.

Duraiappah, A. K., Nakamura, K., Takeuchi, K., Watanabe, M., & Nishi, M. (Eds.). (2012). *Satoyama-Satoumi ecosystems and human well-being: Socio-ecological production landscapes of Japan.* Tokyo: United Nations University Press.

Ekelund, F., & Ronn, R. (1994). Notes on protozoa in agricultural soil with emphasis on heterotrophic flagellates and naked amoebae and their ecology. *FEMS Microbiology Reviews, 15,* 321–353.

Ekelund, F., Ronn, R., & Griffiths, B. S. (2001). Quantitative estimation of flagellate community structure and diversity in soil samples. *Protist, 152,* 301–314.

Firoza, A., Zakaria, Z. A., Tahmida, Z. N., & Mondal, R. (2002). Increase in macro-nutrients of soil by the application of cyanobacteria in rice production. *Pakistan Journal of Biological Sciences, 5,* 19–24.

Foissner, W. (1987). Soil protozoa: Fundamental problems, ecological significance, adaptations in ciliates and testaceans, bioindicators, and guide to the literature. *Proger. Protistology, 2,* 69–212.

Griffiths, B. S. (1994). Soil nutrient flow. In J. F. Darbyshire (Ed.), *Soil protozoa* (pp. 65–91). Wallingford, Oxon: CAB International.

Hibi, N., Yamamoto, T., & Yuma, M. (1998). Life histories of aquatic insects living in man-made water systems located around paddy fields. In T. Ezaki, & T. Tanaka (Eds.), *The preservation in water environments: From the viewpoint of biodiversity* (pp. 111–123). Tokyo: Asakura shoten [in Japanese].

Hidaka, K. (1998). The biodiversity and the preservation in paddy fields. In T. Ezaki, & T. Tanaka (Eds.), *The preservation in water environments: From the viewpoint of biodiversity* (pp. 125–148). Tokyo: Asakura shoten [in Japanese].

Hirano, T. (1958). Studies on blue green algae (part 2). Study on the formation of fumus due to the growth of blue green algae. *Bulletin of the Shikoku Agricultural Experiment Station, 4,* 63–74 [in Japanese with English summary].

Hiroki, S. (Ed.). (2002). *Ecology of Satoyama.* Nagoya: Nagoya University Press.

Horie, T. (2009). Reductionistic and integrated approach in agricultural research. *Japanese Journal of Crop Science, 78,* 399–406.

Ichimura, S. (1954). Ecological studies on the plankton in paddy fields. I. Seasonal fluctuations in the standing crop and productivity on plankton. *Japanese Journal of Botany, 14,* 269–274.

Inui, T. (1996). The history and use of lowland forest in Kanto plain. *Shinrin-Kagaku, 18,* 15–20.

Ishibashi, N., & Choi, D. R. (1991). Biological control of soil pests by mixed application of entomopathogenic and fungivorous nematodes. *Journal of Nematology, 23,* 175–181.

Kamijo, T., Kitayama, K., Sugawara, A., Urushimichi, S., & Aasai, K. (2002). Primary succession of the warm-temperate broad-leaved forest on a volcanic Island, Miyake-jima, Japan 'jointly worked'. *Folia Geobotanica, 37,* 71–91.

Kasahara, Y. (1951). Studies on the number of arable weeds and their distribution in Japan. Part IV. *Nogaku- kenkyu, 39,* 143–154 [in Japanese].

Kimura, M., Murase, J., & Lu, Y. (2004). Carbon cycling in rice field ecosystems in the context of input, decomposition and translocation of organic materials and the fates of their end products (CO_2 and CH_4). *Soil Biology and Biochemistry, 36,* 1399–1416.

Kiritani, K. (2000). Integrated biodiversity management in paddy fields: Shift of paradigm from IPM to IBM. *Integrated Pest Management Reviews, 5,* 175–183.

Koyama, T., Chammek, C., & Niamsrichand, N. (1973). Nitrogen application technology for tropical rice as determined by field experiements using ^{15}N tracer technique. *Technical Bulletin (Tropical Agr. Res. Cen. Tokyo), 3.*

Kobayashi, T., Noguchi, Y., Hiwada, T., Kanayama, K., & Maruoka, N. (1973). Studies on the arthropod associations in paddy fields, with particular reference to insecticidal effect on them. I. General composition of the arthropod fauna in paddy fields revealed by net-sweeping in Tokushima Prefecture. *Kontyu, 41*, 359–373 [in Japanese with English summary].

Kuroda, K. (2001). Response of *Quercus* sapwood to infection with the pathogenic fungus of a new silt disease vectored by the ambrosia beetle *Platypus quercivorus*. *Journal of Wood Science, 47*, 425–429.

Maeda, N., & Matsui, M. (1989). *Frogs and toads of Japan*. Tokyo: Bun-ichi Sogo shuppan [in Japanese with English summary].

MAFF (2010). <http://www.maff.go.jp/e/annual_report/2010/pdf/e_all.pdf>.

MAFF (2012a). <http://www.e-stat.go.jp/SG1/estat/List.do?lid=000001087149>.

MAFF (2012b). <http://www.maff.go.jp/j/press/kanbo/anpo/pdf/120810-01.pdf>.

MAFF (2012c). <http://www.maff.go.jp/j/chikusan/sinko/lin/l_siryo/pdf/haigou_2409.pdf>.

MAFF (2012d). <http://www.maff.go.jp/j/chikusan/kikaku/lin/pdf/megru_1207.pdf>.

Matsuzaki, I., & Itakura, J. (1992). Effect of collembola grazing on the reduction of rhizoctonia damping-off of vegetable seedlings in three types of toils under the laboratory conditions. *Annual Report of the Society of Plant Protection of North Japan, 43*, 133–134.

Moriyama, H. (1997). *What is protecting paddy fields?* Tokyo: Nobunkyo [in Japanese].

Nakamura, Y., Itakura, J., & Matsuzaki, I. (1991). Mycophagous meso soil animals from crop fields in Fukushima Pref. *Edaphologia, 45*, 49–54 [in Japanese].

Nemoto, H. (2010). The status and subjects of rice production for bio-ethanol materials in Japan. *Japanese Journal of Crop Science, 79*, 224–226.

Nishigaki, S., & Shioiri, M. (1959). Nitrogen cycles in the rice field soil: The effect of blue–green algae on the nitrogen fixation of atmospheric nitrogen in the water-logged rice soils. *Soil Plant Food, 5*, 36–39.

Nishimura, S., Sawamoto, T., Akiyama, H., Sudo, S., & Yagi, K. (2004). Mathane and nitrous oxide emissions from a paddy field with Japanese conventional water management and fertilizer application. *Global Biological Cycles, 18*, GB2017.

Ronn, R., Ekelund, F., & Christensen, S. (1995). Optimizing soil extract and broth media for MPN-enumeration of naked amoebae and heterotrophic flagellates in soil. *Pedobiologia, 39*, 10–19.

Saijo, H. (2001). Seasonal prevalence and migration of aquatic insects in paddies and an irrigation pond in Shimane Prefecture. *Japanese Journal of Ecology, 51*, 1–11 [in Japanese with English summary].

Saito, K., Katano, O., & Koizumi, K. (1988). Movement and spawning of several fresh water fishes in temporary waters around paddy fields. *Japanese Journal of Ecology, 38*, 35–47 [in Japanese with English summary].

Sakai, M., Iida, S., Maeda, H., Sunohara, Y., Nemoto, H., & Imbe, T. (2003). New rice varieties for whole crop silage use in Japan. *Breeding Science, 53*, 271–275.

Sato, A., & Seto, M. (2000). Fungal or bacterial biomass in volcanic soil with various land used and soil treatments (in Japanese). *Soil Microorganisms, 54*, 23–30.

Schowalter, T. D. (2000). Decomposition and pedgenesis. In *Insect ecology* (pp. 361–388). New York: Academic Press.

Seastedt, T. R., & Crossley, D. A., Jr. (1984). The influence of arthropods on ecosystems. *Bioscience, 34*, 157–161.

Sekiya, S. (1992). *Function of paddy field*. Tokyo: Ienohikari-kyoukai [in Japanese].

Sharma, P. K., & De Datta, S. K. (1985). Effect of puddling on soil physical properties and processes. In *Soil physics and rice* (pp. 217–234). Los Banos: Philippines. IRRI.

Smith, S. E., & Read, D. J. (1997). *Mycorrhizal symbiosis*. San Diego: Academic Press.

Taguchi, M., & Watanabe, M. (1985). Ecological studies of dragonflies in paddy fields surrounded by hills. IV. Spatial distribution of Sympetrum eroticum eroticum in relation to the seasonal fluctuations of shaded area. *Bulletin of the Educational Faculty of Mie University, 38*, 57–67 [in Japanese].

Takai, Y., & Kamura, T. (1966). The mechanism of reduction in waterlogged paddy soil. *Folia Microbiology, 11*, 304–313.

Takayama, S. (1986). *"Guidebook of chronological science table" Natural history of river.* Maruzen, Tokyo, pp. 160–161 (in Japanese).

Toda, H. (2004). Effects of slash-removal on the nutrients circulation in forest ecosystems. *Shinrin-Kagaku, 40*, 33–38.

Toda, H., Haibara, K., & Arai, M. (1991). Nutrient circulation of a small watershed under an established Sugi (*Cryptomeria japonica*) and Hinoki (*Chamaecyparis obtusa*) stand. *Bulletin of Experiment Forest, Tokyo University of Agriculture and Technology, 28*, 1–22.

Toda, H., Hanaoka, K., Kishimoto, K., Haibara, K., & Kametani, Y. (2007). Mapping and mechanisms of acid buffer effect in volcanic ash on surface soil in Miyake Island. *Journal of Japanese Society for Revegetation Technology, 33*, 21–26.

Toda, H., Hanaoka, K., Ebara, M., Sasaki, R., Haibara, K., Kametani, Y., & Choi, D. S. (2008). Effects of forest soil properties on revegetation after volcano eruption 2000 in Miyake Island. *J. Jpn. Soc. Reveget. Tech., 34*, 21–26.

Trofymow, J. A., & Coleman, D. C. (1982). The role of bacterivorous and fungivorous nematodes in cellulose and chitin decomposition. In D. W. Freckman (Ed.), *Nematodes in soil ecosystems* (pp. 117–138). Austin, TX: University of Texas Press.

Wardle, D. A. (1995). Impacts of disturbance on detritus food webs in agro-ecosystems of contrasting tillage and weed management practices. In (Series Ed.) & M. Begon, & A. H. Fitter (Vol. Eds.). *Advances in ecological research*, Vol. 26 (pp. 105–185). New York: Academic Press.

Whitford, W. G. (2000). Keystone arthropods as webmasters in desert ecosystems. In D. C. Coleman, & P. Hendrix (Eds.), *Invertebrates as webmasters in ecosystems* (pp. 25–42). Wallingford, Oxon: CAB International.

Zhou, S., & Hosomi, M. (2008). Nitrogen transformations and balance in a constructed wetland for nutrient-polluted river water treatment using forage rice in Japan. *Ecological Engineering, 32*, 147–155.

Chapter 4

Production Technology for Bioenergy Crops and Trees

Tadashi Hirasawa, Taiichiro Ookawa, Shinya Kawai, Ryo Funada and Shinya Kajita

Chapter Outline

Research Approaches to Sustainable Biomass Systems. http://dx.doi.org/10.1016/B978-0-12-404609-2.00004-0

4.1. PHOTOSYNTHESIS AND BIOMASS PRODUCTION IN ENERGY CROPS

Tadashi Hirasawa

4.1.1. Introduction

Annual crop plants that contain a large amount of oil, starch, and sugar have the potential to be converted into biomass energy; sunflower and rape seeds, as well as corn grain and sugar cane stem, are being used to produce oil and ethanol. The yield needs to be improved in order to lower the production cost of the biomass energy. The grain yield can be considered as a product of biomass yield and harvest index (i.e. yield/biomass). Modern, semi-dwarf varieties of crops such as rice and wheat generally have very high harvest index values and decent grain yield (Evans, 1993). However, harvest indices for these plants have approached the theoretical maximum level (Mann, 1999). Increasing the total biomass production is currently considered essential for further improving the grain yield of crop plants.

Recently, more crop land has been diverted to growing grain for producing fuel that causes an increase in food price. If non-edible plant parts such as rice or wheat straw, and corn stalk, among many others, can be used for producing ethanol effectively, plants can be used to produce both food and fuel. In other words, all the solar energy stored in plants and the biomass produced can be used more effectively and efficiently. Additionally, any land that is not feasible for growing food crops may be used for harvesting fuel crops so that problems caused by conflicting uses of grain as food and energy can be solved.

4.1.2. The Concept of Biomass Production

More than 90% of plant dry matter is derived from photosynthesis; hence the yield of plant biomass (BY) can be expressed as (Hay and Porter, 2006):

$$BY = Q \times Ic \times \varepsilon \tag{4.1}$$

where Q = the total quantity of incident solar radiation received over the growing period of the crop, Ic = the fraction of Q that is intercepted by the canopy, and ε = the overall photosynthetic efficiency of the crop.

For annual plants, the crop growth rate (CGR), expressed as dry matter accumulated per unit land area per unit time, is low right after planting, and it then increases rapidly because the leaf area index (LAI), defined as the leaf area per unit land area, increases exponentially (Figure 4.1A). It reaches maximum level when the leaf area index is equal to the critical LAI. Such changes of CGR with respect to time to increase biomass production conform to a sigmoid curve (Figure 4.1B). Subsequently, the growth rate decreases in proportion to lessening LAI. Based on equation (4.1) and Figure 4.1, images of the plant canopy and the individual leaves that contribute to biomass production for a given growing season can be constructed as shown in Table 4.1.

a. Ic, the Fraction of Q that is Intercepted by the Canopy

If land is completely covered by a plant canopy, the interception of solar radiation increases because of high LAI (Gardner et al., 1985), resulting in a high crop

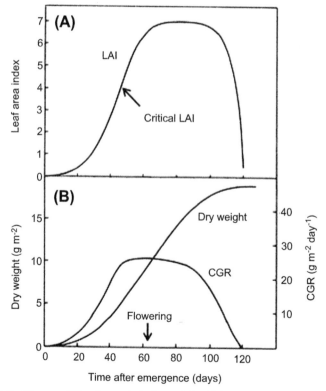

FIGURE 4.1   Diagram of the changes in leaf area index (LAI), crop growth rate (CGR), and above-ground dry weight of annual, determinate grain crop *(Adapted from Gardner et al., 1985.)*

TABLE 4.1 Characteristics Affecting Increased Dry Matter Production in High-Yielding Rice Varieties

Growth stage	Characteristics	Corresponding term in eq. (4.1)	High-yielding cultivars with these characteristics*
Tillering stage	Rapid increase in leaf area	Ic	Nanjing 11 and high-yielding hybrid
After the panicle formation stage	(1) Canopy structure well suited to light penetration into canopy (2) Canopy structure well suited to CO_2 diffusion into canopy	ε	(1) Milyang 23 and Takanari (2) Leaf Star
Late ripening stage	Large leaf area during ripening	Ic	Akenohoshi
Throughout plant growth	Photosynthesis by canopy leaves (1) High photosynthetic capacity (2) High rate of photosynthesis in the afternoon (3) High rate of photosynthesis during ripening	ε	(1) Takanari (2) Akenohoshi and Takanari (3) Akenohoshi

Adapted from Ishihara (1996).

growth rate (Horie and Sakuratani, 1985; Hay and Porter, 2006). The energy plant, i.e. *Miscanthus*, develops a leaf canopy earlier and maintains it later, which contributes to its prolific production of biomass as compared with maize (Dohleman and Long, 2010). The F_1 rice varieties show vigorous growth during the early stages (Ishihara, 1996; Peng et al., 1998) so that biomass production of F_1 is large, although the mechanism of its superior growth is presently unclear. This observation can be used to increase LAI during the early stage by increasing planting density (Box 4.1). However, increasing planting density sometimes causes severe lodging and deteriorations in canopy architecture that lead to reduction of biomass and grain production. Hence, crop plants with lodging resistance must be considered when increased planting density is practiced.

b. ε, the Overall Photosynthetic Efficiency of the Crop

Once the land is completely covered by leaves, the canopy architecture becomes an important factor that affects light penetration and CO_2 diffusion into

> **BOX 4.1 Direct Sowing of Rice with an "Air-Assisted Drill" for Labor-Saving and High-Yielding Rice Growing**
>
> In Japan, direct seeding is under consideration for saving capital and labor costs in rice cultivation. Dry matter production and grain yield can be increased in lodging tolerant rice when it is grown under the conditions of high hill density because of vigorous early growth and the improved light-intercepting characteristics after heading (San-oh et al., 2004). An "air-assisted drill" (AAD) has been developed to improve the efficiency of direct seeding (Chapter 6). Grain yield and dry matter production of rice direct seeded with the AAD (AAD plants) are compared with conventionally transplanted plants (CT plants) and directly hill-seeded plants (HS plants).
>
> AAD plants were planted at inter-row spacings of 30 and 15 cm at a rate of approximately 50 kg ha^{-1}. CT and HS plants were grown at a density of 22.2 hills m^{-2} (30 cm × 15 cm) with three plants per hill for CT and five plants per hill for HA in 33.3 hills m^{-2} (30 cm × 10 cm).
>
> The AAD plants grew in an approximately 12-cm-wide zone along each row, which was rather different from that observed in CT and HS plants. Both AAD and HS plants have higher number of tillers and more rapid interception of solar radiation by the canopy than CT plants. The light extinction coefficient of the canopy is smaller in the AAD and HS plants than in the CT plants after heading. The AAD plants has a larger number of crown roots to accumulate a larger amount of nitrogen at ripening; the net assimilation rate was kept high in spite of the larger LAI in the AAD plants. Finally, AAD plants have more grain yield and dry weight of aboveground parts at harvest than both HS and CT plants (Table 4.2) (Mukouyama et al., 2012).
>
> In conclusion, the broad casting seeding using a lodging resistant cultivar has the potential to reduce capital and labor costs while growing high-yielding rice. The air-assisted drill may be implemented to grow rice crops for energy as well as forage if the reliability in the seeding rate is improved.

the canopy, which in turn influence dry matter production. High rates of individual leaf photosynthesis from the seedling to the ripening stages are considered desirable in order to maximize biomass production. The maximum efficiency of the intercepted radiation is only 5–6% in many crop plants (Hay and Porter, 2006). If each of these parameters can be improved, the efficiency of crop plants to use solar energy will be improved significantly.

(i) Light Penetration into Canopy

The penetration of light into the canopy is quantified as the extinction coefficient k used in the Monsi and Saeki equation (Monsi und Saeki, 1953):

$$I = I_0 e^{-kL} \tag{4.2}$$

where I_0 = the irradiance above the canopy and I = the irradiance at a point in the canopy above which there is a leaf area index of L.

TABLE 4.2 Biomass Production, Grain Yield, and Yield Components in the AAD, HS, and CT Plants (Mukouyama et al., 2012) Means within a single column and in the same year followed by the same letter are not significantly different by LSD test (P<0.05).

Year	Plots	Panicle number (m^{-2})	Spikelet number (per panicle)	Spikelet number (10^3 m^{-2})	Ripened grain (%)	1000 grain weight (g)	Grain yield (g m^{-2})	Dry weight (g m^{-2})	Harvest index (%)
2009	AAD30	358a	178a	71.9a	78.6a	21.3a	984a	2297a	42.8a
	AAD15	380a	185a	64.0a	70.6a	21.0a	971a	2354a	41.3a
	CT	279b	204a	56.8a	76.5a	20.6a	861b	2078b	42.7a
2010	AAD30	257a	212a	54.0ab	88.4a	18.7a	950a	2012a	45.0a
	AAD15	280a	221a	61.2a	87.9a	17.8a	922a	2190ab	42.2a
	HS	270a	182b	49.2b	76.6b	18.8a	776b	1934ab	42.5a
	CT	245a	206a	50.7b	86.9a	18.6a	796b	1823b	43.5a
2011	AAD30	320a	205ab	65.4a	84.6a	19.9a	951a	2048a	49.0a
	AAD15	313a	213a	66.8a	82.1a	19.9a	946a	2104a	48.2a
	HS	326a	188b	61.2ab	71.1b	20.0a	809b	1893ab	41.8a
	CT	274b	189b	51.8b	86.5a	20.1a	768b	1719b	45.0a

Parameter k is a useful indicator of the light-intercepting characteristics of the canopy. The leaf angle is a major factor governing k in a given species. Leaf surface properties affecting reflection, leaf properties including thickness, leaf size and shape, and characteristics affecting the three-dimensional arrangement of leaves also affect k (Hay and Porter, 2006).

(ii) CO_2 Diffusion into Canopy

The CO_2 concentration decreases to some extent in the canopy during daytime on a sunny day (Hay and Porter, 2006). This may induce the reduction in the rate of photosynthesis of a leaf in the canopy because of low ambient CO_2 concentration as compared with the appropriate level for maximum photosynthesis, especially for C_3 plants. When plant populations with similar LAI are compared, CO_2 diffuses more effectively into a canopy consisting of taller plants than shorter plants because the former has smaller leaf area density (leaf area (m^2) per unit volume of air space (m^3)) (Kuroda et al., 1989). Increasing plant height sometimes causes severe lodging so that crop plants with lodging resistance should be used when increasing the height of canopy plants.

(iii) Rates of Individual Leaf Photosynthesis

We can consider the rate of individual leaf photosynthesis from three viewpoints: the capacity of photosynthesis, the rate of photosynthesis under senescence, and the rate of photosynthesis under abiotic stress.

The capacity of leaf photosynthesis is defined as the rate of photosynthesis measured for the fully expanded young leaf at optimum temperature with saturated light intensity and low vapor atmospheric pressure deficit without other abiotic stresses. The capacity of leaf photosynthesis increases with higher nitrogen content of a leaf (Makino, 2011). C_4 plants have a higher capacity of photosynthesis than C_3 plants due to the additional pathway of carbon assimilation in C_4 plants. Various species of C_3 plants have different photosynthetic capacity because of the properties of ribulose-1,5-bisphosphate carboxylase/oxygenase (Rubisco) (Makino, 2011). Varietal differences in photosynthetic capacity are also observed for plants of the same species because of the different leaf nitrogen and Rubisco levels.

The rate of leaf photosynthesis decreases with senescence; the reduction differs among various plant species and varieties. Growth conditions also affect the rate of photosynthesis due to senescence. It is well known that there is a close correlation between the rate of leaf photosynthesis and leaf nitrogen content or Rubisco content during senescence (Hidema et al., 1991). The reduction in the rate of leaf photosynthesis is delayed in plants with a high capacity of cytokinin synthesis (Soejima et al., 1995).

The reduction in the rate of photosynthesis due to abiotic stress is usually remarkable during daytime on a sunny day. Plants sometimes show a midday and afternoon reduction in the rate of leaf photosynthesis even under the conditions of sufficient soil moisture (Hirasawa and Hsiao, 1999). Specific and varietal differences in the reduction in leaf photosynthesis have also been observed.

4.1.3. Rice as a Potential Plant for Energy Crops in Japan

Plants adapting well to the conditions of a region usually grow with relatively high productivity. Plants with high productivity can be considered as a candidate for energy crops. Although rice plants originated in tropical regions, it is one of the highest productive crops in Japan at present. Therefore, rice may become an important potential energy crop in Japan.

In Japan, brown rice yields have increased by approximately 1000 kg ha^{-1} over the 20-year period since the late 1940s; this is equivalent to the increase in the rate of rice yield in Asia during the "green revolution". Since the 1970s, when Japan attained self-sufficiency in rice, further increase in grain yields has decreased in response to a shift in the eating habits of Japanese consumers so that growing rice crops emphasizes quality instead of quantity for consumption. During the same period, several high-yielding rice cultivars were released in Korea to produce more rice for human consumption. In the 1980s and 1990s, similar high-yielding rice cultivars were also released in Japan for various uses other than human consumption. These cultivars have approximately 20–30% higher yields than the regular cultivars grown in Japan for human consumption at that time (Ishihara, 1996). In the latter half of this section, characteristics of crops that can produce heavy biomass and methods to further increase their biomass production will be discussed using case studies.

4.1.4. Characteristics of the Rice Varieties Yielding Heavy Biomass

a. Characteristics of the Improved Commercial Rice

When compared to the yields of several Japanese rice varieties under similar growth conditions (Figure 4.2), the cultivars, which were released about 50 years ago and are still grown for human consumption, have higher yields than the leading varieties that were released more than 100 years ago in the Kanto area. This increase in yield over time is the direct result of breeding. For the cultivars used in the experiment, the increase in yield resulted from an increase in dry matter production rather than from an increase in harvest index. Comparing rice cultivars currently grown in Japan with older cultivars, Kumura (1995) made the following observations:

1. The extinction coefficient of the canopy is smaller in currently grown cultivars than in the older cultivars.
2. With nitrogen top-dressing, the level of leaf nitrogen, and therefore the rate of leaf photosynthesis, increases significantly in currently grown cultivars compared with older cultivars.

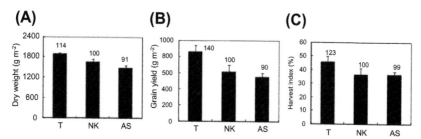

FIGURE 4.2   Comparisons of total dry weight of above-ground plant parts at harvest (A), grain yield (B), and harvest index (C) of varieties from different eras grown in a paddy field. Data are averages of three years (2002, 2005, and 2006); error bars indicate standard deviations. T: Takanari, a high-yielding *indica* rice variety released in 1990. NK: Average values for Nipponbare and Koshihikari, *japonica* varieties currently grown in Japan for human consumption that were released in 1963 and 1956 respectively. AS: Average values for Aikoku and Sekitori, *japonica* varieties released in 1882 and 1848 respectively. The number above each bar represents the value for that trait relative to that of NK, which was set at 100. *(Adapted from Taylaran et al., 2009.)*

3. Currently grown cultivars have a shorter culm and higher lodging resistance than the older cultivars.

b. Characteristics of the Most Productive Varieties

Compared with the varieties currently cultivated in Japan for human consumption, the high-yielding varieties that have been released in Korea and Japan since the 1970s have one or more superior characteristics that relate to canopy photosynthesis (Table 4.1). An *indica* variety "Takanari" is considered to be one of the most productive varieties in Japan, with higher grain yields and dry matter production consistently than any of the new or old commercial *japonica* varieties ever cultivated in Japan (Figure 4.2). Specifically, Takanari can produce 8–9 t grain and 19–21 t total dry matter per ha with the same rate of fertilizer application (San-oh et al., 2004; Taylaran et al., 2009). As shown in Table 4.1, the superior canopy characteristics of Takanari for increasing biomass are considered to be those aspects of canopy structure that affect light penetration into the canopy (Taylaran et al., 2009), photosynthetic capacity of the leaf (Hirasawa et al., 2010; Taylaran et al., 2011), and the rate of leaf photosynthesis at midday and in the afternoon as well. Compared with other high-yielding varieties, Takanari has many superior characteristics concerning the canopy photosyntheses that may explain why this strain can produce relatively higher dry matter than many of the other high-yielding varieties (Ishihara, 1996).

The superiority of Takanari with respect to dry matter production is most apparent after heading, when approximately 70% of the final carbohydrates in the rice grains are derived from the photosynthates. The larger dry matter production after heading will also increase the harvest index of Takanari compared with the *japonica* varieties that have been examined to date (Figure 4.2).

c. Characteristics of the Highest Photosynthesis Capacity in the Most Productive Varieties

Higher levels of leaf nitrogen and larger leaf stomatal conductance in Takanari as compared with other Japanese cultivars can generally be attributed to the relatively higher rate of leaf photosynthesis (Hirasawa et al., 2010).

Takanari tends to accumulate a larger amount of nitrogen, as measured at harvest, even at the same rate of nitrogen application (Figure 4.3A). The currently grown commercial varieties fall in this ranking, whereas the old commercial cultivars have the lowest ranking. A large portion of nitrogen accumulates in Takanari during the period from heading to harvest (Figure 4.3B); other varieties after heading have not shown any larger relative partitioning of nitrogen to leaves as Takanari. These results indicate that the higher level of leaf nitrogen in Takanari may result from its ability to accumulate nitrogen more efficiently after heading than other varieties.

The rate of leaf photosyntheses at an ambient CO_2 concentration of $370 \ \mu mol \ mol^{-1}$ is closely correlated to leaf nitrogen content for all varieties (Figure 4.4A). However, Takanari is shown to have a higher rate of leaf photosynthesis than other current varieties based on the examination of leaf nitrogen content. The stomatal conductance at an ambient CO_2 concentration of $370 \ \mu mol \ mol^{-1}$ in Takanari is also greater than that in other current varieties (Figure 4.4B). For both Takanari and other varieties, the rates of photosynthesis at an intercellular CO_2 concentration of $260 \ \mu mol \ mol^{-1}$ in terms of leaf Rucisco content are quite similar, implying that these species have similar leaf photosynthetic activities with almost identical Rubisco contents in leaves (Figure 4.4C).

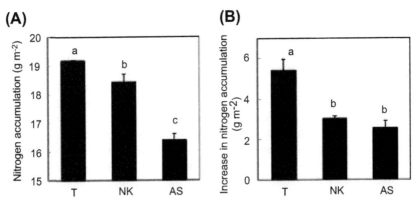

FIGURE 4.3 Comparisons of accumulated nitrogen at harvest (A) and from full heading to harvest (B). T: Takanari, a high-yielding rice variety released in 1990. NK: Average values for Nipponbare and Koshihikari varieties currently grown in Japan for human consumption, released in 1963 and 1956 respectively. AS: Average values for Aikoku and Sekitori varieties released in 1882 and 1848 respectively. The same letters are not significantly different at the 5% level. *(Adapted from Taylaran et al., 2009.)*

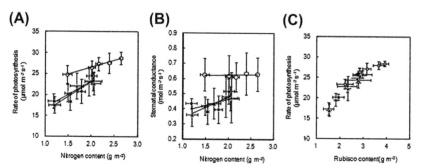

FIGURE 4.4   Relationships between leaf nitrogen content and the rate of photosynthesis at an ambient CO_2 concentration of 370 μmol mol^{-1} (A), between leaf nitrogen content and stomatal conductance (B), and between leaf Rubisco content and the rate of photosynthesis at an intercellular CO_2 concentration of 260 μmol mol^{-1} (C) of a flag leaf at full heading. Open circles represent Takanari variety and filled symbols varieties currently grown in Japan for human consumption. *(Adapted from Hirasawa et al., 2010.)*

The water potential of the flag leaf in the currently grown varieties decreases significantly when compared with that in Takanari despite the fact that plants of all varieties are growing in submerged soil (Taylaran et al., 2011). This causes the larger stomatal conductance in Takanari (Figure 4.4B). Takanari has a far larger root surface area than other varieties. The hydraulic conductance from roots to leaves is thus much higher in Takanari whereas the hydraulic conductivity, defined as hydraulic conductance per unit root surface area, is not different among all varieties. Hence, the larger root surface area is suggested as the reason why Takanari has higher hydraulic conductance and higher leaf water potential.

The higher rate of photosynthesis in Takanari appears to result from both higher leaf nitrogen content and stomatal conductance than those in the currently grown cultivars even under similar nitrogen applications and even at the same levels of leaf nitrogen. The characteristics such as increased nitrogen uptake and hydraulic conductance may be related to the larger root surface area that leads to a higher rate of leaf photosynthesis in Takanari (Figure 4.5).

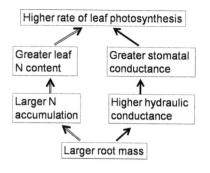

FIGURE 4.5   A schematic to illustrate how the high-yielding *indica* variety, Takanari, achieves a higher rate of leaf photosynthesis than other *japonica* varieties. *(Adapted from Taylaran et al., 2011.)*

4.1.5. Further Increasing Production of Rice Biomass for Energy and Food

Increasing parameters Ic and ε in equation (4.1) may be the key to increasing biomass production. Takanari is not superior to other high-yielding cultivars in all of the characteristics listed in Table 4.1. Consequently, if these non-superior characteristics in Takanari can be improved, its biomass production will be further enhanced.

a. Light-Intercepting Characteristics of the Canopy

Compared with Japanese cultivars currently cultivated for human consumption, both Takanari and another high-yielding variety, Milyang 23, have erect leaves and small canopy extinction coefficients at the heading stage. However, because of its large and downward-pointing panicles, Takanari has a larger canopy extinction coefficient during the ripening stage than in Milyang 23.

b. Rate of Photosynthesis in Fully Expanded Young Leaves

The maximum rate of rice leaf photosynthesis ranges from approximately 20 to 30 μmol m^{-2} s^{-1} when the ambient CO_2 is between 370 and 400 μmol mol^{-1}. The highest recorded rate of leaf photosynthesis is approximately 30–33 μmol m^{-2} s^{-1} as observed in Takanari, which has leaves with higher nitrogen content and larger stomatal conductance as mentioned above. Conversely, Koshihikari, the most popular rice variety in Japan, has a relatively low photosynthetic rate of 25–28 μmol m^{-2} s^{-1}. Among backcrossed inbred lines derived from a cross between Takanari and Koshihikari (Koshihikari/Takanari//Takanari), rice lines with leaf photosynthesis values approximately 20% higher than that observed in Takanari have been identified (Adachi et al., 2013). These lines had mesophyll cells with large surface areas and unprecedented rates of leaf photosynthesis achieved by increasing CO_2 diffusion from intercellular air spaces to chloroplasts, and from atmospheric CO_2 to these intercellular spaces through high levels of stomatal conductance as well.

c. Reduced Rates of Leaf Photosynthesis Associated with Senescence

Compared with Nipponbare, the high-yielding variety, Akenohoshi that maintains high rates of leaf photosynthesis during ripening is not observed in Takanari.

d. Lodging Resistance and CO_2 Diffusion into the Canopy

When above-ground biomass increases, the bending moment increases as well. Increasing the bending moment of the basal internode at breaking is considered to be important for lodging resistance in transplanted rice. This value is significantly larger in the high-biomass-producing variety Leaf Star than in

Takanari owing to the larger section modulus in Leaf Star (Ookawa et al., 2010a). Additionally, because Takanari is shorter than Leaf Star, canopy leaf area density in Takanari is relatively larger. Consequently, shorter Takanari experiences the effect of decreasing CO_2 diffusion into the canopy.

4.1.6. Genetic Analysis of the Traits Responsible for Biomass Production: Concluding Remarks

If the characteristics that affect biomass production as listed in Table 4.1 could be improved, biomass production in Takanari and other high-yielding varieties would increase in addition to enhancing biomass production in commercial varieties that are currently cultivated for human consumption. The major concern is how to improve these characteristics effectively. A marker-assisted approach is likely to become one of the most effective approaches for improving the traits as discussed in above sections. Indeed, the identification of the loci for numerous important quantitative traits that is currently being undertaken will be of great importance in this regard (Yamamoto et al., 2009; Ookawa et al., 2010b; Adachi et al., 2011). The capacity of biomass production for both food and fuel crops can be improved significantly in the near future.

4.2. AGRONOMY AND BREEDING TECHNOLOGY FOR BIOENERGY CROPS

Taiichiro Ookawa

Biofuel crops used as raw materials for producing bioenergy are grouped into herbaceous bioenergy crops and woody bioenergy crops. Herbaceous bioenergy crops grow faster than woody bioenergy crops in addition to having higher environmental adaptability, so that they grow in most environments on earth. In this section, the definition of herbaceous bioenergy crops as well as agronomy and breeding technology for these crops will be discussed.

4.2.1. Types of Herbaceous Energy Crops

Bioenergy crops can be classified into three development stages, i.e. the first generation, the second generation, and the third generation (Karp and Shield, 2008). The multiple use of these crops as feedstock often confuses their classification as bioenergy crops; the lignocellulose of the first-generation bioenergy crops (corn stover, bagasse, rice straw, and wheat straw) are considered as the second-generation crop.

a. The First Generation

Energy conversion technologies have already been established in dealing with the first generation, such as sugar crops (sugar cane, sweet sorghum, sugar beet, etc.), starch crops (rice, wheat, barley, corn, sweet potato, potato, etc.), and oil crops (rapeseed, sunflower, soybean, etc.).

b. The Second Generation

The process is to switch the raw materials for biofuels from food to non-food lignocelluloses consisting of the polysaccharides cellulose, hemicellulose and lignin; the energy conversion technologies are still under development.

c. The Third Generation

Future technologies such as genetically modified crops or microbes, and conversion of organic matter into hydrogen gas, are proposed for third-generation biofuels.

4.2.2. First-Generation Bioenergy Crops

a. Sugar Crops

(i) Energy Cane and Sugar Cane

Sugar cane (*Saccharum officinarum*) is a C_4 perennial grass. Brazil is the leading sugar producer in the world, followed by India, China, and Thailand (Lichts, 2010). In Brazil, about half of the sugar cane crop is used for producing bioethanol.

Modern sugar cane varieties are interspecific hybrids among the thick culm species, *S. officinarum*, a high yielding plant originating in New Guinea, and fine culm species such as *S. barberi* or *S. sinensis* originating from India. Sugar cane and energy sugar cane are the same plant species; energy cane indicates that the sugar cane is used for producing energy like other forms of organic materials such as feedstock, and plant leaves and stems.

In first-generation energy crops, the accumulation of sucrose in the culm is used for producing fuel; energy sugar cane has the most efficient energy conversion in the process of biomass production and energy conversion to result in the highest output with the lowest input energy among all plants.

The plant height of sugar cane reaches 3–6 m with sugar accumulating in parenchyma cells of culms. The cane juice is squeezed from culms within one day after harvest to prevent decomposition. The juice is then heated and concentrated with the bagasse used as fuel to power the distillation of alcohol.

Cultivation and Breeding

Brazil. The sugar cane growing area in Brazil has been increased and expanded to about 500 million ha, with 79 t ha^{-1} average yield (FAOSTAT, 2010); about half of the cultivated area is used for growing the energy sugar cane.

Sugar cane can grow vigorously on fertile soils, hence applying the appropriate amount of fertilizer is important in getting high yields. Nutrient recycling is made possible by applying liquid fertilizers obtained from residues of bioethanol production. Crop rotation by growing soybean has been practiced in order to conserve the soil fertility of the field for growing energy sugar cane.

In Brazil, a number of private seed companies have bred and released many sugar cane varieties. The main breeding objective is to develop sugar cane with high biomass production, high sugar content and yield, pest resistance, and drought resistance.

The Brazilian government promotes the research on improving sugar yield by funding several biofuel projects (Ministry of Agriculture, 2006). Specific objectives of these studies include the maintenance of high sugar content from March to September through the entire growing period, high nutrient use efficiency, sugar cane genome research, and the development of DNA markers (Waclawovsky et al., 2010).

Japan. In Japan, sugar cane is grown in lowland and gently sloping alluvial fans in Kagoshima prefecture and Okinawa prefecture, where sandy soil is suitable for growing sugar cane because sugar cane roots need to be well ventilated.

In Japan, Sugimoto and other researchers of the National Agricultural Research Center in Kyushu Okinawa Region (NARO) have bred "monster cane" by crossing common sugar cane with *Erianthus* that is closely related to wild species of sugar cane (Figure 4.6). The new breed has higher biomass production than common sugar cane varieties. Bioethanol plants have been established for the demonstration of practical experiments in Ie Island, Okinawa prefecture (Kawamitsu et al., 2003), in collaboration with Asahi Beer Company in Japan.

(ii) Sugar Beet (*Beta vulgaris* L.)

Sugar beet is a C_3 plant belonging to the Chenopodiaceae family; the cultivated species are F_1 hybrid. France is a leading producer of sugar beet, and has been

FIGURE 4.6   Monster cane.

producing bioethanol using beet as the raw material just as sugar cane is used for biofuel elsewhere. In Japan, sugar beet has been grown in Hokkaido that is located in the northern part of Japan using the paper pot transplanting practice. Sugar beet has been introduced into the crop rotation with wheat, legumes, potato, and corn.

(iii) Sweet Sorghum

Sorghum is a C_4 annual grass, and is classified in (*Sorghum bicolor* L.) the Poaceae family. Grain sorghum is mainly used as livestock feed concentrate, whereas sweet sorghum is used as a raw material for bioethanol in the USA, the EU, and China.

b. Starch Crops

(i) Maize (*Zea mays* L.)

Maize is an annual C_4 grass belonging to the Poaceae family. Similar to wheat and rice, maize is a major food crop and may also be utilized as a forage crop.

The USA is a major producer of maize, with harvest accounting for about one-half of the total production in the world (Lichts, 2010). In recent years, maize has been used as a raw material to produce bioethanol in the USA. The production of bioethanol from maize seed in the USA is the highest in the world. F_1 hybrid maize varieties using hybrid vigor are easy to process for biomass and foodstuffs.

(ii) Rice (*Oryza sativa* L.)

Rice is an annual C_3 grass belonging to the *Oryza* genus. *Oryza sativa*, which is the cultivated Asian native species, has two ecotypes: *Indica* and *Japonica*. *Japonica* is grown mainly in East Asia and *Indica* is distributed in the sub-tropical regions of Southeast Asia.

Cultivation and Breeding In Asia, the International Rice Research Institute began the dwarf rice breeding program in 1962, and a dwarf selection from Dee-geo-woo-gen/Peta cross was released in 1966 as IR8. These short-culm varieties have a semi-dwarf gene *sd1*, which is recessive and inhibits the production of gibberellin biosynthetic enzyme, introduced into rice to reduce lodging at higher rates of fertilizer use.

In China, the F_1 hybrid variety using hybrid vigor has been introduced in the southern parts of China and contributed to increased yields.

Since the 1960s in Japan, rice crops with high yield have been achieved by using some technologies with important innovations, i.e. controlling application of chemical fertilizer, mechanical transplanting, and planting semi-dwarf varieties. In recent years, direct seeding, which has been practiced in the USA and Australia, and has been widespread in Southeast Asia, has been introduced as a practical technology for reducing cultivation costs in Japan.

In Japan, the Ministry of Agriculture, Forestry and Fisheries started a research project in the 1980s to utilize rice as animal feed. One of the many objectives is to develop rice varieties with high yield and low cost production for forage use. High-yielding varieties have been developed, e.g. *Indica* type variety, and Takanari with the highest grain yield over $10\,\mathrm{t\,ha^{-1}}$. In recent years, rice varieties for biofuel have also been developed; bioethanol demonstration plants in Hokkaido and Niigata prefecture have been put into operation using the new high-yielding varieties developed for biofuel.

c. Oil Crops

(i) Rapeseed (*Brassica* spp.)

Among the six species belonging to rapeseed *Brassica* (Brassicaceae), two species, *Brassica napus* L. and *B. rapa* L., are generally referred to as rapeseed.

The order of magnitude for nations with yield of rapeseed is China, Canada, India, France, and Germany. France and Germany in Europe have encouraged the use of rapeseed oil to replace biodiesel for automobiles. In Japan, the national "Nanohana Project" to process recycled cooking oil from rapeseed into synthesized biodiesel fuel has been implemented in many local towns.

4.2.3. Second-Generation Bioenergy Crops

a. Current Status of the Technology for Producing Second-Generation Bioenergy Crops

(i) USA

The US Department of Energy (USDOE) and Department of Agriculture (USDA) started a joint development and demonstration project in 2005, to appropriate funds for USDOE to promote basic research on bioenergy. The research results were published in a report entitled "Breaking the Biological Barriers to Cellulosic Ethanol" that shows the roadmap for using cellulosic biomass biofuel (Houghton et al., 2006). The information on producing cellulosic biomass is taken from this report as follows.

Road map of research involved in the production of cellulosic biomass: Technology development phase within 10 years, 10–15 years individual system integration phase:

1. Development of sustainable biomass production technology:
   - Development of next-generation energy crops with a high yield in sustainable agriculture (see Figure 4.7).
   - Low-input and no tillage cultivation using perennial crops.
   - Introduction of biological diversity by crop rotation and the mixing of a wide variety of plant species.

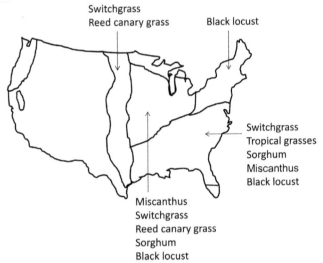

Switchgrass
Reed canary grass   Black locust

Switchgrass
Tropical grasses
Sorghum
Miscanthus
Black locust

Miscanthus
Switchgrass
Reed canary grass
Sorghum
Black locust

FIGURE 4.7 Geographic distribution of energy crops in the USA (US Department of Energy, 2006).

- Effects of the removal of crop residues on nitrogen and carbon recycling and soil microorganisms.
2. Low-cost harvest and transportation technology:
   - Developments of the harvester and the packaging machinery for corn stover, etc.
3. Development of next-generation energy crops with a high conversion efficiency:
   - Improvements of the lignin composition and the structure of cellulose involved in the conversion efficiency, cell wall-related genes and genome analysis.
4. Technology development of the bioenergy system that is applicable to the characteristics of each local region.
   - System integration of the sustainable agriculture in local regions, the conversion technology of bioenergy, the evaluation of economic values of bioenergy.

In the USA, three companies have operated bioethanol plants using lignocellulose such as corn stover, corn cobs and bagasse of energy cane, and another 16 companies have made plans to use switchgrass and wheat straw as raw materials for bioenergy (Jessup, 2009).

(ii) EU

In the EU, the Renewable Energy Road Map was proposed in 2006. The bioethanol derived from wheat grain (and straw) and the BDF from rapeseed will

Leaf Star and its parents.   Lodging resistant variety,
                                               Leaf Star.

Gold hull phenotype in Leaf Star,   Gold internode phenotype in Leaf
Chugoku 117 and gh2 mutant   Star.

FIGURE 4.8   Gold hull phenotype in Leaf Star (Ookawa et al., 2010a).

be produced in order to increase the proportion of bioenergy in energy consumption, whereas bioethanol derived from energy cane and soybean and the BDF from oil palm are scheduled to be imported from overseas.

(iii) China

China established the Renewable Energy Law in 2005. Because productions of biomass from corn stover, rice straw, and wheat straw are enormous, direct combustion for power generation and indirect use as bioethanol from these lignocelluloses have been proposed (Wang Q., 2011). China is divided into the following five regions of "green oil fields": Northeast (sweet sorghum), Northwest (sweet sorghum, shrub), North (sweet sorghum), Southwest (sweet sorghum, shrub), and Southeast (trees, grass).

(iv) Japan

Japan is a highly industrialized country that depends on almost 100% of imported energy with high total GHG emissions. Six ministries collaborated to start the "Biomass Nippon Strategy" program in 2002 with the object to alleviate global warming. The Biomass Nippon Strategy is expected to develop a low-carbon and "circular" economy that utilizes renewable resources in a highly efficient manner. The motivations behind the program include: (1) preventing climate change; (2) creating a recycling-oriented society to use limited resources in an effective manner by utilizing renewable biomass; (3) fostering new strategic industries; (4) activating agriculture, forestry, and fishery, as well as uniting rural communities. In this plan, the utilization of unused portions of farm crops such as rice straw or husk for biofuel will become visible by around 2010. By around 2020, energy crops will be widely cultivated so that they can be utilized as an energy source, and by around 2050 newly developed crops such as marine plants and genetically modified crops will contribute to an increased production of biomass.

In 2007, the Biomass Research Center was founded by the Japan Ministry of Agriculture, Forestry and Fisheries as a virtual research organization to start the project "Development of biomass utilization technology for regional revitalization". In this project, the following tasks will be undertaken: (1) Development of biofuel production technology using the first-generation energy crops (e.g. sugar cane, sugar beet, potato, sweet potato, and sorghum); and (2) Development of technology to efficiently convert lignocellulose such as rice straw to bioethanol.

b. Second-Generation Energy Crops

(i) Miscanthus (*Miscanthus* spp.)

Miscanthus is a perennial C_4 grass originating in East Asia; *Miscanthus sacchariflorus* ($2n = 76$) genotypes are more adapted to warmer climates whereas *Miscanthus sinensis* ($2n = 38$) genotypes grow well in winter. *Miscanthus* × *giganteus*, which is an indeterminate type in triploid ($2n = 57$) (Karp and Shield, 2008), has been considered as a model plant for second-generation bioenergy crops in the EU. It was introduced to Denmark from Yokohama, Japan in 1935, and has now spread throughout Europe. This plant is a natural hybrid of *M. sacchariflorus* and *M. sinensis* (Lewandowski et al., 2000). The details of its origin are unclear, but the possibility of its origination from "Ogisusuki" is high (Adati, 1958). These two species are of interest for breeding genotypes for bioenergy. Interspecific hybrids show more vigorous growth than parental plants.

In Europe except Scandinavia, *Miscanthus* × *giganteus* can be grown in most countries, and the biomass yield is higher in warm climates such as Southern European countries, Italy and Greece.

Miscanthus × *giganteus* is exceptional among C_4 species for its high productivity in cold climates because it can maintain photosynthetically active leaves at temperatures below the minimum for maize (*Zea mays*) so that it has a longer growing season in cool climates than most other plants (Naidu et al., 2003; Wang D. et al., 2008).

Cultivation and Breeding of Miscanthus × **giganteus** The sterile hybrid *Miscanthus × giganteus* has to be propagated vegetatively using rhizome (macro-propagation) or tissue culture (micro-propagation).

Because *Miscanthus × giganteus* is a perennial grass, weed control during the first two years is critical but the nutrient requirements are minimal for its growth; 50 kg nitrogen fertilizer, 20 kg phosphate, and 100 kg of potassium oxide per ha are sufficient for a satisfactory yield. In Europe, the harvest is carried out in February or March when large amounts of nutrients are stored in rhizomes to promote spring regeneration.

The biomass production in miscanthus is high even in low nitrogen conditions (Danalatos et al., 2007) with no changes of biomass production even if the nitrogen fertilizer were reduced by half to 50 kg-N ha^{-1}. The highest biomass production was also obtained at a sparse planting density of 1 plant m^{-2}.

In Europe, many genotypes of *Mischantus × giganteus* have been bred and widely used for productivity trails since its introduction in Denmark in 1930. Lewandowski et al. (2000) reported the following current situations and problems for cultivating and breeding miscanthus:

1. To improve biomass productivity and cold tolerance, many strains have been developed by crossing genetic resources of *M. sinensis* and *M. sacchariflorus* in the European Mischanthus Improvement Project (EMI).
2. From reports in European countries, the yields are 30 t ha^{-1} for irrigated fields and 10–25 t ha^{-1} for fields without irrigation in Southern Europe.
3. Results of case studies show that 30–35 cm young seedling or 10 cm cut rhizome is transplanted to fields in Denmark and Germany, whereas in the UK seedling or tissue culture using a piece of rhizome is transplanted in March or April.
4. High costs to construct and operate nursery facilities have become a serious problem.
5. Mischanthus cannot overwinter in Denmark, Ireland, and Germany.
6. In low fertilized fields, the retranslocation of nutrients to the rhizome becomes significant (N: 21–46%).
7. Weed control should be carried out thoroughly in the first year.
8. The yield limiting factors are soil type and soil moisture.

In the USA, strains from the *Miscanthus × giganteus* "Illinois" clone are selected for their high biomass yield based on studies conducted at Illinois State University. The amounts of bioethanol production and biomass production of this energy crop were estimated as shown in Table 4.3 (Heaton et al., 2008). In this estimation, mischanthus yields 260% ethanol production per unit land area compared with that from corn grain.

(ii) Switchgrass (*Panicum virgatum* L.)

Switchgrass is a perennial C_4 grass originating from central Mexico and has spread to 55° north latitude; it is classified into two ecotypes of lowland and

TABLE 4.3 Biomass and Bioethanol Production of Energy Crops in the Illinois State University, USA

	Biomass (Mg ha^{-1})	Bioethanol (gal ha^{-1})
Corn grain	10.2	1127
Corn stover	7.4	741
Corn total	17.6	1868
LIHD*	3.8	380
Switchgrass	10.4	1040
Miscanthus	29.6	2960

*Low input high diversity.

upland. The upland ecotype, which is octoploid ($2n = 72$) or hexaploid ($2n = 54$), has adapted to dry land. On the other hand, the lowland ecotype is tetraploid ($2n = 36$), and has adapted to swamps (Karp and Shield, 2008).

The breeding of switchgrass as a livestock feed was started in the 1930s, and the technology for breeding and cultivating switchgrass as feed has been established (Bassam, 1998). In recent years, evaluating switchgrass as energy crops has become a model US DOE project for using herbaceous crops in the Biofuel Feedstock Development Program undertaken by Oak Ridge National Laboratory.

For the quantitative trait locus (QTL) analysis of yield traits, the DOE Great Lakes Bioenergy Research Center and the University of Wisconsin have jointly established a research center. The breeding program has been carried out to select switchgrass of low lignin content by using marker-assisted selection (MAS). In addition, the syntenies of sorghum genome is expected to be applied to the same grasses. Heterosis of 32% in biomass production has been obtained from the F_1 in both upland and lowland types (Vogel et al., 2002).

Switchgrass reproduces by seed propagation so that the biomass yield can be maintained even when grown in fields of less fertile soil. In order to reuse the carbon and nitrogen for growing next year's crop, resources are recycled into rhizomes, thus the fertilizer requirement can be reduced (McLaughlin and Walsh, 1998). Typically, the application of P and K is not necessarily as frequent as once a year in the cropping system (McLaughlin and Adams, 2005). Although switchgrass has lower biomass yield than miscanthus, it requires less fertilizer; 50 kg-N ha^{-1} of nitrogen fertilizer is sufficient for switchgrass with other advantages by keeping the low cost of chemical fertilizer application in swithchgrass cultivation.

> **BOX 4.2 New Rice Variety "Leaf Star" as a Raw Material for Bioenergy**
>
> Lignin modification in feed and bioenergy crops has been the main objective for breeding new plants with improved energy yield (Chen and Dixon, 2007). Improving energy yield due to reduced lignin content is often accompanied by reduced lodging resistance. The rice species "*gold hull and internode 2 (gh2)*" was identified as a lignin-deficient mutant (Zhang et al., 2006) with the *GH2* gene mapped to the short arm region on chromosome 2 to encode a cinnamyl-alcohol dehydrogenase. In the *gh2* mutant, the conifenyl alcohol dehydrogenase (CAD) activity is reduced and sinapyl alcohol dehydrogenase (SAD) activity is not detectable. The new rice variety "Leaf Star" exhibits high lodging resistance, whereas the similar reddish-brown pigmentation in the hull and internode is observed with its parent "Chugoku 117" and *gh* mutants (Ookawa et al., 2010b). Leaf Star, its parents, and *gh* mutants have been compared based on the location of the *gh* locus, the properties of CAD and SAD activities, the lignin content and the culm strength. By conducting QTL analyses using parents, the *gh* locus is located in the same region on chr.2 as *gh2*. Leaf Star exhibits similar substrate specificity for CAD and SAD activities with Chugoku 117 and *gh2* mutant. The lignin content of stems in Leaf Star is reduced by 20% when compared with its parent species "Koshihikari". The *gh2* mutant "SG0207" with low lignin content is susceptible to lodging due to weak and fine culms. By contrast, Leaf Star has large culm strength due to thick and strong culms. These observations suggest that the improvement of energy yield is compatible with improving the high lodging resistance by utilizing the rice's genetic resources with thick and strong culm, such as the *gh2* cultivar Leaf Star.

(iii) Rice (Whole Crop, Straw)

In Japan, fallow and abandoned paddy fields have been caused by the over-production of rice. On the other hand, the proportion of self-sufficient food supply is only 40% on a calorie basis in Japan. Furthermore, 75% of the domestic demand of feed for livestock is imported from overseas. To increase the self-sufficient feed supply and establish the recycling system between rice cultivation and the livestock industry, the cultivation of forage crops in fallow and abandoned paddy fields has been conducted since the 1970s. Whole crop silage (WCS) is one of the ways to achieve self-efficiency in livestock feed, and breeding rice varieties for WCS have been developed for feed.

High biomass production, lodging resistance and digestibility are the main targets for the breeding of WCS rice varieties. These varieties can be applied as a raw material for second-generation biofuel (Box 4.2).

4.2.4. Prospects: Future Research for the Development of Energy Crop Production Technologies

Many researchers have developed new technology for second- and third-generation bioenergy (Sarath et al., 2008; Vega-Sanchez and Ronald, 2010).

In the USA, the research has been promoted toward using energy crops such as switchgrass, which can be grown widely in this country, to expand biofuel production from lignocellulose in the future. To utilize the biomass for both bioethanol and BDF, new GM graminaceous crops that can accumulate sugar and oil in straw are proposed. On the other hand, Tilman et al. (2006) suggested that achieving sustainable bioenergy production is important to maintain biodiversity in a form close to natural vegetation from the viewpoint of plant ecology.

In countries with limited land space such as Japan, improving the self-sufficiency of food and feed and recycling natural resources depends on the development of a recycling-based high biomass production system. The ecophysiological properties associated with sustainable biomass production, such as expansion of leaf area, photosynthetic rate, nitrogen use efficiency, lodging resistance, etc., need to be improved in order to breed new energy crops capable of adapting to this system. DNA marker-assisted selection using genome information and crop physiology is a powerful tool for breeding new bioenergy crops (Ookawa et al., 2010b). Further research is required to develop the sustainable production technology of energy crops.

4.3. PLANT MOLECULAR BREEDING TO ENERGY CROPS AS GENETIC IMPROVEMENTS OF BIOMASS SACCHARIFICATION

Shinya Kawai

4.3.1. Importance of Plant Molecular Breeding

This review is written based on prior excellent reviews carried out by previous researchers (Sticklen, 2006; Torney et al., 2007; Weng et al., 2008; Simmons et al., 2010) and recent research. Most bioethanol produced in the world is derived from starch of maize seeds and fermentable sugars (e.g. sucrose) of molasses from sugar canes. But starch and sucrose are also important sources of nutrients for both humankind and livestock. Therefore, the development of biofuels including bioethanol from non-food crops and agricultural residues has been strongly promoted.

Projects of fermentable sugars for ethanol production made from non-utilized biomass (such as agricultural residues, timber, switchgrass, and *Miscanthus*) have been undertaken in various parts of the world. However, saccharification (decomposition of polysaccharides into monosaccharides or fermentable sugars) of lignocellulosic materials contained in the plant cell wall is expensive, and the converted products are difficult to convert into ethanol due to its rigidity, complexity, and tolerance to cellulolytic enzymes. Addtionally, lignin restricts the availability of polysaccharides to enzymes, therefore limiting the enzymatic degradability and digestibility of biomass. In brief, lignin is an effective barrier against enzymatic saccharification. The improvement in pulping and bioconversion efficiencies of the wood seems useful to improve biomass for saccharification (Leple et al., 2007). Pretreating the lignocellulosic materials prior to the enzymatic saccharification process with cell wall degradation enzymes, microbial ligninases, cellulases, hemicellulases, etc. becomes necessary. Although efficient delignification is accomplished by treating lignocellulosic materials with organic solvents, ionic liquids, and Kraft pulping, these methods are costly. Therefore, other less expensive methods such as treating the fibers with dilute acid, ammonia fiber expansion, or heat are also applied. Although dilute acid and heat pretreatments are effective in decomposing the cell wall matrix, allowing enzymes to access cellulose, these methods require high temperatures (Hamelinck et al., 2005).

Accordingly, developing plant materials that are suitable for converting the contained cellulose into glucose cost-effectively provide a better solution to this problem. Recent plant molecular breeding research has aimed to increase the digestibility of the cellulose by modifying the lignin content and composition, and the accumulation of "redesign" lignin in the plant cell wall because the structures of lignin contribute to the resistance to degradation. Unlike cellulose and amylose, lignin is not a linear polymer of identical and repetitive monomers; it is a three-dimensional phenolic polymer linked by several types of carbon–carbon and ether bonds. Lignin monomers consist of three major monolignols: (1) coniferyl alcohol, (2) sinapyl alcohol and p-coumaryl alcohol, and (3) minor monolignols (Figures 4.9 and 4.10A). Lignin is mainly present only in certain types of mature plant cells such as xylem and fibers.

Several of the energy crops, including rice, maize, wheat, sugar cane, sorghum, switchgrass, and sugar beet, can be practically transformed. A direct method for altering saccharification efficiencies of plant materials is through these transformation technologies to modify the expression of the enzymes and transcription factors involved in lignin biosynthesis, and to introduce lignin degradation enzymes.

This review covers the recent research on improving plant biomass characteristics through plant genetic engineering for bioethanol production. The following strategies for modifying the lignin to breed transgenic plants with characteristics of easy saccharification have been examined by researchers using many plant species.

4.3.2. Reduction of Lignin Contents

Firstly, transgenic plants with reduced lignin have been developed. Lignin is the second most abundant biomass after cellulose on Earth. For example, maize stover (leaves and stalks) constitutes a large portion of agricultural biomass. Lignocellulosic materials of maize stover are composed of 30% hemicellulose, 44% cellulose, and 26% lignin (US Department of Energy, 2006). Lignocellulosic materials of plant cell walls consist of crystalline cellulose embedded in the lignocarbohydrate complex (LCC), which is a complicated compound of high molecular weight made of lignin and hemicellulose. Pretreatments, such as acid and heat, prior to the saccharification process disrupt the structures of

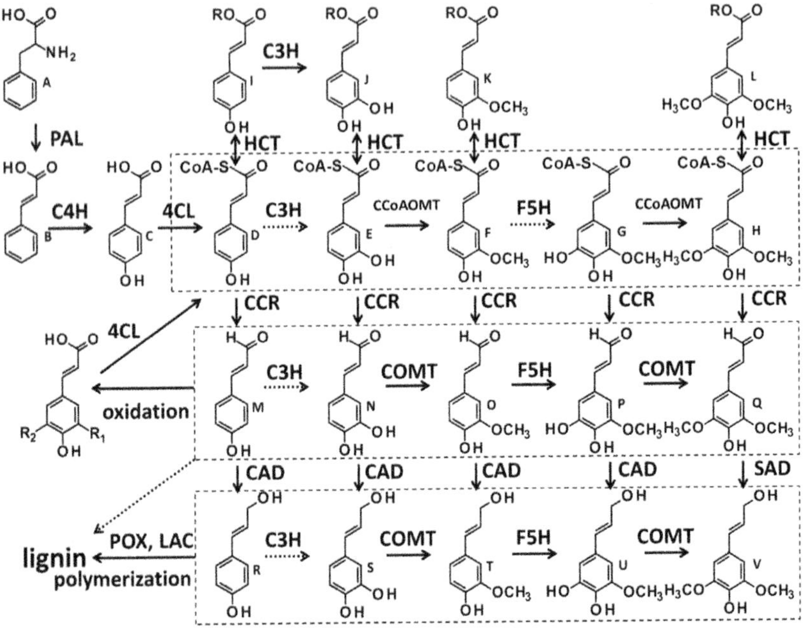

FIGURE 4.9 The grid model of the partial pathway for lignin biosynthesis. Enzymes: 4CL, 4-coumarate:coenzyme A (CoA) ligase; C3H, coumarate 3-hydroxylase; C4H, cinnamate 4-hydroxylase; CAD, hydroxycinnamyl alcohol dehydrogenase; CCoAOMT, S-adenosyl-methionine caffeoyl-CoA/5-hydroxyferuloyl-CoA O-methyltransferase; CCR, hydroxycinnamoyl-CoA reductase; COMT, caffeate/5-hydroxyferulate O-methyltransferase; F5H, ferulate 5-hydroxylase; HCT, hydroxycinnamoyl-CoA shikimate/quinate hydroxycinnamoyl transferase; LAC, laccase; PAL, phenylalanine ammonia-lyase; POX, peroxidase; SAD, sinapyl alcohol dehydrogenase. **Substrates**: A, phenylalanine; B, *trans*-cinnamic acid; C, *p*-coumaric acid; (*p*-hydroxycinnamoyl-CoAs: D, E, F, G, and H); D, *p*-coumaroyl-CoA; E, caffeoyl-CoA; F, feruloyl-CoA; G, 5-hydroxyferuloyl-CoA; H, sinapoyl-CoA; I, *p*-coumaroyl shikimic acid/quinic acid; J, caffeoyl shikimic acid/quinic acid; K, feruloyl shikimic acid/quinic acid; L, sinapoyl shikimic acid/quinic acid; (*p*-hydroxycinnamyl aldehydes: M, N, O, P, and Q); M, *p*-coumaraldehyde; N, caffeylaldehyde; O, coniferaldehyde; P, 5-hydroxyconiferaldehyde; Q, sinapaldehyde; (*p*-hydroxycinnamyl alcohols: R, S, T, U, and V; major monolignols: R, T, and V); R, *p*-coumaryl alcohol; S, caffeyl alcohol; T, coniferyl alcohol; U, 5-hydroxyconiferyl alcohol; V, sinapyl alcohol. –ORs of I, J, K, and L represent shikimate/quinate ester bonds. *p*-Hydroxycinnamyl aldehydes: M, N, O, P, and Q are highly oxidizable, and they are often oxidized to *p*-hydroxycinnamic acids: C, caffeic acid, ferulic acid, 5-hydroxyferulic acid, and sinapic acid respectively. These *p*-hydroxycinnamic acids are substrates of 4CL and are recovered to *p*-hydroxycinnamoyl-CoAs: D, E, F, G, and H respectively.

the lignocellulosic materials of the cell wall and remove lignin to allow easy access to the cellulose by cellulases (Hamelinck et al., 2005). Plant molecular breeding has been adapted to decrease the lignin contents and/or change the compositions of lignin. In previous studies, reducing lignin content was examined as a tool for enhancing sugar recovery.

Decreases in lignin contents through down-regulation of several genes for lignin biosynthetic enzymes (Figure 4.9) and transcriptional factors have been

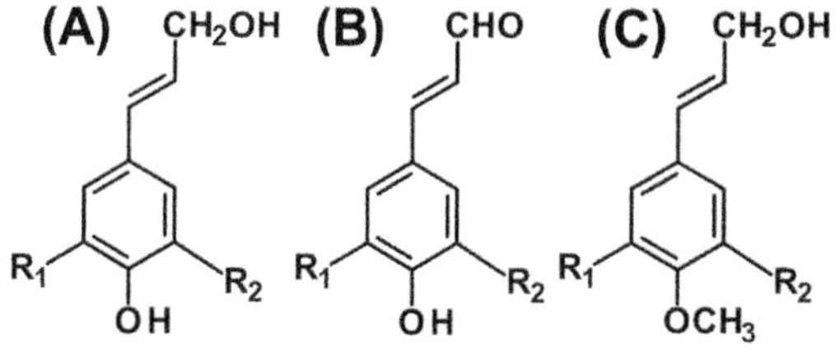

FIGURE 4.10   Monolignols and related molecules. (A) p-Hydroxycinnamyl alcohols: $R_1 = H$, $R_2 = H$, p-coumaryl alcohol; $R_1 = H$, $R_2 = OCH_3$, coniferyl alcohol; $R_1 = OCH_3$, $R_2 = OCH_3$, sinapyl alcohol; $R_1 = H$, $R_2 = OH$, caffeyl alcohol; $R_1 = OCH_3$, $R_2 = OH$, 5-hydroxyconiferyl alcohol. (B) p-Hydroxycinnamyl aldehydes: $R_1 = H$, $R_2 = H$, p-coumaryl aldehyde; $R_1 = H$, $R_2 = OCH_3$, coniferaldehyde; $R_1 = OCH_3$, $R_2 = OCH_3$, sinapaldehyde; $R_1 = H$, $R_2 = OH$, caffeyl aldehyde; $R_1 = OCH_3$, $R_2 = OH$, 5-hydroxyconiferaldehyde. (C) 4-O-methylated monolignols: $R_1 = H$, $R_2 = H$, 4-O-methylated p-coumaryl alcohol; $R_1 = H$, $R_2 = OCH_3$, 4-O-methylated coniferyl alcohol; $R_1 = OCH_3$, $R_2 = OCH_3$, 4-O-methylated sinapyl alcohol.

reported. These enzymes are involved in the biosynthesis of monolignols, which are the direct precursors of lignin, e.g. phenylalanine ammonia-lyase (PAL), cinnamate 4-hydroxylase (C4H), 4-coumarate:coenzyme A ligase (4CL), hydroxycinnamoyl-CoA shikimate/quinate hydroxycinnamoyl transferase (HCT), coumarate 3-hydroxylase (C3H), S-adenosyl-methionine caffeoyl-CoA/ 5-hydroxyferuloyl-CoA O-methyltransferase (CCoAOMT), hydroxycinnamoyl-CoA reductase (CCR), hydroxycinnamyl alcohol dehydrogenase (CAD, including sinapyl alcohol dehydrogenase (SAD)), ferulate 5-hydroxylase (F5H), and caffeic acid/5-hydroxyferulic acid O-methyltransferase (COMT). Furthermore, monolignol polymerizing enzymes, peroxidase and laccase during the last step of lignin biosynthesis have also been manipulated. The genes for these enzymes have often been adapted to alter the lignin content and/or composition in the transgenic plants. It is very important to note that, according to the types of artificially regulated enzymes, the adopted specific genes in the gene family for the enzyme, plant species and transgenic lines, and various phenotypes in transgenic plants appear in either independent or combinational manner such as reduced lignin content, altered lignin composition, dwarfism, collapsed vessels, etc. Therefore, in some cases, the down-regulation of one gene led to the reduction of lignin content, but in some other cases the same manipulation caused different phenotypes.

Transgenic aspen (*Populus tremuloides* Michx.) trees in which expression of *Pt4CL1* (encoding 4-coumarate:coenzyme A ligase) had been down-regulated were bred (Hu et al., 1999). Trees with suppressed *Pt4CL1* expression exhibit up to 45% reduction of lignin, but this reduction is offset by a 15% increase in cellulose. As a result, the total lignin–cellulose mass remains essentially unchanged. The growth of leaf, root, and stem is substantially enhanced, and the structural integrity is maintained both at the cellular and whole-plant levels in the transgenic lines. These results indicate that lignin and cellulose deposition can be regulated in a compensatory fashion that may contribute to metabolic flexibility and a growth advantage for sustaining the long-term structural integrity of woody perennials.

There are similar examples of 4CL down-regulation in grasses. *Pv4CL1* encodes 4CL in switchgrass (*Panicum virgatum*). RNA interference (RNAi) of *Pv4CL1* reduces extractable 4CL activity by 80%, leading to a reduction in lignin content with decreased guaiacyl (G) unit composition (Xu et al., 2011). Altered lignification patterns in the stems of RNAi transgenic plants were observed with phloroglucinol-HCl staining. The dilute acid pretreatment significantly increases the saccharification efficiency of the low-lignin transgenic biomass. In spite of 4CL down-regulated aspen, the transgenic switchgrass plants also had uncompromised biomass yields. Additionally, transgenic plants with a genetically modified lignin biosynthetic pathway often show dwarfism (Reddy et al., 2005). Therefore, transgenic plants with undesirable agronomic phenotypes such as dwarfism, and the collapse of vessel elements in the xylem, must be excluded.

Other genes for lignin biosynthesis have also been down-regulated. COMT catalyzes the methylation of 5-hydroxyconiferaldehyde, which is converted from coniferaldehyde by F5H. Maize *brown midrib* (*bm*) mutations are known for their naturally reduced lignin content and higher digestibility (Vignols et al., 1995). *bm3*, one of the *bm* mutations, is the recessive mutation of a gene for COMT. Therefore, the COMT antisense gene construct was introduced into maize, and the transgenic maize plants had decreased COMT activity and lignin content (He et al., 2003). Similar results using a sorghum COMT antisense construct in maize were also obtained (He et al., 2003); comparable experiments have also been done using many other plant species. A COMT down-regulated alfalfa (*Medicago sativa* L.) with reduced lignin content and altered lignin composition has been developed. Down-regulation of COMT brings about the decrease of lignin content by losing syringyl (S) units, whereas down-regulation of CCoAOMT caused a decrease of lignin content without reducing S units (Guo et al., 2001). The digestibility of forage from COMT down-regulated alfalfa plants is better than that of wild alfalfa plants, but increasing digestibility is also observed in CCoAOMT down-regulated alfalfa plants (Guo et al., 2001). The results indicate that both lignin content and composition affect the digestibility of alfalfa forage that is closely related to the saccharification efficiency (Guo et al., 2001). The CCoAOMT down-regulated transgenic alfalfa with low lignin content has been on the market with the proprietary name KK179 or the registration name OECD UI:MON-ØØ179-5.

To modify lignification more efficiently in plants, gene expressions for transcriptional factors related to lignin biosynthesis have been altered. Several transcription factors, which control genes for enzymes involved in lignin biosynthesis, have been identified; they belong to members of the NAC or MYB families. For example, several R2R3-MYB transcription factors have been identified to control lignin biosynthesis, such as *Arabidopsis thaliana* MYB61 (AtMYB61), *Pinus taeda* MYB4 (PtMYB4), *Antirrhinum majus* MYB308

(AmMYB308), and *Eucalyptus gunnii* MYB2 (EgMYB2). The down-regulation of these transcription factors may lead to the reduction of carbon flux into the lignin biosynthesis in the down-regulated transgenic plants (Besseau et al., 2007).

4.3.3. Alternation of Lignin Composition

Although reduction of lignin content is an important factor in improving digestibility, other alternatives may also achieve the same objective. The modification of lignin composition can significantly alter its degradation tendency but also leads to undesired agronomic features (Pedersen et al., 2005; Reddy et al., 2005) or undesirable phenotypes for bioenergy crops such as dwarfing, the collapse of vessel elements in the xylem, and increased susceptibility to fungal pathogens. Hence, another strategy has been suggested to alter lignin composition but not the content.

Lignin principally consists of *p*-hydroxyphenyl (H), guaiacyl (G), and syringyl (S) units, which are derived from *p*-hydroxycinnamyl alcohols, and the three major monolignols (*p*-coumaryl alcohol, coniferyl alcohol, and sinapyl alcohol) (Figure 4.10A). Their distributions are different among species, individual plants, and even cell types. In gymnosperms, the lignin is constructed with only G and H units; in dicotyledons, G and S units are predominant. Grass lignin of monocotyledons contains all three units. Monolignols are secreted to the extracellular space, then radicalized by plant peroxidases and laccases within the secondary plant cell walls, and then polymerized.

A higher G unit content creates a highly condensed lignin composed of a greater portion of biphenyl and other carbon–carbon linkages, whereas S units are commonly linked through more labile ether bonds at the 4-hydroxyl position (Figure 4.11). S units assist in lignin degradation; therefore, materials and energy production from angiosperms are more efficient than those from gymnosperm wood. To ensure the efficient degradation of the cell wall biomass in agriculture, paper making, and biofuel production, the lignin content should be lowered or the ratio of the more chemically labile S units in lignin should be increased. However, in some case studies, increasing G units in lignin was observed to cause more efficient digestion of cell wall materials.

Enrichment of the G unit composition without causing total lignin content alteration has been reported to increase alfalfa digestibility (Chen and Dixon, 2007). G units can be enriched by reducing the enzyme activities specific for S unit biosynthesis such as COMT, F5H, and SAD. SAD is a group of CAD that prefers sinapaldehyde and sinapyl alcohol to coniferaldehyde using coniferyl alcohol as the substrate. Therefore, both F5H and COMT are required to supply carbon flux from coniferyl alcohol precursors to sinapyl alcohol, and these hydroxylation and methylation steps determine the ratio of G and S units (S/G) of lignin (Humphreys et al., 1999; Osakabe et al., 1999).

Down-regulation of COMT leads to a dramatic reduction of S unit content in transgenic alfalfa plants that are related to wild-type alfalfa plants but

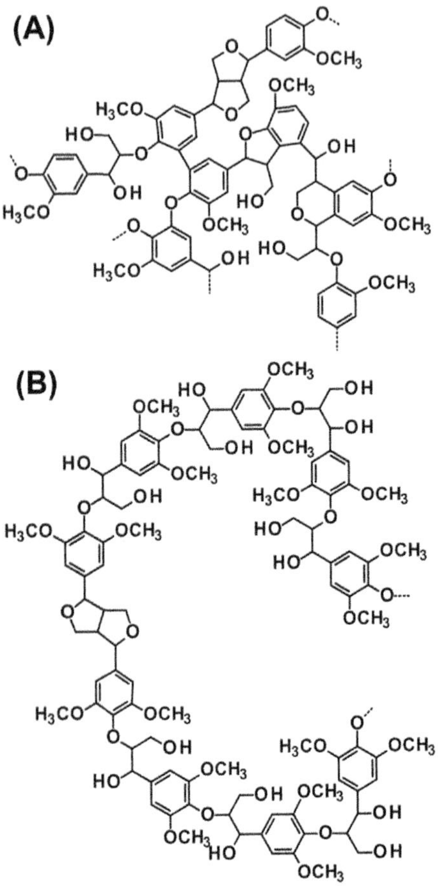

(A)

(B)

FIGURE 4.11 Partial structures of G lignin and S lignin. (A) In the G lignin, there are many kinds of bonds, carbon–carbon bonds and ether bonds, among the coniferyl alcohol residues. The condensed interunit linkages, which are 5–5 bonds, β-5 bonds, β-1 and β–β bonds, are abundant. The G lignin is a random, three-dimensional network polymer. (B) The S lignin basically contains β-O-4 ether bonds and β–β bonds that link between sinapyl alcohol residues, and it has a linear structure.

without changing the G unit content or reducing biomass yields (Guo et al., 2001). Similar results were obtained in the transgenic sugarcane (Jung et al., 2012). Down-regulation of COMT brings about another alteration of lignin composition. In COMT down-regulated plants, 5-hydroxyconiferaldehyde is difficult to convert to sinapaldehyde due to the reduced COMT activity; it is a substrate of CAD and can be reduced to 5-hydroxyconiferyl alcohol by CAD. In the presence of peroxidases and/or laccases, 5-hydroxyconiferyl alcohol can also be oxidized as the three major monolignols, and can be linked with major monolignols to be incorporated into the lignin polymer (Marita et al., 2003). As a result, lignin in the COMT down-regulated plants has novel characteristics that partially contribute to the improvement of digestibility for COMT down-regulated alfalfa in which the G unit composition is increased without altering the total lignin content (Chen and Dixon, 2007).

Conversely, the increase of S unit composition may improve the digestibility of lignin. The overexpression of F5H was examined in order to breed

S-unit-rich plants that have improved degradability and better saccharification efficiency. In transgenic poplar plants, a significant increase of the S/G value in lignin has been found (Stewart et al., 2009). Lignin with a high S/G value is more flexible than lignin with a low S/G value.

Reduction of the CAD gene expression in lignifying tissues improves pulping efficiency (Baucher et al., 1996) and digestibility (Baucher et al., 1999) but sometimes leads to reduction of lignin or plant growth, whereas CAD reduction in some plants does not cause lignin reduction. CAD reduction, CAD-deficient spontaneous, and artificial mutants in some plant species such as *gh2* in rice (Zhang et al., 2006), *bm1* in maize, *bmr6* in sorghum and *cad-n1* in loblolly pine have been reported (MacKay et al., 1997). Lignin composition in these mutants shows dramatic modifications, including increased incorporation of *p*-hydroxycinnamyl aldehydes (Figure 4.10B) as the substrates of CAD/SAD. These observations indicate that CAD/SAD may modulate lignin composition (Ralph et al., 2001). The modification is seen as a reddish-brown color in the modified lignin that causes visual phenotypes in mutant plants. Rice plants with *gh2* show the phenotype of gold hull and internode; maize with *bm1* and sorghum with *bmr6* have characteristic reddish-brown to tan colored midribs of mutant leaf blades from accumulating reddish-brown to yellow pigment in stalks and roots that is in contrast to the pale green midrib of wild-type leaf blades. The heart wood of CAD-deficient loblolly pine with *cad-n1* is brown whereas the wood of wild types is nearly white (MacKay et al., 1997). These similar phenotypes have also been found in some transgenic plants with reduced CAD activity. While no difference in lignin content was reported in maize *bm1* lines, *bmr-6* mutations in sorghum cause both a decrease in lignin and an increase in cinnamyl aldehydes (Pillonel et al., 1991). Similarly, the rice *gh2* mutant and the loblolly pine *cad-n1* mutant have decreased lignin contents (MacKay et al., 1997; Zhang et al., 2006). Artificial CAD reduced switchgrass transformants that are saccharified effectively have already been developed (Saathoff et al., 2011).

Furthermore, down-regulation of C3H in alfalfa changes lignin compositions and structures dramatically, and the accumulated lignin in the transgenic plants significantly enhances the digestibility (Reddy et al., 2005).

Results of catalyzing the last step of the monolignol biosynthesis with CCR show that down-regulation of CCR in transgenic poplar (*Populus tremula* × *P. alba*) is associated with up to 50% reduced lignin content to result in an orange–brown, often patchy, coloration of the outer xylem (Leple et al., 2007). In the reduced lignin of the transgenic poplar plants, S units are relatively more reduced than G units. The cohesion of the walls is affected particularly at sites that are generally richer in the S units contained in wild-type poplar. Ferulic acid residues incorporated into the lignin via ether bonds have been detected from the lignin in the transgenic poplar, and the xylem coloration is due to the presence of ferulic acid residues in the lignin. The reduced lignin and hemi-celluloses levels are associated with an increased proportion of cellulose.

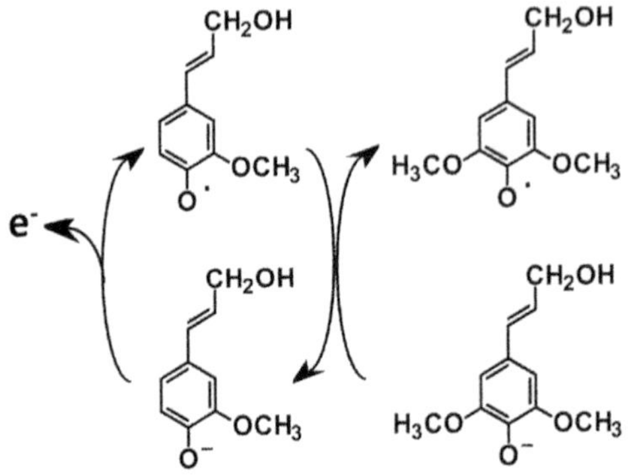

FIGURE 4.12 Coniferyl alcohol radical as a radical mediator to sinapyl alcohol. $O^{\bullet}$ represents the radical atom.

Finally, chemical pulping of wood derived from 5-year-old field-grown transgenic lines shows improved pulping characteristics, but the growth is affected in all transgenic lines tested (Leple et al., 2007).

Another strategy proposed for altering lignin composition is to target the plant peroxidase that catalyzes the polymerization of monolignols along with laccases. These enzymes consist of many isozymes and the substrate specificities of plant peroxidases related to lignin biosynthesis are generally broad. Most plant peroxidase can efficiently polymerize coniferyl alcohol but not sinapyl alcohol. However, many S units are contained in natural angiosperm lignin. It is suggested that coniferyl alcohol radicals can oxidize sinapyl alcohol molecules to sinapyl alcohol radicals through a radical mediation mechanism (Sasaki et al., 2004) (Figure 4.12).

On the other hand, several plant peroxidases, e.g. cationic cell wall-bound peroxidase (CWPO-C) from poplar, can also catalyze oxidization of sinapyl alcohol as a preferable substrate (Sasaki et al., 2004). Where coniferyl alcohol and sinapyl alcohol coexist, sinapyl alcohol molecules are also polymerized through coniferyl alcohol radicals as the radical mediators. Sinapyl alcohol-specific peroxidases are needed to improve lignin degradation through F5H up-regulation. Therefore, these enzymes would be potential targets for lignin engineering in order to control the S/G value of lignin. Peroxidases, which can efficiently oxidize sinapyl alcohol, are thought to have additional substrate-oxidizing sites on the protein surfaces except the heme pockets. PrxA3a, an anionic peroxidase from the hybrid aspen, *Populus kitakamiensis*, is one such typical plant peroxidase (Osakabe et al., 1995) that can polymerize coniferyl alcohol efficiently but not sinapyl alcohol. Down-regulation of PrxA3a results in decreased G unit content whereas the S unit content is not affected (Li et al., 2003). However, PrxA3a with substituted amino acid residues has additional substrate-oxidizing sites on the protein surface (Yoshinaka and Kawai, 2012). The mutated enzymes of PrxA3a can oxidize and polymerize sinapyl alcohol into high-molecular-weight compounds (see Box 4.3). Therefore, manipulating monolignol-specific peroxidases will lead to new approaches for genetic engineering to modify lignin content and composition.

4.3.4. Decrease in the Degree of Polymerization of Lignin and Addition of Easily Hydrolyzable Linkages into the Lignin Polymer

In addition to down-regulation and up-regulation of lignin biosynthetic enzymes and transcription factors, the incorporation of altered monolignols into lignin biosynthesis has been shown to be a more effective method to create "redesigned" lignin with better saccharification of the plant cell wall.

Polymerization of lignin is initiated by the dehydrogenation of monolignols by peroxidases and/or laccases at the *para*-hydroxyl (4-OH) site of the aromatic rings (Figure 4.13). By subsequent coupling of the phenoxy radicals to one another or to the growing polymers, low-molecular-weight polymers become high-molecular-weight lignin. In brief, the generation of radical intermediates and formation of ether linkages between the phenylpropane units require the 4-hydroxyl groups of monolignols. Therefore, methylation of the 4-hydroxyl groups of monolignols may lead to low-molecular-weight lignin. To reduce

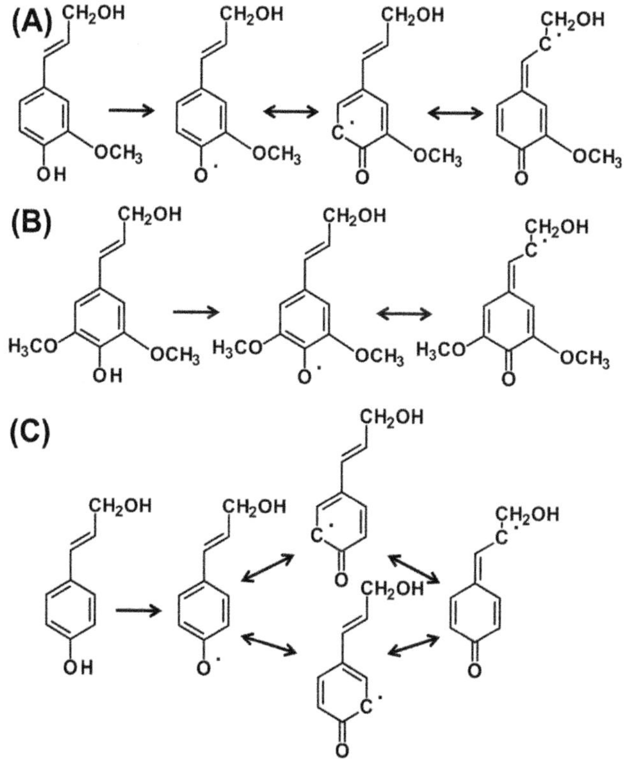

FIGURE 4.13   Radicalization of monolignols. (A) Dehydrogenation of coniferyl alcohol and the resulted radicals. (B) Dehydrogenation of sinapyl alcohol and the resulted radicals. (C) Dehydrogenation of *p*-coumaryl alcohol and the resulted radicals. O• and C• represent the radical atoms.

the degree of polymerization in lignin, a novel monolignol 4-O-methyltransferase is created through amino acid residue substitutions on the active site of the original enzyme, i.e. isoeugenol 4-O-methyltransferase (Bhuiya and Liu, 2010). The mutated enzyme can also methylate the 4-hydroxy residue of monolignols, and the resulting 4-O-methylated monolignols (Figure 4.10C) cannot be catalyzed by peroxidases and laccases so that carbon flux to lignin polymerization is prevented. This enzyme will be a useful tool to reduce lignin content and the degree of polymerization of lignin (Bhuiya and Liu, 2010).

To increase linkages that are easy to hydrolyze in the lignin polymer, amide and ester interunit linkages are introduced into the lignin polymer. The number of amide interunit linkages in a lignin structure can be increased by up-regulating the synthesis of minor monomer units of lignin, e.g. hydroxycinnamic acid amides (HAAs) (Figure 4.14A). The synthesis of HAAs is initiated when plants are wounded or subject to pathogen attack. HAAs are then secreted into the extracellular space, and radicalized like monolignols by peroxidases and laccases (McLusky et al., 1999), and then cross-coupled to lignin in the cell wall. The amide bonds derived from HAAs in the HAA-incorporated lignin become easily

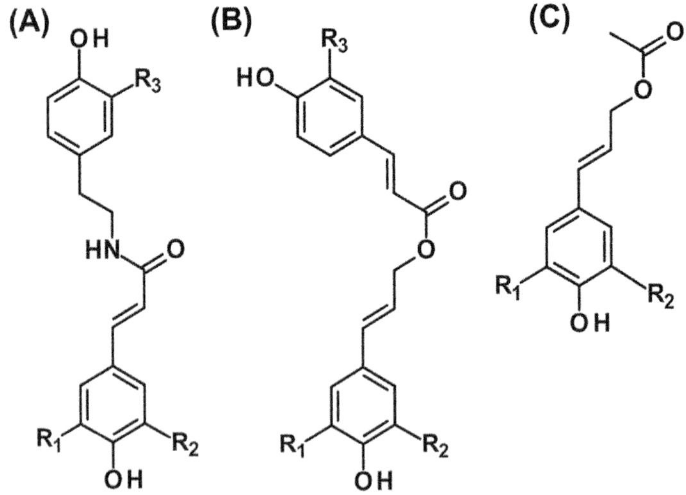

FIGURE 4.14   Monolignol analogs containing easily hydrolyzable bonds. (A) Hydroxycinnamic acid amides (HAAs): R_1 = H, R_2 = H, R_3 = H, coumaroyltyramine; R_1 = H, R_2 = H, R_3 = OCH$_3$, coumaroyl-3'-methoxytyramine; R_1 = OCH$_3$, R_2 = H, R_3 = H, feruloyltyramine; R_1 = OCH$_3$, R_2 = H, R_3 = OCH$_3$, feruloyl-3'-methoxytyramine; R_1 = OCH$_3$, R_2 = OCH$_3$, R_3 = H, feruloyltyramine; R_1 = OCH$_3$, R_2 = OCH$_3$, R_3 = OCH$_3$, feruloyl-3'-methoxytyramine. (B) Acylated monolignols: R_1 = H, R_2 = H, R_3 = H, coumaryl p-coumarate; R_1 = H, R_2 = H, R_3 = OCH$_3$, p-coumaryl ferulate; R_1 = OCH$_3$, R_2 = H, R_3 = H, coniferyl p-coumarate; R_1 = OCH$_3$, R_2 = H, R_3 = OCH$_3$, coniferyl ferulate; R_1 = OCH$_3$, R_2 = OCH$_3$, R_3 = H, sinapyl p-coumarate; R_1 = OCH$_3$, R_2 = OCH$_3$, R_3 = OCH$_3$, sinapyl ferulate. (C) Acetylated monolignols: R_1 = H, R_2 = H, p-coumaryl acetate; R_1 = H, R_2 = OCH$_3$, coniferyl acetate; R_1 = OCH$_3$, R_2 = OCH$_3$, sinapyl acetate.

hydrolyzable sites, thus improving lignin degradability. The enzyme involved in the last step of HAA biosynthesis is hydroxycinnamoyl-CoA:tyramine N-(hydroxycinnamoyl) transferase (THT; EC 2.3.1.110) (Figure 4.15). The genes for synthesizing THTs have been isolated from various plant species (McLusky et al., 1999). These genes are related to tyrosine decarboxylase (TYDC; EC 4.1.1.25), which has been isolated; tyrosine decaroxylase is involved in another step of HAA biosynthesis, to catalyze tyrosine to tyramine. Transgenic tobacco plants with higher THT and TYDC activities than wild-type plants have been characterized (Hagel and Facchini, 2005). For the addition of ester interunit linkages to the lignin polymer, coniferyl ferulate (Figure 4.14B), a methoxylated analog of coniferyl p-coumarate, is introduced into the maize cell walls (Grabber et al., 2008). Coniferyl ferulate moderately reduces lignification and cell wall ferulate copolymerization with monolignols; cell walls lignified with coniferyl ferulate are easier to hydrolyze with cellulase with and without alkaline pre-treatment (Grabber et al., 2008).

Acylated lignin units including mono- and eudicotyledons were found in the milled wood lignins of angiosperms, but were absent in the gymnosperms analyzed (Del Río et al., 2008). The structure of the lignins from sisal (*Agave sisalana*), kenaf (*Hibiscus cannabinus*), abaca (*Musa textilis*) and curaua (*Ananas erectifolius*) is remarkable; they are extensively acylated at the γ-carbon of the lignin side chain (up to 80% acylation) with acetate and/or p-coumarate groups and preferentially over S units. Whereas lignins from sisal and kenaf are γ-acylated exclusively with acetate groups, lignins from abaca and curaua are esterified with acetate and p-coumarate groups. The structures of all these highly acylated lignins are characterized by a very high S/G ratio, a large predominance of β-O-4 linkages (up to 94% of all linkages), and a

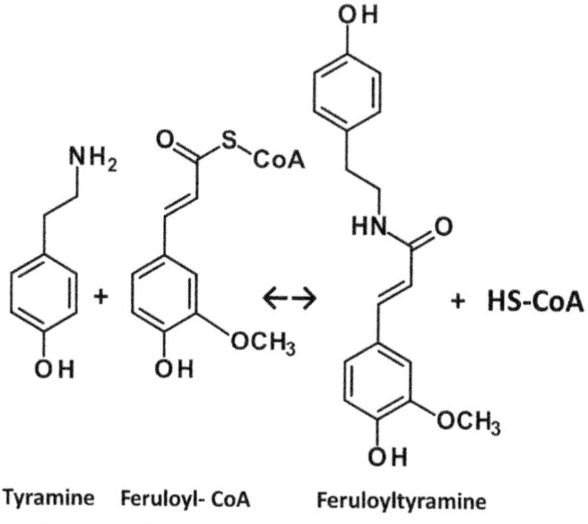

Tyramine   Feruloyl- CoA   Feruloyltyramine

FIGURE 4.15   Reaction of feruloyl-CoA:tyramine N-(hydroxycinnamoyl) transferase.

strikingly low proportion of traditional β–β linkages, which indeed are completely absent in lignins from abaca and curaua. The occurrence of β–β homocoupling and cross-coupling products of sinapyl acetate (Figure 4.14C) in the lignins from sisal and kenaf indicates that sinapyl alcohol is acetylated at the monomer stage, and that sinapyl acetate should be considered as a real monolignol involved in the lignification reactions (Del Río et al., 2008). β-O-4 linkages in lignin can be easily cleaved, thereby the highly acylated lignin is thought to be more digestible than the non-acylated lignin.

4.3.5. Production of Cell Wall Degradation Enzymes in Plants

a. Production of Lignin Degradation Enzymes in Plants

The introduction of genes for producing microbial lignin degradation enzymes in biomass crops in order to enhance the decomposition of cell walls during and before the saccharification process has been studied. White-rot fungi can oxidize lignin to carbon dioxide because they contain many peroxidases, laccases, and other oxidases responsible for generating reactive molecules to degrade lignin. The genome sequence of a white-rot fungus (*Phanerochaete chrysosporium*) has been determined that provides much insightful information on biofuel production (Vanden Wymelenberg et al., 2006).

b. Cellulase Production in Plants

Another approach is the heterogeneous expression of microbial cellulose degradation enzymes in plants. Cellulases are classified into three groups: endoglucanases, exoglucanases, and β-glucosidases. One suggested method is to produce microbial cellulases in plant cells. The catalytic domain of *Acidothermus cellulolyticus* thermostable endoglucanase gene (encoding for endo-1,4-β-glucanase enzyme or E1) is constitutively expressed in rice, tobacco, and maize (Biswas et al., 2006; Oraby et al., 2007) that are healthy and develop

---

BOX 4.3 Addition of Substrate-Oxidizing Sites to the Protein Surface of PrxA3a — Substrate-Specific Alteration of a Peroxidase to Monolignols

Mutations were introduced into *prxA3a*, a peroxidase gene of the hybrid aspen, *Populus kitakamiensis*, to substitute the amino acid residues at the surface of the protein, and the mutated enzymes were then analyzed for their substrate specificities. PrxA3a and mutated enzyme heterogeneous genes expressed in *Saccharomyces cerevisiae* were purified using Ni affinity chromatography, hydrolysis of sugar chain (Endoglycosidase Hf), and gel filtration. The substrate specificities are altered by substituted amino acid residues. PrxA3a F77Y and PrxA3a F77YA165W can polymerize sinapyl alcohol (Figure 4.16). In addition, PrxA3a A165W, F77Y, and F77YA165W improve the cytochrome c oxidizing activity. These substituted amino acid residues should function as a catalytic site outside of the heme pocket (Yoshinaka and Kawai, 2012).

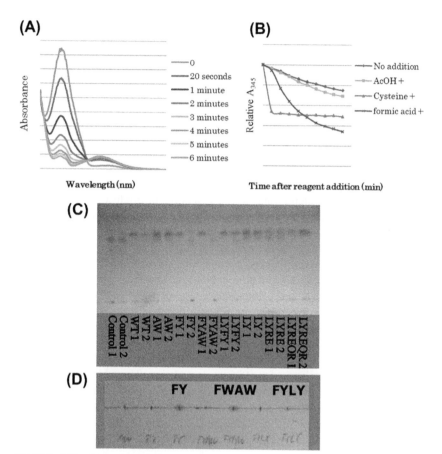

FIGURE 4.16 Sinapyl alcohol polymerizing activities. Sugar chains of enzymes were removed. Abbreviations [FYAW; PrxA3a F77Y A165W, LY; PrxA3a L182Y, FYLY; PrxA3a F77Y L182Y, LYRE; PrxA3a L182Y R245E, LYREQR; PrxA3a L182Y R245E S178Q N181R]. (A) The peak of A_{285} decreased rapidly and the peak of A_{345} increased temporally when the PrxA3a excessive amount (2 µg mL^{-1}) was added. (B) The peak of A_{345} was derived from syringyl-type quinone methide intermediate because it decreased rapidly by adding the acid, the oxidant, and the reducing agent. (C) 1: After a 6-minute reaction the A_{285} decrease was not measured. 2: Reaction mixtures incubated overnight were developed on TLC. Spots from reaction mixtures of PrxA3a F77Y, PrxA3a F77Y A165W, and PrxA3a F77Y L182Y were different from the others because those acquired sinapyl alcohol polymerizing activity. The sinapyl alcohol polymerizing activity of PrxA3a F77YL182Y is weaker than those of PrxA3a F77Y and PrxA3a F77Y A165W. (D) Expanded picture for the reaction mixture spotted points of PrxA3a F77Y, PrxA3a F77Y A165W, and PrxA3a F77Y L182Y in (C).

normally as compared with the wild-type (WT) plants. After thermochemical pretreatment and enzyme digestion, the transformed plants are clearly more digestible than WT and require lower pretreatment severity to achieve comparable conversion levels. Furthermore, the decreased recalcitrance is not due to post-pretreatment residual E1 activity and cannot be reproduced by the addition of exogenous E1 to the biomass prior to pretreatment. This indicates that the expression of E1 during cell wall construction alters the inherent recalcitrance of the cell wall (Oraby et al., 2007). Other solutions have also been suggested. For example, the use of expansin, a hydrogen bond-breaking protein, to loosen the cell wall during plant growth, as a candidate for plant endogenous enzymes to promote cell wall degradation, is being studied.

4.3.6. Conclusions

Recently, breeding of renewable lignocellulosic plant materials that can be saccharified more efficiently and less expensively to produce bioethanol has been studied by altering the content and structure of lignin in the cell walls of these energy crops so that the lignocellulosic materials can be easily decomposed into glucose and pentose. Several approaches have been attempted, and some mutant and transgenic plants are available on the market. Most research efforts have been directed toward up-regulating or down-regulating the enzymes involved in lignin biosynthesis. Some success has been obtained in altering both lignin content and composition to ease lignin decomposition, but many transgenic plants have shown undesirable characteristics such as dwarfing or vascular collapse. Therefore, recent research on the transcriptional factors involved in lignin biosynthesis has focused on controlling the lignin biosynthetic system in addition to modifying lignin for efficient degradation. The addition of new or rare linkages that are easier to hydrolyze to the interunits of lignin polymer has been attempted. Also, there have been several approaches to use *in planta* production of microbial lignin degradation enzymes in the energy crops to decompose the cell walls during and before the saccharification process.

4.4. IMPROVEMENT OF WOODY BIOMASS

Ryo Funada and Shinya Kajita

4.4.1. Wood and Cell Formation

a. Wood Formation by Cambial Activity

Wood has been used for thousands of years as a raw material for timber, furniture, pulp and paper, chemicals, and fuels. In addition, since wood is a major carbon sink, it is expected to play an important role in removing the excess atmospheric CO_2 that is generated by burning fossil fuels. Moreover, wood has recently been used as a resource for bioethanol. Therefore, there is still great demand for woody biomass as a renewable source of biomaterial and bioenergy.

Wood is produced by the vascular cambium of trees (Catesson, 1994; Larson, 1994; Funada, 2000, 2008). Cambium is defined as the active layer of active and delicate meristematic tissue between the inner bark or phloem and the wood or xylem that produces the secondary phloem on the outside and the secondary xylem on the inside (Figure 4.17). Cambium consists of fusiform cambial cells and ray cambial cells. The length of fusiform cambial cells varies depending on the species and the age of the cambium; it ranges from 1100 to 4000 μm in conifers and from 170 to 940 μm in hardwoods (Larson, 1994). The periclinal division of cambial cells leads to an increase in the stem diameter of trees. The amount of secondary xylem cells produced is usually much higher than the amount of secondary phloem cells. Thus, mature xylem cells are usually used as wood.

The cambial activity of trees exhibits seasonal cycles of activity and dormancy, which are known as annual periodicity, in temperate and cool zones. This periodicity plays an important role in the formation of wood and reflects the environmental adaptivity of trees, i.e. their tolerance to cold in winter in cool and temperate zones. The quantity and quality of wood depend on the division of cambial cells and the differentiation of cambial derivatives. Therefore, details of the cell biological and physiological aspects of the regulation of cambial activity in trees are of considerable interest.

Cambial activity ceases in autumn or winter and the cambial dormancy can be considered to consist of two stages, i.e. rest followed by quiescence (Catesson, 1994; Larson, 1994). The resting stage is maintained by conditions within the tree whereas the quiescent stage is controlled by environmental conditions. The resting stage of dormancy is a physiological state wherein the cambium cannot divide even under favorable growth conditions. By contrast, during the quiescent stage of cambial dormancy, the cambium is able to divide when exposed to appropriate environmental conditions. The transition from rest to quiescence involves structural, histochemical, and functional changes in cambial cells (Lachaud et al., 1999; Samuels et al., 2006).

Cambial activity generally resumes in the spring with a change from the quiescent dormant state to the active state (cambial reactivation). It is regulated by internal factors, such as plant hormones, and environmental factors including temperature, rainfall, and photoperiod. However, with respect to plant hormone auxins, no increase in the level of indole-3-acetic acid (IAA) is detected in the cambial region of conifers at the onset of cambial reactivation.

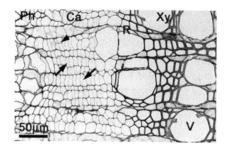

FIGURE 4.17   Light micrograph of a transverse section showing active cambium and differentiating secondary xylem of the main stem in hybrid poplar (*Populus sieboldii* × *P. grandidentata*). Arrows: cell division; Ca: cambium; Ph: secondary phloem; R: ray parenchyma cell; V: vessel element; Xy: secondary xylem. (*Courtesy of Dr S. Begum.*)

This suggests the absence of a clear relationship between the timing of cambial reactivation and endogenous levels of IAA (Sundberg et al., 1991; Funada et al., 2001a, 2002). Therefore, other factors appear to be involved in cambial reactivation. Recent studies have demonstrated that the timing of cambial reactivation is controlled by ambient temperature, which influences both the quantity and quality of wood (Oribe and Kubo, 1997; Oribe et al., 2001, 2003; Gričar et al., 2006; Begum et al., 2007, 2008, 2010, 2013).

b. Process of Cell Wall Formation

As soon as a cambial cell loses the ability to divide, it will start to differentiate into secondary phloem or xylem cells. The stages in the development of secondary xylem cells can be categorized into the following steps: cambial cell division, cell enlargement, cell wall thickening, cell wall sculpturing (formation of modified structure), lignification, and cell death (Panshin and de Zeeuw, 1980; Thomas 1991; Funada, 2000). Fusiform cambial cells differentiate into longitudinal tracheids (tracheids), vessel elements, wood fibers, and axial parenchyma cells, while ray cambial cells differentiate into ray parenchyma cells. In some conifers, such as *Pinus*, ray cambial cells differentiate into ray tracheids. Cells derived from fusiform cambial cells increase in length and diameter as they approach their final shape during differentiation. For example, tracheids in conifers increase only slightly in length but they increase considerably in radial diameter. Vessel elements do not increase significantly in length but their increases in radial and tangential diameter are conspicuous.

The major component of the very thin (often less than 0.1 μm thick) and plastic cell wall that is characteristic of the stage of cell enlargement is called the primary wall; it is cellulose that is highly crystalline and has very high tensile strength. Thus, cellulose microfibrils form a framework in the cell wall, and the primary wall is considered to consist of loose aggregates of cellulose microfibrils (Harada, 1965; Harada and Côté, 1985). This structure allows relatively unimpeded expansion of the cells derived from the cambium. In addition, the primary wall of cambial cells has a characteristic histochemical composition that allows for considerable extensibility (Catesson, 1990, 1994).

When cell expansion is almost complete, the well-ordered cellulose microfibrils are deposited on the inner surface of the primary wall to form the so-called secondary wall (Harada, 1965). Once the formation of the secondary wall has begun, no further expansion of cells will occur. The secondary xylem cells of woody plants, such as tracheids and wood fibers, have cell walls with a highly organized structure. Continuous deposition of the secondary wall increases the thickness of the cell wall. The thickness of the cell wall varies depending on cell function, cambial age, and the season in which the cell is formed (early wood or late wood; Figure 4.18). In general, cells that function to support the tree form thick secondary walls. The cell wall supports the heavy weight of the tree itself and allows the transport of water from roots to leaves, which can sometimes reach more than 100 m in height. In addition, the cell wall prevents microbial

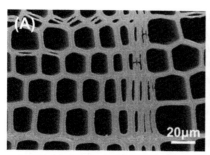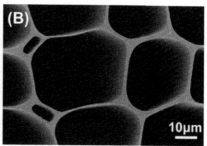

FIGURE 4.18   Scanning electron micrographs of transverse section showing tracheids of *Chamaecyparis obtusa* (A) and wood fibers of *Ochroma lagopus* (B). *(Courtesy of Dr Y. Sano.)*

and insect attack, thereby protecting the tree during its very long life that in some cases can exceed thousands of years. The thickness of the cell wall of wood fibers varies depending on species, showing different amounts of fixation of CO_2 in cell walls (Figure 4.18). Thus, the ultrastructure of tracheids and wood fibers is of great importance to define the mechanical properties of wood.

During formation of the secondary wall in tracheids or wood fibers, the cellulose microfibrils change their orientation progressively from a flat helix to a steep Z-helix in a clockwise rotation when viewed from the lumen side. They are oriented at about 5–20° with respect to the cell axis. No cellulose microfibrils with an S-helix are observed during formation of the S_2 layer. This shift in the angles of cellulose microfibrils is considered to generate a semi-helicoidal structure (Prodhan et al., 1995a; Abe and Funada, 2005). The concept of a helicoidal pattern has been proposed in the cell walls of numerous plants (Roland and Vian, 1979; Roland et al., 1987). This pattern consists of a series of planes in which the direction of cellulose microfibrils changes progressively. The arc-shaped or bow-shaped patterns observed by transmission electron microscopy in oblique ultrathin sections of tracheids or wood fibers correspond to the helicoidal structure (Roland and Mosiniak, 1983; Prodhan et al., 1995b).

The cellulose microfibrils of the S_2 layer are closely aligned, with a high degree of parallelism. The continuous deposition of cellulose microfibrils in one direction produces a thick cell wall layer with a consistent texture. When the rotational change in the orientation of cellulose microfibrils has been arrested, a thick cell wall layer is formed as a result of the repeated deposition of cellulose microfibrils (Roland and Mosiniak, 1983; Roland et al., 1987; Abe et al., 1991; Prodhan et al., 1995a). The thickness of the secondary wall is important in terms of the properties of wood because it is closely related to the specific gravity of wood (Panshin and de Zeeuw, 1980; Zobel and van Buijitenen, 1989). The duration of the arrest in the orientation of cellulose microfibrils seems to determine the thickness of the S_2 layer and, thus, the thickness of the secondary wall.

The average microfibril angle in the S_2 layer differs among species, and between the radial and tangential wall; it also depends on the time of cell formation. For example, the angles in differentiating early-wood tracheids with respect to the cell axis are 3–14° for *Abies sachalinensis*, 9–21° for *Larix kaempferi*, and 17–32° for *Picea jezoensis* (Abe and Funada, 2005). In addition, the microfibril angles in the S_2 layer can vary within the stem; they are usually large in the growth ring near the pith (juvenile wood), and decrease outwards to the bark side with increases in the age of cambium. Moreover, tracheids of compression wood, which is a reaction wood of conifers formed on the lower side of inclined stems in conifers, have large microfibril angles of about 45° with respect to the cell axis (Timell, 1986; Yoshizawa, 1987). These variations in microfibril angles are due to differences in the pattern of orientation of cellulose microfibrils. For example, while the direction of cellulose microfibrils in normal wood tracheids rearranges into a steep Z-helix oriented at about 5–20° from the tracheid axis, the orientation of cellulose microfibrils in compression wood tracheids becomes oblique until the cellulose microfibrils are oriented in a Z-helix about 45° from the tracheid axis. Such a rotation of cellulose microfibrils controls the microfibril angle in the S_2 layer, and thus it is one of the most important factors in determining the wood's properties.

At the final stage of the formation of secondary wall, the orientation of newly deposited cellulose microfibrils changes from a steep Z-helix to a flat helix, with counterclockwise rotation from the outer towards the inner part of the layer when viewed from the lumen side. This corresponds to a directional switch in the orientation of the cellulose microfibrils from clockwise to counterclockwise (when viewed from the lumen side) during formation of the secondary wall. The deposition of cellulose microfibrils in a flat helix results in the formation of an S_3 layer to bundle the deposited cellulose microfibrils (Abe et al., 1991, 1992, 1994). This texture differs from that of the S_2 layer in which the cellulose microfibrils have a high degree of parallelism. When transverse sections are observed with polarization microscopy, the S_3 layer exhibits birefringence, as does the S_1 layer. The shift in angles of cellulose microfibrils is more abrupt during the transition from the S_2 to the S_3 layer as compared to the transition from the S_1 to the S_2 layer (Harada and Côté, 1985; Abe et al., 1991). The rate of change in the orientation of cellulose microfibrils determines the structure of the cell wall layer.

A model of the orientation of newly deposited cellulose microfibrils in a tracheid is shown schematically in Figure 4.19. The direction of orientation of cellulose microfibrils changes progressively with changing speed of rotation during the formation of the secondary wall (Funada, 2000, 2008).

4.4.2. Effect of Tree Breeding on Wood Quality

a. Targets of Tree Breeding on Wood Quality

As mentioned above, wood is produced by the vascular cambium of living trees. Therefore, the wood quality among various species or even within a single species varies significantly depending on environmental conditions such as

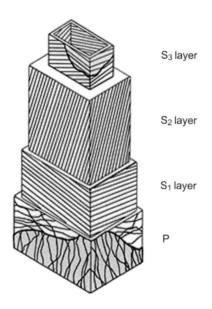

S₃ layer

S₂ layer

S₁ layer

P

FIGURE 4.19   A schematic model of cell wall structure in a tracheid. The orientation of newly deposited cellulose microfibrils in primary wall (P) and secondary wall (S).

climate, soil, and the spacing of growing trees, as well as cambial age, stem position, and distance from the crown. The tree-to-tree variability within members of a species under identical growth conditions shows that genetic factors also influence the wood structure. The wood quality as indicated by its mechanical properties is largely due to differences in wood structure. These observations indicate the strong possibility that the wood quality may be improved not only by silvicultural treatments but also by breeding to select genetically desirable trees (Panshin and de Zeeuw, 1980; Zobel and van Buijtenen, 1989; Zobel and Jett, 1995). The molecular biology also has the potential to improve wood quality (Mellerowicz et al., 2001).

The secondary xylem cells of woody plants, such as tracheids, vessel elements, and wood fibers, have cell walls with a highly organized structure. The orientation of cellulose microfibrils in the primary wall determines the direction of cell elongation and expansion, thereby controlling the shape and size of xylem cells. In addition, the orientation of cellulose microfibrils is one of the most important characteristics that determine the physical properties of wood. In particular, microfibril angles of the S_2 layer have a significant influence on the strength of wood, and it is possible to control the microfibril angles genetically. The microfibril angles differ significantly among clones of different origin. Thus, detailed screens to select plus trees with low microfibril angles should be included in future breeding programs. In addition, cortical microtubules, which are part of the cytoskeleton, are parallel to the newly deposited cellulose microfibrils and change their orientation progressively during the formation of secondary wall in a similar manner as the formation of cellulose microfibrils (Funada, 2000, 2002, 2008; Funada et al., 2001b; Chaffey, 2002). Therefore, cortical microtubules control the orientation of cellulose microfibrils in the secondary wall of tracheids or wood fibers. The manipulation of cortical microtubules might control microfibril angles of cell walls.

b. Woody Biomass is One of the Present Energy Sources

The total forest area of the world is over 4 billion hectares, covering about 30% of the world's land. Although the total carbon stock of the forest biomass in the world decreased by about 10 Gt from 1990 to 2010, it still amounts to 600 Gt (Global Forest Resources Assessment, 2010). According to the report from the Food and Agriculture Organization of the United Nations (FAO), the global wood production in 2005 was estimated to reach 3.4 billion m^3 (Figure 4.20). Fuel wood and industrial roundwood each account for nearly 50% of the total wood production, with about 70% and 90% of wood used as fuel wood in Asia and Africa respectively. In these developing regions, wood is either burned directly or processed into charcoal for fuel. Although burning fuel wood or charcoal directly is the most widespread and typical application of wood, the total heating value stored in fuel wood is not fully and efficiently used. Thus, alternative strategies for using woody biomass more efficiently such as gasification, pyrolysis, and cellulosic ethanol are desirable for the sustainable development of the future world.

c. Lignocellulose: The Main Part of Woody Biomass

In woody plants, about 60–80% of the biomass is made up of cell walls of woody tissue of stems. For example, approximately 75% of the biomass contained in poplars is present in terrestrial parts of stems and branches. Secondary cell wall composed of cellulose, hemicelluloses, and lignin constitutes the main part of the woody biomass. Cellulose consists of polymer with β-O-4-linked glucose units that are bonded to one another via hydrogen bonds to form

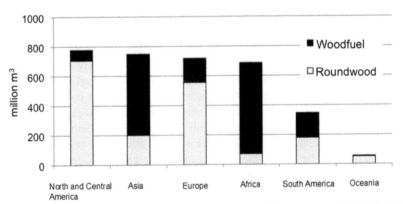

FIGURE 4.20 Wood production in the world. Values are 5-year averages for 2003–2007 (Global Forest Resources Assessment 2010, FAO).

microfibrils with crystalline properties. Cellulose microfibrils consist of bundles of around 36 cellulose chains that are embedded in a matrix of hemicelluloses and lignin. This chemical complex of cellulose, hemicellulose, and lignin is commonly designated as "lignocellulose". In recent years, lignocellulose is considered as an alternative source for heat, electricity, fuels, and chemicals in the future. Because the carbon contained in wood has been derived from atmospheric carbon dioxide during the growth of trees, using lignocelluloses as fuel will cycle the carbon between the atmosphere and wood. In other words, using forest resources to replace fossil fuels as energy sources will not increase atmospheric carbon dioxide as does burning fossil fuels. Thus, there are currently significant efforts to improve the productivity and characteristics of woody biomass (trees) by using emerging silvicultural systems and new biotechnology.

4.4.3. Molecular Breeding for Tailoring Lignocellulose

a. Strategy for Selection of Desirable Woody Biomass

Since woody biomass has huge potential as raw material for cellulosic ethanol, its saccharification efficiency is becoming one of the target traits for new tree breeding. Unlike corn grains, rice and sugar cane juice, which are typical sources of fermentable sugars for bioethanol production, lignocellulose in woody biomass consists of a larger quantity of lignin and highly crystallized cellulose that are difficult to convert into fermentable sugars (Table 4.4). Although these characteristics favor woody biomasses for combustion, pyrolysis, and gasification, they are less favorable as feedstocks for producing bioethanol. They have recalcitrant characteristics that resist enzymatic, chemical, and physical breakdown. Thus, pretreatments of the woody mass before the saccharification process are necessary in order to enhance the recovery of fermentable sugars from lignocelluloses. Toxic acids, alkalis, and high temperature along with mechanical size reduction are applied to pretreat the biomass to make it susceptible to enzymatic hydrolyses for recovery of fermentable sugars. However, the pretreatment step is the major bottleneck for reducing the cost of bioethanol production. Thus, new types of woody biomass, which are easy to break down into simple components (simple sugars and phenolics), should be developed.

In addition to hydrolysis conditions, efficiency of the saccharification is affected by many other biomass-related factors including cellulose content, degree of cellulose polymerization, chemical heterogeneity of hemicelluloses, content and chemical compositions of lignin. Substantial variations in chemical composition such as cellulose and lignin have been identified within the species of woody plants (Brereton et al., 2010; Studer et al., 2011). These results suggest that identification of low-recalcitrance variants of lignocelluloses for different end-uses can be selected from natural and cultivated forests in the near future. In order to breed improved woody biomasses, several technologies to assist in the selection by using markers such as DNA/RNA, transcripts, and

TABLE 4.4 Chemical Composition of Biomasses

Biomass	Cellulose	Hemicellulose	Lignin	Extractives	Ash
Japanese cedar*	38.4	24.5	33.4	3.4	0.3
Japanese beech*	43.9	29.3	24.2	1.9	0.6
Poplar[†]	54.5	16.1	18.8	nd	1.1
Bamboo*	34.6	30.1	30.8	3.3	1.1
Rice[‡]					
Whole plant	30.4	32.3	8.6	nd	6.3
Corn[§]					
Stalk rind	39.4	25.2	20.1	2.5	9.9
Stalk pith	31.6	27.3	17.0	6.6	10.0
Oil palm*					
Trunk	31.8	30.5	29.7	3.7	4.3

Values are all expressed as percentages (w/w). nd: not determined.
*Rabemanolontsoa et al. (2011).
[†]Guidi et al. (2009).
[‡]Jin & Chen. (2006).
[§]Li et al. (2012).

metabolite variations have been developed recently. Mapping populations of willow (138 different genotypes) were used to characterize their variation in enzymatic glucose release from stem and to identify quantitative trait loci (QTL) associated with the saccharification yield (Brereton et al., 2010). Glucose yields after enzymatic hydrolysis varied significantly among the population, and the four QTLs that influence this trait have been identified. Another study examined the effect of lignin characteristics on sugar release from 1100 individual undomesticated *Populus trichocarpa* trees and concluded that the sugar release is positively affected by lignin content but not by lignin composition (ratio of syringyl to guaiacyl residues) (Studer et al., 2011). These results indicate that there are opportunities to improve tree breeding programs for increasing enzymatic saccharification yields and production of bioethanol.

b. Molecular Breeding is an Attractive Tool for Tailoring Lignocellulose to Fitting Saccharification

The approach to modify gene expression such as antisense RNA, RNA interference, and overexpression of any genes will provide easier ways for genetic modification of cell wall biosynthetic pathway to develop new species of woody plants with higher saccharification characteristics. Three main cell wall components, i.e. cellulose, hemicelluloses, and lignin, should be targeted for the modification of the cell wall architecture.

Cellulose microfibrils have high crystallinity; therefore, their surface area is restricted for hydrolysis enzymes to act upon. On the other hand, the cellulose microfibrils also have paracrystalline regions that have a greater surface area exposed to degrading enzymes. Thus, increasing the portion of paracrystalline regions in cellulose microfibrils may improve the efficiency of saccharification of the woody biomass. Kaida et al. (2009a) overexpressed a gene for poplar cellulase in a woody species, sengon (*Paraserianthes falcataria*). Although the crystallinity of cellulose molecules had not been characterized, one of the resultant transgenic plants exhibits 1.4-fold more efficiency in cellulose hydrolysis than a wild-type plant. This modification also contributes to increasing the subsequent ethanol production from the transgenic plants.

Modification of the hemicellulose is another option for improving the saccharification characteristics of woody biomass. This cell wall component plays a key role in the attachment of cellulose and lignin to the surface of the microfibrils for building up the massive complex. Kaida et al. (2009b) introduced three different genes for hemicellulases, xyloglucanase, xylanase, and galactanase independently into poplar plants. Among the genes tested, the expression of the genes for xyloglucanase and xylanase exhibits a positive effect of saccharification in the transgenic poplar plants. The amount of glucose released from the transgenic biomasses increases significantly to reach 1.8-(xyloglucanase) and 1.7-fold (xylanase) higher than the wild-type plant. Increased susceptibility of cellulose in the xyloglucanase-expressing poplar may be due, at least in part, to enlargement of the cellulose microfibrils in the cell walls. Recently, 24–44% reduction of above-ground biomass of the xyloglucanase-expressing poplar has been revealed in field trials carried out in Japan. Cell wall modification without any detrimental effects to plant growth and development should be future tasks to be achieved for the genetic engineering of cell wall polysaccharides.

Lignin, the third component of the secondary cell wall, is a hydrophobic biopolymer that provides a barrier for enzymes to access cellulose. Reduction of lignin content and modification of its molecular structure is expected to increase the capability of cellulytic enzymes to attack the cell wall structure. Numerous studies on the increment of saccharification efficiency of lignocelluloses via the lignin modification have already been reported. However, most of these studies concern grass and herbaceous plants (see Section 4.3). A successful example of the transgenic woody biomass with modification of lignin for improving saccharification efficiency has been reported by Wang H. et al. (2012). In this study, the biomass characteristics were analyzed with stem prepared from 5-year-old transgenic poplar plants cultivated in a field. Transgenic plants with down-regulation of caffeoyl CoA 3-*O*-methyltransferase (CCoAOMT) gene exhibit 7–10% reduction in lignin content as compared to wild-type poplar.

The yield of total sugar released from the transgenic biomasses by enzyme saccharification shows significant increase after pretreatments both with and without 1% sodium hydroxide solution. These results suggest that lignin modification contributes to both increase of sugar recovery from woody biomass and decrease of chemical consumption in the pretreatments.

REFERENCES

Abe, H., & Funada, R. (2005). The orientation of cellulose microfibrils in the cell walls of tracheids in conifer: A model based on observations by field emission-scanning electron microscopy. *IAWA Journal, 26*, 161–174.

Abe, H., Ohtani, J., & Fukazawa, K. (1991). FE-SEM observation on the microfibrillar orientation in the secondary wall of tracheids. *IAWA Bull. New Series, 12*, 431–438.

Abe, H., Ohtani, J., & Fukazawa, K. (1992). Microfibrillar orientation of the innermost surface of conifer tracheid walls. *IAWA Bull. New Series, 13*, 411–417.

Abe, H., Ohtani, J., & Fukazawa, K. (1994). A scanning electron microscopic study of changes in microtubule distributions during secondary wall formation in tracheids. *IAWA Journal, 15*, 185–189.

Adachi, S., Nito, N., Kondo, M., Yamamoto, T., Arai-Sanoh, Y., Ando, T., Ookawa, T., Yano, M., & Hirasawa, T. (2011). Identification of chromosomal regions involving in the rate of photosynthesis of rice leaves by using a progeny from *japonica* and high-yielding *indica* varieties. *Plant Production Science, 14*, 118–127.

Adachi, S., Nakae, T., Uchida, M., Soda, K., Takai, T., Oi, T., Yamamoto, T., Ookawa, T., Miyake, H., Yano, M., & Hirasawa, T. (2013). The mesophyll anatomy enhancing CO_2 diffusion is a key trait for improving rice photosynthesis. *Journal of Experimental Botany, 64*(4), 1061–1072.

Adati, S. (1958). Studies on the genus *Miscanthus* with special reference to the Japanese species for breeding purpose as fodder crops. *Bulletin of the Faculty of Agriculture, Mie University, 17*, 1–112.

Bassam, N. E. I. (1998). *Enery plant species*. London: Science Publishers.

Baucher, M., Chabbert, B., Pilate, G., Van Doorsselaere, J., Tollier, M. T., Petit-Conil, M., Cornu, D., Monties, B., Van Montagu, M., Inze, D., Jouanin, L., & Boerjan, W. (1996). Red xylem and higher lignin extractability by down-regulating a cinnamyl alcohol dehydrogenase in poplar. *Plant Physiology, 112*, 1479–1490.

Baucher, M., Bernard-Vailhe, M. A., Chabbert, B., Besle, J.-M., Opsomer, C., Van Montagu, M., & Botterman, J. (1999). Down-regulation of cinnamyl alcohol dehydrogenase in transgenic alfalfa (*Medicago sativa* L.) and the effect on lignin composition and digestibility. *Plant Molecular Biology, 39*, 437–447.

Begum, S., Nakaba, S., Oribe, Y., Kubo, T., & Funada, R. (2007). Induction of cambial reactivation by localized heating in a deciduous hardwood hybrid poplar (*Populus sieboldii* × *P. grandidentata*). *Annals of Botany, 100*, 439–447.

Begum, S., Nakaba, S., Bayramzadeh, V., Oribe, Y., Kubo, T., & Funada, R. (2008). Responses to ambient temperature of cambial reactivation and xylem differentiation in hybrid poplar (*Populus sieboldii* × *P. grandidentata*) under natural conditions. *Tree Physiology, 28*, 1813–1819.

Begum, S., Nakaba, S., Oribe, Y., Kubo, T., & Funada, R. (2010). Changes in the localization and levels of starch and lipids in cambium and phloem during cambial reactivation by artificial heating of main stems of *Cryptomeria japonica* trees. *Annals of Botany, 106*, 885–895.

Begum, S., Nakaba, S., Yamagishi, Y., Oribe, Y., & Funada, R. (2013). Regulation of cambial activity in relation to environmental conditions: Understanding the role of temperature in wood formation of trees. *Physiologica Plantarum, 147*(1), 46–54.

Besseau, S., Hoffmann, L., Geoffroy, P., Lapierre, C., Pollet, B., & Legrand, M. (2007). Flavonoid accumulation in Arabidopsis repressed in lignin synthesis affects auxin transport and plant growth. *Plant Cell, 19*, 148–162.

Bhuiya, M. W., & Liu, C. J. (2010). Engineering monolignol 4-*O*-methyltransferases to modulate lignin biosynthesis. *The Journal of Biological Chemistry, 285*, 277–285.

Biswas, G. C. G., Ransom, C., & Sticklen, M. (2006). Expression of biologically active *Acidothermus* cellulolyticus endoglucanase in transgenic maize plants. *Plant Science, 171*, 617–623.

Brereton, N. J. B., Pitre, F. E., Hanley, S. J., Ray, M. J., Karp, A., & Murphy, R. J. (2010). QTL mapping of enzymatic saccharification in short rotation coppice willow and its independence from biomass yield. *Bioenergy Research, 3*, 251–261.

Catesson, A. M. (1990). Cambial cytology and biochemistry. In M. Iqbal (Ed.), *The vascular cambium* (pp. 63–112). Taunton: Research Studies Press.

Catesson, A. M. (1994). Cambial ultrastructure and biochemistry: changes in relation to vascular tissue differentiation and the seasonal cycle. *International Journal of Plant Sciences, 155*, 251–261.

Chaffey, N. (2002). Immunolocalisation of the cytoskeleton in the secondary vascular system of angiosperm trees and its visualization using epifluorescence microscopy. In N. Chaffey (Ed.), *Wood formation in trees: Cell and molecular biology techniques* (pp. 113–142). London: Taylor & Francis.

Chen, F., & Dixon, R. A. (2007). Lignin modification improves fermentable sugar yield for biofuel production. *Nature Biotechnology, 25*, 759–761.

Danalatos, N. G., Archontoulis, S. V., & Mitsios, I. (2007). Potential growth and biomass productivity of *Mischanthus × giganteus* as affected by plant density and N-fertilization in central Greece. *Biomass and Bioenergy, 31*, 145–152.

Del Río, J. C., Rencoret, J., Marques, G., Gutiérrez, A., Ibarra, D., Santos, J. I., Jiménez-Barbero, J., Zhang, L., & Martínez, A. T. (2008). Highly acylated (acetylated and/or *p*-coumaroylated) native lignins from diverse herbaceous plants. *Journal of Agricultural and Food Chemistry, 56*, 9525–9534.

Dohleman, F. G., & Long, S. P. (2010). More productive than maize in the midwest: How does Miscanthus do it? *Plant Physiology, 150*, 2104–2115.

Evans, L. T. (1993). *Crop evolution, adaptation and yield.* Cambridge: Cambridge University Press.

FAOSTAT. (2010). *FAO.* <http://faostat.fao.org/site/567/default.aspx#ancor>.

Funada, R. (2000). Control of wood structure. In P. Nick (Ed.), *Plant microtubules* (pp. 51–81). Berlin: Springer.

Funada, R. (2002). Immunolocalisation and visualisation of the cytoskeleton in gymnosperms using confocal laser scanning microscopy (CLSM). In N. Chaffey (Ed.), *Wood formation in trees: Cell and molecular biology techniques* (pp. 143–157). London: Taylor & Francis.

Funada, R. (2008). Microtubules and the control of wood formation. In P. Nick (Ed.), *Plant microtubules* (pp. 83–119). Berlin: Springer.

Funada, R., Kubo, T., Tabuchi, M., Sugiyama, T., & Fushitani, M. (2001a). Seasonal variations in endogenous indole-3-acetic acid and abscisic acid in the cambial region of *Pinus densiflora* stems in relation to earlywood–latewood transition and cessation of tracheid production. *Holzforschung, 55*, 128–134.

Funada, R., Miura, H., Shibagaki, M., Furusawa, O., Miura, T., Fukatsu, E., & Kitin, P. (2001b). Involvement of localized cortical microtubules in the formation of modified structure of wood. *Journal of Plant Research, 114*, 491–497.

Funada, R., Kubo, T., Sugiyama, T., & Fushitani, M. (2002). Changes in levels of endogenous plant hormones in cambial regions of stems of *Larix kaempferi* at the onset of cambial activity in springtime. *Journal of Wood Science, 48*, 75–80.

Gardner, F. P., Pearce, R. B., & Mitchell, R. L. (1985). *Physiology of crop plants*. Ames: Iowa State University Press.

Global Forest Resources Assessment. (2010). *Food and Agriculture Organization of the United Nations (FAO)*. <http://www.fao.org/forestry/fra/fra2010/en/>.

Grabber, J. H., Hatfield, R. D., Lu, F., & Ralph, J. (2008). Coniferyl ferulate incorporation into lignin enhances the alkaline delignification and enzymatic degradation of cell walls. *Biomacromolecules, 9*, 2510–2516.

Gričar, J., Zupančič, M., Čufar, K., Koch, G., Schmitt, U., & Oven, P. (2006). Effect of local heating and cooling on cambial activity and cell differentiation in the stem of Norway spruce (*Picea abies*). *Annals of Botany, 97*, 943–951.

Guidi, W., Tozzini, C., & Bonari, E. (2009). Estimation of chemical traits in poplar short-rotation coppice at stand level. *Biomass and Bioenergy, 33*, 1703–1709.

Guo, D., Chen, F., Wheeler, J., Winder, J., Selman, S., Peterson, M., & Dixon, R. A. (2001). Improvement of in-rumen digestibility of alfalfa forage by genetic manipulation of lignin *O*-methyltransferases. *Transgenic Research, 10*, 457–464.

Hagel, J. M., & Facchini, P. J. (2005). Elevated tyrosine decarboxylase and tyramine hydroxycinnamoyl transferase levels increase wound-induced tyramine-derived hydroxycinnamic acid amide accumulation in transgenic tobacco leaves. *Planta, 221*, 904–914.

Hamelinck, C. N., van Hooijdonk, G., & Faaij, A. P. C. (2005). Ethanol from lignocellulosic biomass: techno-economic performance in short-, middle- and long-term. *Biomass and Bioenergy, 28*, 384–410.

Harada, H. (1965). Ultrastructure and organization of gymnosperm cell walls. In W. A. Côté Jr (Ed.), *Cellular ultrastructure of woody plants* (pp. 215–233). New York: Syracuse University Press.

Harada, H., & Côté, W. A., Jr. (1985). Structure of wood. In T. Higuchi (Ed.), *Biosynthesis and biodegradation of wood components* (pp. 1–42). Orlando, FL: Academic Press.

Harris, J. M., & Meylan, B. A. (1965). The influence of microfibril angle on longitudinal and tangential shrinkage in *Pinus radiata*. *Holzforschung, 19*, 144–153.

Hay, R., & Porter, J. (2006). *The physiology of crop yield*. Oxford: Blackwell Publishing.

He, X., Hall, M. B., Gallo-Meagher, M., & Smith, R. L. (2003). Improvement of forage quality by downregulation of maize *O*-methyltransferase. *Crop Science, 43*, 2240–2251.

Heaton, E. A., Dohleman, F. G., & Long, S. P. (2008). Meeting US biofuel goals with less land: The potential of Miscanthus. *Global Change Biology, 14*, 2000–2014.

Hidema, J., Makino, A., Mae, T., & Ojima, K. (1991). Photosynthetic characteristics of rice leaves aged under different irradiances full expansion through senescence. *Plant Physiology, 97*, 1287–1293.

Hirasawa, T., & Hsiao, T. H. (1999). Some characteristics of reduced leaf photosynthesis at midday in maize growing in the field. *Field Crops Research, 62*, 53–62.

Hirasawa, T., Ozawa, S., Taylaran, R. D., & Ookawa, T. (2010). Varietal differences in rates of leaf photosynthesis in rice plants, with special reference to the nitrogen content of leaves. *Plant Production Science, 13*, 53–57.

Horie, T., & Sakuratani, T. (1985). Studies on crop-weather relationship model in rice. (1) Relation between absorbed solar radiation by the crop and the dry matter production. *Journal of Agricultural Meteorology, 40*, 331–342.

Houghton, J., Weatherwax, S., & Ferrell, J. (Eds.). (2006). *Breaking the biological barriers to cellulosic ethanol: a joint research agenda. A Research Roadmap Resulting from the Biomass to Biofuels Workshop*. Rockville, MD: US Department of Energy Office of Science and Office of Energy. December 7–9, 2005.

Hu, W. J., Harding, S. A., Lung, J., Popko, J. L., Ralph, J., Stokke, D. D., Tsai, C. J., & Chiang, V. L. (1999). Repression of lignin biosynthesis promotes cellulose accumulation and growth in transgenic trees. *Nature Biotechnology, 17*, 808–812.

Humphreys, J. M., Hemm, M. R., & Chapple, C. (1999). New routes for lignin biosynthesis defined by biochemical characterization of recombinant ferulate 5-hydroxylase, a multi-functional cytochrome P450-dependent monooxygenase. *Proceedings of the National Academy of Sciences USA, 96*, 10045–10050.

Ishihara, K. (1996). Eco-physiological characteristics of high yielding cultivars in crop plants – a case study of the rice plant. *Japanese Journal of Crop Science, 65i*(Extra issue 2), 321–326.

Jessup, R. W. (2009). Development and status of dedicated energy crops in the United States. *In Vitro Cellular & Developmental Biology, 45*, 282–290.

Jin, S. Y., & Chen, H. Z. (2006). Structural properties and enzymatic hydrolysis of rice straw. *Process Biochemistry, 41*, 1261–1264.

Jung, J. H., Fouad, W. M., Vermerris, W., Gallo, M., & Altpeter, F. (2012). RNAi suppression of lignin biosynthesis in sugarcane reduces recalcitrance for biofuel production from lignocellulosic biomass. *Plant Biotechnology Journal, 10*, 1067–1076.

Kaida, R., Kaku, T., Baba, K., Hartati, S., Sudarmonowati, E., & Hayashi, T. (2009a). Enhancement of saccharification by overexpression of poplar cellulase in sengon. *Journal of Wood Science, 55*, 435–440.

Kaida, R., Kaku, T., Baba, K., Oyadomari, M., Watanabe, T., Nishida, K., Kanaya, T., Shani, Z., Shoseyov, O., & Hayashi, T. (2009b). Loosening xyloglucan accelerates the enzymatic degradation of cellulose in wood. *Molecular Plant, 2*, 904–909.

Karp, A., & Shield, I. (2008). Bioenergy from plants and the sustainable yield challenge. *New Phytologist, 179*, 15–32.

Kawamitsu, Y., Fukuzawa, Y., Ueno, M., Komiya, Y., Sugimoto, A., & Matsuoka, M. (2003). Nutrient uptake characteristics and photosynthetic rate in Monster cane. *Japanese Journal of Tropical Agriculture, 47*(Extra 2), 49–50.

Kumura, A. (1995). Physiology of high-yielding rice plants from the viewpoint of dry matter production and its partitioning. In T. Matsuo, K. Kumazawa, R. Ishii, K. Ishihara, & H. Hirata (Eds.), *Science of the rice plant. Vol. 2. Physiology* (pp. 691–696). Tokyo: Food and Agriculture Policy Research Center.

Kuroda, E., Ookawa, T., & Ishihara, K. (1989). Analysis on difference of dry matter production between rice cultivars with different plant height in relation to gas diffusion inside stands. *Japanese Journal of Crop Science, 58*, 374–382.

Lachaud, S., Catesson, A. M., & Bonnemain, J. L. (1999). Structure and functions of the vascular cambium. *Comptes Rendus de l Acadmie des Sciences – Series III – Sciences Vie (Paris), 322*, 633–724.

Larson, P. R. (1994). *The vascular cambium: development and structure*. Heidelberg: Springer.

Leple, J. C., Dauwe, R., Morreel, K., Storme, V., Lapierre, C., Pollet, B., Naumann, A., Kang, K. Y., Kim, H., Ruel, K., Lefebvre, A., Joseleau, J. P., Grima-Pettenati, J., De Rycke, R., Andersson-Gunneras, S., Erban, A., Fehrle, I., Petit-Conil, M., Kopka, J., Polle, A.,

Messens, E., Sundberg, B., Mansfield, S. D., Ralph, J., Pilate, G., & Boerjan, W. (2007). Downregulation of cinnamoyl-coenzyme A reductase in poplar: Multiple-level phenotyping reveals effects on cell wall polymer metabolism and structure. *Plant Cell, 19*, 3669–3691.

Lewandowski, I., Clifton-Brown, J. C., Scurlock, J. M. O., & Huisman, W. (2000). Mischanthus: European experience with a novel energy crop. *Biomass and Bioenergy, 19*, 209–227.

Li, Y., Kajita, S., Kawai, S., Katayama, Y., & Morohoshi, N. (2003). Downregulation of an anionic peroxidase in transgenic aspen and its effect on lignin characteristics. *Journal of Plant Research, 116*, 175–182.

Li, Z., Zhai, H., Zhang, Y., & Yu, L. (2012). Cell morphology and chemical characteristics of corn stover fractions. *Industrial Crops and Products, 37*, 130–136.

Lichts, F. O. (2010). *Industry statistics: 2010 world fuel ethanol production.* Renewable Fuels Association.

MacKay, J. J., O'Malley, D. M., Presnell, T., Booker, F. L., Campbell, M. M., Whetten, R. W., & Sederoff, R. R. (1997). Inheritance, gene expression, and lignin characterization in a mutant pine deficient in cinnamyl alcohol dehydrogenase. *Proceedings of the National Academy of Sciences of the United States of America, 94*(15), 8255–8260.

Makino, A. (2011). Photosynthesis, grain yield, and nitrogen utilization in rice and wheat. *Plant Physiology, 155*, 125–129.

Mann, C. C. (1999). Crop scientists seek a new revolution. *Science, 283*, 310–314.

Marita, J. M., Ralph, J., Hatfield, R. D., Guo, D., Chen, F., & Dixon, R. A. (2003). Structural and compositional modifications in lignin of transgenic alfalfa down-regulated in caffeic acid 3-*O*-methyltransferase and caffeoyl coenzyme A 3-*O*-methyltransferase. *Phytochemistry, 62*, 53–65.

McLaughlin, B., Samuel, B., & Kszos, L. A. (2005). Development of switchgrass (Panicum virgatum) as a bioenergy feed stock in the United States. *Biomass and Bioenergy, 28*, 515–535.

McLaughlin, S. B., & Walsh, M. E. (1998). Evaluating environmental consequences of producing herbaceous crops for bioenergy. *Biomass and Bioenergy, 14*, 317–324.

McLusky, S. R., Bennett, M. H., Beale, M. H., Lewis, M. J., Gaskin, P., & Mansfield, J. W. (1999). Cell wall alterations and localized accumulation of feruloyl-3'-methoxytyramine in onion epidermis at sites of attempted penetration by *Botrytis allii* are associated with actin polarisation, peroxidase activity and suppression of flavonoid biosynthesis. *Plant Journal, 17*, 523–534.

Mellerowicz, E. J., Baucher, M., Sundberg, B., & Boerjan, W. (2001). Unravelling cell wall formation in the woody dicot stem. *Plant Molecular Biology, 47*, 239–274.

Ministry of Agriculture. (2006). *Brazilian Agroenergy Plan 2006–2011.* Brasilia: Embrapa Publishing House.

Monsi, M., & Saeki, T. (1953). Über den Lichtfaktor in den Pflanzengesellschaften und seine Bedeutung für die Stoffproduktion. *The Journal of Japanese Botany, 14*, 22–52.

Mukouyama, T., Motobayashi, T., Chosa, T., Ookawa, T., Furuhata, M., Tojo, S., & Hirasawa, T. (2012). Dry matter production and grain yield of high yielding rice cultivar, Takanari, directly seeded in the paddy field with an "Air-assisted Drill": Effects of planting pattern on ecophysiology of direct seeded rice. *Japanese Journal of Crop Science, 81*, 414–423.

Naidu, S. L., Moose, S. P., Al-Shoaibi, A. K., Raines, C. A., & Long, S. P. (2003). Cold tolerance of C4 photosynthesis in Miscanthus × giganteus: Adaptation in amounts and sequence of C4 photosynthetic enzymes. *Plant Physiology, 132*, 1688–1697.

Ookawa, T., Yasuda, K., Seto, M., Sunaga, K., Kato, H., Sakai, M., Motobayashi, T., Tojo, S., & Hirasawa, T. (2010a). Biomass production and lodging resistance in 'Leaf Star', a new long-culm rice forage cultivar. *Plant Production Science, 13*, 58–66.

Ookawa, T., Hobo, T., Yano, M., Murata, K., Ando, T., Miura, H., Asano, K., Ochiai, Y., Ikeda, M., Nishitani, R., Ebitani, T., Ozaki, H., Angeles, E. R., Hirasawa, T., & Matusoka, M. (2010b). New approach for rice improvement using a pleiotropic QTL gene for lodging resistance and yield. *Nature Communication, 1*, 1–11.

Oraby, H., Venkatesh, B., Dale, B., Ahmad, R., Ransom, C., Oehmke, J., & Sticklen, M. (2007). Enhanced conversion of plant biomass into glucose using transgenic rice-produced endoglucanase for cellulosic ethanol. *Transgenic Research, 16*, 739–749.

Oribe, Y., & Kubo, T. (1997). Effect of heat on cambial reactivation during winter dormancy in evergreen and deciduous conifers. *Tree Physiology, 17*, 81–87.

Oribe, Y., Funada, R., Shibagaki, M., & Kubo, T. (2001). Cambial reactivation in the partially heated stem in an evergreen conifer. *Abies sachalinensis. Planta, 212*, 684–691.

Oribe, Y., Funada, R., & Kubo, T. (2003). Relationships between cambial activity, cell differentiation and the localization of starch in storage tissues around the cambium in locally heated stems of *Abies sachalinensis* (Schmidt) Masters. *Trees, 17*, 185–192.

Osakabe, K., Koyama, H., Kawai, S., Katayama, Y., & Morohoshi, N. (1995). Molecular cloning of two tandemly arranged peroxidase genes from *Populus kitakamiensis* and their differential regulation in the stem. *Plant Molecular Biology, 28*, 677–689.

Osakabe, K., Tsao, C. C., Li, L., Popko, J. L., Umezawa, T., Carraway, D. T., Smeltzer, R. H., Joshi, C. P., & Chiang, V. L. (1999). Coniferyl aldehyde 5-hydroxylation and methylation direct syringyl lignin biosynthesis in angiosperms. *Proceedings of the National Academy of Sciences of the United States of America, 96*, 8955–8960.

Panshin, A. J., & de Zeeuw, C. (1980). *Textbook of wood technology* (4th ed.). New York: McGraw-Hill.

Pedersen, J. F., Vogel, K. P., & Funnell, D. L. (2005). Impact of reduced lignin on plant fitness. *Crop Science, 45*, 812–819.

Peng, S., Yang, J., Garcia, F. V., Laza, R. C., Visperas, R. M., Sanico, A. L., Chavez, A. Q., & Virmani, S. S. (1998). Physiology-based crop management for yield maximization of hybrid rice. In S. S. Virmani, E. A.Siddiq, & Muralidharan (Eds.), *Advances in hybrid rice technology*. Los Banos: IRRI (pp. 157–176).

Pillonel, C., Mulder, M., Boon, J., Forster, B., & Binder, A. (1991). Involvement of cinnamylalcohol dehydrogenase in the control of lignin formation in *Sorghum bicolor* (L.) Moench. *Planta, 185*, 538–544.

Prodhan, A. K. M. A., Funada, R., Ohtani, J., Abe, H., & Fukazawa, K. (1995a). Orientation of microfibrils and microtubules in developing tension-wood fibers of Japanese ash (*Fraxinus mandshurica* var. *japonica*). *Planta, 196*, 577–585.

Prodhan, A. K. M. A., Ohtani, J., Funada, R., Abe, H., & Fukazawa, K. (1995b). Ultrastructural investigation of tension wood fibre in *Fraxinus mandshurica* Rupr. var. *japonica* Maxim. *Annals of Botany, 75*, 311–317.

Rabemanolontsoa, H., Ayada, S., & Saka, S. (2011). Quantitative method applicable for various biomass species to determine their chemical composition. *Biomass and Bioenergy, 35*, 4630–4635.

Ralph, J., Lapierre, C., Marita, J. M., Kim, H., Lu, F., Hatfield, R. D., Ralph, S., Chapple, C., Franke, R., Hemm, M. R., Van Doorsselaere, J., Sederoff, R. R., O'Malley, D. M., Scott, J. T., MacKay, J. J., Yahiaoui, N., Boudet, A., Pean, M., Pilate, G., Jouanin, L., & Boerjan, W. (2001). Elucidation of new structures in lignins of CAD- and COMT-deficient plants by NMR. *Phytochemistry, 57*, 993–1003.

Reddy, M. S. S., Chen, F., Shadle, G., Jackson, L., Aljoe, H., & Dixon, R. A. (2005). Targeted down-regulation of cytochrome P450 enzymes for forage quality improvement in alfalfa

(*Medicago sativa* L.). *Proceedings of the National Academy of Sciences of the United States of America, 102*, 16573–16578.

Roland, J. C., & Mosiniak, M. (1983). On the twisting pattern, texture and layering of the secondary cell walls of lime wood. proposal of an unifying model. *IAWA Bull. New Series, 4*, 15–26.

Roland, J. C., & Vian, B. (1979). The wall of the growing plant cell: Its three dimensional organization. *International Review of Cytology, 61*, 129–166.

Roland, J. C., Reis, D., Vian, B., Satiat-Jeunemaitre, B., & Mosiniak, M. (1987). Morphogenesis of plant cell walls at the supermolecular level: Internal geometry and versatility of helicoidal expression. *Protoplasma, 140*, 75–91.

Saathoff, A. J., Sarath, G., Chow, E. K., Dien, B. S., & Tobias, C. M. (2011). Tobias: Down-regulation of cinnamyl-alcohol dehydrogenase in switchgrass by RNA silencing results in enhanced glucose release after cellulase treatment. *PLoS One, 6*, e16416.

Samuels, A. L., Kaneda, M., & Rensing, K. H. (2006). The cell biology of wood formation: From cambial divisions to mature secondary xylem. *Canadian Journal of Botany, 84*, 631–639.

San-oh, Y., Mano, Y., Ookawa, T., & Hirasawa, T. (2004). Comparison of dry matter production and associated characteristics between direct-sown and transplanted rice plants in a submerged paddy field and relationships to planting patterns. *Field Crops Research, 87*, 43–58.

Sarath, G., Mitchell, R. B., Sattler, S. E., Funnell, D., Pedersen, J. F., Graybosch, R. A., & Vogel, K. P. (2008). Opportunities and roadblocks in utilizing forages and small grains for liquid fuels. *Journal of Industrial Microbiology and Biotechnology, 35*, 343–354.

Sasaki, S., Nishida, T., Tsutsumi, Y., & Kondo, R. (2004). Lignin dehydrogenative polymerization mechanism: a poplar cell wall peroxidase directly oxidizes polymer lignin and produces *in vitro* dehydrogenative polymer rich in β-*O*-4 linkage. *FEBS Letters, 562*, 197–201.

Simmons, B. A., Loqué, D., & Ralph, J. (2010). Advances in modifying lignin for enhanced biofuel production. *Current Opinion in Plant Biology, 13*, 313–320.

Soejima, H., Sugiyama, T., & Ishihara, K. (1995). Changes in the chlorophyll contents of leaves and in levels of cytokinins in root exudates during ripening of rice cultivars Nipponbare and Akenohoshi. *Plant Physiology, 36*, 1105–1114.

Stewart, J. J., Akiyama, T., Chapple, C., Ralph, J., & Mansfield, S. D. (2009). The effects on lignin structure of overexpression of ferulate 5-hydroxylase in hybrid poplar. *Plant Physiology, 150*, 621–635.

Sticklen, M. (2006). Plant genetic engineering to improve biomass characteristics for biofuels. *Current Opinion in Biotechnology, 17*, 315–319.

Studer, M. H., DeMartini, J. D., Davis, M. F., Sykes, R. W., Davison, B., Keller, M., Tuskan, G. A., & Wyman, C. E. (2011). Lignin content in natural Populus variants affects sugar release. *Proceedings of the National Academy of Sciences of the United States of America, 108*, 6300–6305.

Sundberg, B., Little, C. H. A., Cui, K., & Sandberg, G. (1991). Level of endogenous indole-3-acetic acid in the stem of *Pinus sylvestris* in relation to the seasonal variation of cambial activity. *Plant, Cell & Environment, 14*, 241–246.

Taylaran, R. D., Ozawa, S., Miyamoto, N., Ookawa, T., & Hirasawa, T. (2009). Performance of a high-yielding modern rice cultivar Takanari and several old and new cultivars grown with and without chemical fertilizer in a submerged paddy field. *Plant Production Science, 12*, 365–380.

Taylaran, R. D., Adachi, S., Ookawa, T., Usuda, H., & Hirasawa, T. (2011). Hydraulic conductance as well as nitrogen accumulation plays a role in the higher rate of leaf photosynthesis of the most productive variety of rice in Japan. *Journal of Experimental Botany, 62*, 4067–4077.

Thomas, R. J. (1991). Wood: formation and morphology. In M. Lewin, & I. S. Goldstein (Eds.), *Wood structure and composition* (pp. 7–47). New York: Marcel Dekker.

Tilman, D., Hill, J., & Lehman, C. (2006). Carbon-negative biofuels from low-input high-diversity grassland biomass. *Science, 314*, 1598–1600.

Timell, T. E. (1986). *Compression Wood in Gymnosperms, (Vol. 1).* Berlin: Springer.

Torney, F., Moeller, L., Scarpa, A., & Wang, K. (2007). Genetic engineering approaches to improve bioethanol production from maize. *Current Opinion in Biotechnology, 18*, 193–199.

US Department of Energy. (2006). *Breaking the biological barriers to cellulosic ethanol. A research roadmap resulting from the biomass to biofuels workshop.* Rockville, MD: US Department of Energy.

Vanden Wymelenberg, A., Sabat, G., Mozuch, M., Kersten, P. J., Cullen, D., & Blanchette, R. A. (2006). Structure, organization, and transcriptional regulation of a family of copper radical oxidase genes in the lignin-degrading basidiomycete *Phanerochaete chrysosporium*. *Applied and Environmental Microbiology, 72*, 4871–4877.

Vega-Sanchez, M. E., & Ronald, P. C. (2010). Genetic and biotechnological approaches for biofuel crop improvement. *Current Opinion in Biotechnology, 21*, 218–224.

Vignols, F., Rigau, J., Torres, M. A., Capellades, M., & Puigdomènech, P. (1995). The *brown midrib3 (bm3)* mutation in maize occurs in the gene encoding caffeic acid *O*-methyltransferase. *Plant Cell, 7*, 407–416.

Vogel, K. P., Bredja, J. J., Walters, D. T., & Buxton, D. R. (2002). Switchgrass biomass production in the Midwest USA: Harvest and nitrogen management. *Agronomy Journal, 94*, 413–420.

Waclawovsky, A. J., Sato, P. M., Lembke, C. G., Moore, P. H., & Souza, G. M. (2010). Sugarcane for bioenergy production: An assessment of yield and regulation of sucrose content. *Plant Biotechnology Journal, 8*, 263–276.

Wang, D., Portis, A. R., Jr., Moose, S. P., & Long, S. P. (2008). Cool C_4 photosynthesis: Pyruvate Pi dikinase expression and activity corresponds to the exceptional cold tolerance of carbon assimilation in *Miscanthus* × *giganteus*. *Plant Physiology, 148*, 557–567.

Wang, H., Xue, Y., Chen, Y., Li, R., & Wei, J. (2012). Lignin modification improves the biofuel production potential in transgenic *Populus tomentosa*. *Industrial Crops and Products, 37*, 170–177.

Wang, Q. (2011). Time for commercializing non-food biofuel in China. *Renewable and Sustainable Energy Reviews, 15*, 621–629.

Weng, J. K., Li, X., Bonawitz, N. D., & Chapple, C. (2008). Emerging strategies of lignin engineering and degradation for cellulosic biofuel production. *Current Opinion in Biotechnology, 19*, 166–172.

Xu, B., Escamilla-Treviño, L. L., Sathitsuksanoh, N., Shen, Z., Shen, H., Zhang, Y. H., Dixon, R. A., & Zhao, B. (2011). Silencing of 4-coumarate:coenzyme A ligase in switchgrass leads to reduced lignin content and improved fermentable sugar yields for biofuel production. *New Phytologist, 192*, 611–625.

Yamamoto, T., Yonemaru, J., & Yano, M. (2009). Towards the understanding of complex traits in rice: Substantially or superficially? *DNA Research, 16*, 141–154.

Yoshinaka, K., & Kawai, S. (2012). Mutagenesis, heterogeneous gene expression, and purification and amino acid substitution analyses of plant peroxidase, PrxA3a. *Journal of Wood Science, 58*, 231–242.

Yoshizawa, N. (1987). Cambial responses to the stimulus of inclination and structural variation of compression wood tracheids in gymnosperms. *Bulletin of the Utsunomiya University Forests, 23*, 23–141.

Zhang, K., Qian, Q., Huang, Z., Wang, Y., Li, M., Hong, L., Zeng, D., Gu, M., Chu, C., & Cheng, Z. (2006). *GOLD HULL AND INTERNODE2* encodes a primarily multifunctional cinnamyl-alcohol dehydrogenase in rice. *Plant Physiology, 140*, 972–983.

Zobel, B. J., & Jett, J. B. (1995). *Genetics of wood production*. Berlin: Springer.

Zobel, B. J., & van Buijtenen, J. P. (1989). *Wood variation: Its causes and control*. Berlin: Springer.

Chapter 5

Soil Fertility and Soil Microorganisms

Haruo Tanaka, Akane Katsuta, Koki Toyota and Kozue Sawada

Research Approaches to Sustainable Biomass Systems. http://dx.doi.org/10.1016/B978-0-12-404609-2.00005-2

5.1. SOIL FERTILITY

Haruo Tanaka

5.1.1. Definition of Soil Fertility

Soil fertility is defined as "the quality of a soil that enables it to provide nutrients in adequate amounts and in proper balance for the growth of specified plants or crops" (Soil Science Glossary Terms Committee, 2008). It is not only based on the natural conditions or peculiar property that the soil has but also the human activities such as growing various crops by applying different cultivation methods. Soil fertility is the most important factor to affect the production of biomass in a field. Usually, sustainable high yields of biomass can be expected from fertilized soil. For land that is not fertile, application of a large quantity of natural or chemical fertilizer with high labor demand is necessary to maintain high biomass production.

An agro-ecosystem is viewed as a subset of a natural ecosystem. Traditionally, the agro-ecosystem is characterized as having a simpler species composition and simpler energy and nutrient flows than a natural ecosystem. The soil has the function as the decomposer in the agro-ecosystem to decompose the organic matter such as composts or plant residues applied to the soil into inorganic matters biologically by animals and microorganisms inhabiting the soil. However, excessive application of organic matter causes environmental pollution such as the groundwater contamination with ammonium. In addition, as a producer, the soil supports the growth of plants that absorb nutrients and water from the soil to grow under solar irradiation. Hence, "soil productivity" is a synonym of "soil fertility".

5.1.2. Soil Potential Productivity Classification

In Japan, "Soil Potential Productivity Classification" (National Conference of Soil Conservation and Survey Project, 1979) is used to evaluate soil fertility; the soil potential productivity classification is presented based on the report by Hamazaki and Micosa (1991).

Classification of soil potential productivity is a form of interpreting the results of soil surveys. It is a practical method to grade or group soils based upon their limitations or hazards for crop production and/or risk of soil damage to jeopardize crop production; all these concerns are closely related to soil's physical and chemical properties. The objective of land capability classification is to eliminate limitations for increasing crop productivity. There are four soil capacity classifications, i.e. I, II, III, and IV, defined as follows:

- Class I – Land has almost no limitations or hazards for crop production and/or risks of soil damage.
- Class II – Land has some limitations or hazards and/or risks of soil damage, and some improvement practices are required for normal crop production.
- Class III – Land has many limitations or hazards and/or risks of soil damage, and fairly intensive improvement practices are required.
- Class IV – Land has great natural limitations, so it is difficult to use as arable land.

5.1.3. Factors Affecting the Capability Classification

The classification standard varies according to types of crops, such as paddy, upland, orchard, and grassland. In Japan, the Soil Potential Productivity Classification is carried out by evaluating each of the standard factors, or the inherent soil characteristics; some of these factors are determined by a combination of dependent factors, or supplementary soil characteristics. The standard and dependent factors are described in sections (a)–(m) that follow.

a. Thickness of Topsoil (t: code in simplified formula; Table 5.1)

Topsoil is the surface horizon where plant roots can easily penetrate; this horizon corresponds to the plowed layer in general arable land. It is grouped according to the thickness of topsoil as shown in Table 5.1. For upland or orchard, more than 25 cm of topsoil is classified as class I, and less than 25 cm is classified as class II or III. On the other hand, for the paddy or grassland, more than 15 cm is classified as class I, because paddy or grassland crops are capable of growing in thinner topsoil than upland or orchard crops.

TABLE 5.1 Thickness of Topsoil (t)

| Thickness | Class | | | |
t (cm)	Paddy	Upland	Orchard	Grassland
>25	I	I	I	I
25–15	I	II	II	I
<15	II	III–IV*	III–IV*	II–III[†]

*When effective depth (d) is placed in class IV, (t) factor is also placed in class IV.
[†]When effective depth (d) is placed in class III or IV, then (t) factor is also placed in class III.

TABLE 5.2 Effective Depth of Soil (*d*)

Depth	Class			
d (cm)	Paddy	Upland	Orchard	Grassland
>100	I	I	I	I
100–50	I	II	II	I
50–25	II	III	III	I–II
25–15	III	III	IV	II–III
<15	IV	IV	IV	III–IV

b. Effective Depth of Soil (d; Table 5.2)

Effective depth is the maximum depth beyond which the soil's physical conditions become unfavorable for the downward root to develop for normal crops, for example bedrock with hard pan more than 29 mm of soil hardness as determined using Yamanaka's core penetration, and more than 10 cm of thick gravel layer.

c. Gravel Content of Topsoil (g; Table 5.3)

Gravel (rock and mineral fragments) contents in the topsoil are expressed as percentages of exposed surface area of gravel in the soil profile.

d. Ease of Plowing (p; Table 5.4)

Estimation of plowing is based on the evaluation of resistance against agricultural machinery and friability of soil clods. Both properties are largely dependent upon the soil moisture condition as well as the quantity of clay and organic matter. Classes of this factor are evaluated using four dependent factors

TABLE 5.3 Gravel Content of Topsoil (*g*)

Gravel	Class			
g (%)	Paddy	Upland	Orchard	Grassland
<5	I	I	I	I
5–10	I	II	I	I
10–20	I–II	II–III	I–II	II
20–50	II–III	III–IV	II–III	III–IV
>50	IV	IV	IV	IV

TABLE 5.4 Ease of Plowing (p)

a	b	c	d	Class	Criteria
Dependent factors					
1	1	(2)	1	I	
2	2	2	1	I	Easy to slightly difficult
2	2	2	2	I	
2	2	3	2	II	Moderately difficult
3	3	3	1	II	
2	2	3	3	III	Very difficult
3	3	3	2	III	

[a] Texture of top soil (Coarse to medium: 1; Fine: 2; Very fine: 3).
[b] Stickiness of topsoil (Non-sticky to slightly sticky: 1; Sticky: 2; Very sticky: 3).
[c] Consistence of topsoil when dry (Loose to soft: (2); Slightly hard: 1; Hard: 2; Very hard to extremely hard: 3).
[d] Moisture condition of topsoil (Dry to moderately dry: (2); Moist: 1; Wet: 2; Very wet: 3). Moisture condition refers to the major part of the year and/or 2–3 days after considerable rainfall.

and their combinations. For example, soil texture is based on the classification used in the ISSS system; textures of S, LS, SL, FSL, L, and SiL are classified as "Coarse to medium"; SCL, CL, and SiCL are classified as "Fine"; and SC, LiC, SiC, and HC are classified as "Very fine".

e. Permeability Under Submerged Conditions (l; only for paddy fields; Table 5.5)

This soil permeability affects the movement of water in the soil, soil temperature, and leaching of nutrients or development of reduced condition of the soil. It is evaluated mainly by the combination of soil texture and the presence of a compact layer within 50 cm of the surface as dependent factors. The measurement of the water permeability coefficient and/or water requirement in depth has been widely used, and hence the permeability data are handy for carrying out the soil permeability classification. The dependent factors that are relevant to classifying the soil are as shown in Table 5.5. This table shows similar classifications of finest soil texture as in Table 5.4, but the ratings are different. To determine the maximum compactness within 50 cm of the surface, Yamanaka's core penetrometer is used. Value ranges of >24, 24–11, and <11 are classified as "Very compact to compact", "Medium to loose", and "Very loose" respectively.

f. State of Oxidation–Reduction Potentiality (r; only for paddy fields; Table 5.6)

This standard factor indicates the risk of root damage due to strong reduction of soil, resulting in low rice production. Three dependent factors, i.e. "Content

TABLE 5.5 Permeability under Submerged Conditions (l; only for paddy fields)

Dependent factors		Class	Criteria
a	b	(Paddy)	
1	1	I	Poorly to imperfectly permeable
1	2	I	
2	2	II	Moderately to well permeable
3	2	II	
3	3	III	Well to excessively permeable

[a]Finest soil texture within 50 cm of the surface (Very fine: 1; Fine: 2; Medium to coarse: 3).
[b]Maximum compactness within 50 cm of the surface (Very compact to compact: 1; Medium to loose: 2; Very loose: 3).

TABLE 5.6 State of Oxidation–Reduction Potentiality (r; only for paddy fields)

Dependent factors			Class	Criteria
a	b	c	(Paddy)	(Risk of root damage)
1	1	2	I	None to weak
1	3	2	I	
2	1	2	I	
1	1–2	3	II	Moderate to strong
1	3	3	II	
2	1–2	3	II	
3	1	2	II	
2	3	3	III	Very strong
3	2	2	III	
3	1	3	III	
3	3	2	III	

[a]Contents of easily decomposable organic matter in topsoil (Low: 1; Medium: 2; High: 3).
[b]Contents of free iron oxides in topsoil (High: 1; Medium: 2; Low: 3).
[c]Degree of gleyzation (Weak: 1; Moderate: 2; Strong: 3).

of easily decomposable organic matter in topsoil", "Contents of free iron oxides in topsoil", and "Degree of gleyzation" are used. Organic matter that is easily decomposed is represented by NH_4-N cg kg^{-1} in air-dried soil (A) and NH_4-N cg kg^{-1} in soil after the soil sample has been incubated at 30°C for 4 weeks (B). The easily decomposable organic matter is classified as "Low"

($A < 10$, $B < 10$), "Medium" (A: 10–20; B: 10–15), and "High" (A: >20; B: >15). If based on the content of free iron oxides (cg kg^{-1}), the topsoil can be classified as "High" (>1.5), "Medium" (1.5–0.8), and "Low" (<0.8). For the degree of gleyzation, the depth of the top gley horizon is used, and the soil is classified as "Weak" (≥50 cm), "Moderate" (50 cm to bottom of plow layer), and "Strong" (above the bottom of plow layer). Peat and muck are regarded as the gley horizon.

g. Wetness of Land (w; wet condition, (w) dry condition, for upland, orchard, and grasslands; Table 5.7)

This standard factor is used for estimating of the risk of drought or wetness of upland crops, trees, and grasslands; the classes and criteria are as follows: (Class-criteria) (IV) – High possibility of drought; (III) – Possibility of drought; (II) – Low possibility of drought; I – None; II – Low possibility of over-wetness; III – Possibility of over-wetness; IV – High possibility of over-wetness. These classes are evaluated by the combination of three dependent factors, as shown in Table 5.7. One of the dependent factors, water-holding ability of surface and subsurface horizons, is evaluated based on the quantity of water between field capacity (pF 1.5) and permanent wilting point (pF 4.2). A value of available water (g L^{-1}) >200 is "High", 200–100 is "Medium", and <100 is "Low". The above classification is for upland, orchard, and grassland; different criteria are used for paddy fields.

TABLE 5.7 High Possibility of Over-Wetness (w; for upland, orchard, and grassland field)

Dependent factors			Class		
a	b	c	Upland	Orchard	Grassland
1	3	(2)	(IV)	(III)–(IV)	(III)–(IV)
1	3	1	(III)	(II)–(III)	I–(II)
1	2	1	(II)	I–(II)	I
1	1	1	I	I	I
2	2	2	II	II–III	I
1–3	1	3	III	III–IV	II
3	2	3	IV	IV	III–IV

[a]Permeability of solum (High: 1; Medium: 2; Low: 3).
[b]Water-holding ability of surface and subsurface horizons (High: 1; Medium: 2; Low: 3, determined by available water).
[c]Moisture condition (Dry to moderately dry: (2); Moist: 1; Wet: 2; Very wet: 3).

TABLE 5.8 Inherent Fertility of the Soil for Upland, Orchard, and Grassland (f)

| Dependent factors | | | | |
a	b	c	Class	Criteria
1	2	1	I	Fertile
2	1	2	I	
1	2	3	II	Medium
2	1	3	II	
1	3	1	II	
1	3	2	II	
1	3	3	III–II	Infertile–medium
3	1	1	III	Infertile
2	4	2	III–II	Infertile–medium

[a] Nutrient holding capacity (High: 1; Medium: 2; Low: 3).
[b] Nutrient fixation power (Very low: 1; Low: 2; Medium: 3; High: 4).
[c] Base status of the soil (Good: 1; Medium: 2; Poor: 3).

h. Inherent Fertility of the Soil (f; Table 5.8)

The soil's inherent fertility is evaluated by using combinations of the following three dependent factors: nutrient-holding capacity, nutrient fixation power, and base status. Nutrient-holding capacity is evaluated based on cation exchanged capacity (CEC; $cmol_c$ kg^{-1} soil) as "High" (>20), "Medium" (20–6), and "Low" (<6). Classification of nutrient fixation power is based on the phosphate absorption coefficient of the topsoil evaluated using the Blakemore method, and the classifications are "Very low" (value <35%), "Low" (35–80%), "Medium" (80–90%), and "High" (>90%). Base on the pH (H_2O) value, the soil can be classified as "good" (pH > 5.5), "medium" (pH between 5.5 and 5.0), and "poor" (pH < 5.0). Table 5.8 shows the classification for upland, orchard, and grassland, while other tables are used for paddy fields.

i. Content of Available Nutrients (n; Table 5.9)

Available nutrients in topsoil are closely related to the inherent soil fertility, but evidently influenced by the combination of the following dependent factors: contents of exchangeable Ca, exchangeable Mg, exchangeable K, content available P (determined by using the Truog method), available N and Si (for paddy), and micro-elements (evaluated by using the risk of deficiency), as well as soil acidity as indicated by the pH (H_2O) value.

TABLE 5.9(a) Content of Available Nutrients (n)

Rating	Class	Criteria
1	I	High
2	II	Medium
3	III	Low

TABLE 5.9(b) Rating of Dependent Factors of Available Nutrients

Dependent factors	Rating			
	1	2	3	4
Content of exchanged Ca (cmol$_c$ kg^{-1})	>7.1	7.1–3.6	<3.6	–
Content of exchanged Mg (cmol$_c$ kg^{-1})	>1.2	1.2–0.5	<0.5	–
Content of exchanged K (cmol$_c$ kg^{-1})	>0.32	0.32–0.17	<0.17	–
Content of available P (mg kg^{-1})	>44	44–9	<9	–
Content of available N (cg NH$_4$-N kg^{-1})	>20	20–10	<10	–
Content of available Si (cg SiO$_2$ kg^{-1})	>15	15–5	<5	–
Content of micro-element	None–slight	Moderate	Severe	–
Acidity pH (H$_2$O)	>6	6–5	5–4.5	<4.5

j. Hazard (i; Table 5.10)

This standard factor refers mainly to the limitation caused by the presence of excessive quantities of hazardous substances such as sulfur compounds, soluble salts, and heavy metals, among many others. Dependent factors for this standard factor include: (1) Harmful sulfur compounds (None, Slight, Moderate, and Severe), salt content (None to slight, Moderate, and Severe, evaluated by using chlorine as the indicator), heavy metals including Cr, Ni, Cu, Zn, As, Mn, Mo, etc. (None, Slight, Moderate, and Severe); (2) irrigation water quality based on temperature, pH (H$_2$O), total N, salts, and heavy metal (none to Slight, Moderate, and Severe), and (3) physical hazards that involve the difficulty of removal of bedrock, pan, compact layer or gravel layer being within 50 cm of the surface (None, Difficult, and Very difficult).

k. Frequency of Accidents (a; Table 5.11)

This standard factor is mainly influenced by natural environmental conditions rather than by soil characters. Classification of this standard factor is

TABLE 5.10 Hazard (*i*)

Class	Criteria
I	None
II	Slight
III	Moderate
IV	Severe

TABLE 5.11 Frequency of Accidents (*a*)

Class	Criteria
I	None
II	Moderate
III	Frequent

determined based on the following two independent factors: risk of overhead flooding inundation (None to slight, Moderate, and Frequent), and risk of land creep (None to slight, Moderate, and Frequent).

l. Slope of the Field (s; for upland and orchard; Table 5.12)

The natural slope is the main dependent factor; its classification is decided by a combination of natural slope, direction of slope, and artificial slope.

TABLE 5.12 Slope of the Field (*s*; for upland and orchard)

Slope (%)	Rating	Upland	Orchard
<5	1	I	I
5–14	2	II	I
14–27	3	III	I–II
27–47	4	IV	II–III
>47	5	IV	IV

TABLE 5.13 Erosion (e; for upland orchard)	
Class	Criteria
I	None or very slight
II	Slight
III	Serious
IV	Very severe

m. Erosion (e; for upland orchard; Table 5.13)

The degree of erosion or occurrence of rill or gully (Very slight, Slight, Moderate, and Severe), the power to resist water erosion for topsoil determined using the dispersion ratio (Strong, Moderate, and Weak), and the resisting power wind erosion for topsoil determined using the soil bulk density (Strong and Weak) are pertinent here.

5.1.4. Expression of Productive Capability Class

The evaluation of these standard and dependent factors is expressed as either a simplified code formula or a detailed code formula. When a simplified code formula, e.g. "II *plrn*", is used, the productive capability class is placed in the lowest class of factors. This code formula arranges briefly the information regarding the kind and the degree of limitations with each class of land. The code "II *plrn*" means that the land is classified class II because the factors *p* (ease of plowing), *l* (permeability under submerged conditions), *r* (state of oxidation–reduction potentiality), and *n* (content of available nutrient) are grouped in class II. If the soil fertility of this land is to be improved, improving all of these factors together is necessary.

5.1.5. Improvement of Potential Productivity Classification

Soil Potential Productivity Classification, which is focused on crop production, represents the inhibitory factors of the crop production clearly with a simple code formula. Now that environmental conservation has become an important issue, these issues may include prevention of nitrogen eluviation to the groundwater, and prevention of the emission of greenhouse gases including methane and nitrous oxide. The classification must include environmental conservation functions. In addition, as far as the current situation is concerned, the classification must consider not only chemical and physical factors, but also biological factors such as activity or variety of the soil organisms.

5.2. SOIL MANAGEMENT AND SOIL ORGANIC MATTER

5.2.1. Carbon Dynamics on the Earth Scale

In recent years, the importance of soil organic carbon (SOC) in the carbon cycle on a global scale has attracted the attention of public and private sectors, and the general public as well. Increasing greenhouse gases such as CO_2 in the atmosphere causes adverse effects on the climate system. For the whole Earth, about 750 Pg of carbon exist in the atmosphere as CO_2 and about 500 Pg of carbon accumulates in the land plant biomass. On the other hand, about 1500 Pg carbon exists as SOC in soil organic matter (SOM), and the quantity is equivalent to approximately twice the total atmospheric CO_2 and approximately three times the total land plant biomass (IPCC, 2001). Hence, it is certain that the slightly increasing or decreasing SOC quantity influences global carbon greatly because of the huge quantity of SOC existing in the soil. It is estimated that about 2000 Pg of carbon used to exist in soil as SOC during the prehistoric age, and the amount SOC presently is estimated to be about 1500 Pg. Human activities to degrade soil and reduce SOM contained in the soil lead to the loss of about 500 Pg SOC. The amount of carbon released as a result of soil degradation is equivalent to double the amount of 230 Pg, which is the total quantity of carbon released by consuming fossil fuels (Hakamata et al., 2000). If the soil degradation is restored using appropriate management practices, soil can become a huge sink of atmospheric carbon through carbon sequestration. Much research has been undertaken to study how to predict the quantitative change of SOC, and estimate the carbon balance more accurately. This is also emphasized in the Kyoto Protocol in order to prevent the expected global warming.

5.2.2. Factors in the Increase and Decrease of SOC

The SOC decomposition rate is affected by environmental factors as discussed in the following paragraphs:

1. In the range of natural temperature, both the microbial activity and the SOC decomposition rate increase at higher temperature. Hence, tropical zones have less SOC than frigid zones due to the temperature difference.
2. Either higher or lower water content than the level for maximum microbial growth in the soil causes the SOC decomposition rate to slow down, so that the amount of SOC increases. If the soil water content is appropriate, microbial activity reaches the highest level so that more SOC is decomposed to result in less SOC in the soil. In paddy fields, SOC accumulates because the paddy soil is fully submerged in irrigation water when rice is cultivated.
3. The quantity of SOC increases when the soil is clayey.
4. Soil with extreme acidity or alkalinity inhibits the activity of soil microorganisms, so that the SOC decomposition rate becomes slow, and SOC accumulates in the soil.

5. The amount of SOC increases in soil that is not tilled for growing crops.
6. The amount of SOC increases if more organic matter such as plant residues or composts is applied.
7. The amount of SOC increases when the organic matter applied to the soil contains a higher C/N ratio.
8. If the ratio of organic matter that is refractory to microbial decomposition, e.g. lignin, is applied, the quantity of SOC increases.

5.2.3. Principle of Rothamsted Carbon Model

Among the many SOM turnover models proposed by international researchers, the Rothamsted Carbon model (RothC) developed for non-waterlogged soil will be reviewed. RothC was developed by Jenkinson and Rayner (1977) based on data from long-term experiments on soils at the Rothamsted Experimental Station (currently known as Rothamsted Research) in the UK. The model has been improved several times, and the current version is known as RothC-26 (Coleman and Jenkinson, 1996).

RothC is one of the multicompartmental models in which SOC is split into one non-active compartment and four active compartments with different decomposition rates. The calculation is carried out on a monthly basis. Incoming plant carbon is split between decomposable plant material (DPM) and resistant plant material (RPM) depending on the DPM/RPM ratio of the carbon. Both DPM and RPM decompose to form microbial biomass (BIO), humified organic matter (HUM), and partly CO_2 that is lost from the system. The proportion that goes to CO_2 and to BIO + HUM is determined by the clay content of the soil; clayey soil has a slower SOC decomposition rate than sandy soil. Both BIO and HUM undergo further decomposition repeatedly to produce more CO_2 as well as other species of BIO and HUM. The model also includes inert organic matter (IOM) that does not decompose so that it remains in the soil with the quantity unchanged. Decomposition of SOCs in the four active compartments except IOM follows first-order kinetics with different characteristic decomposition rate constants. Both soil temperature and/or water content influence the magnitude of this rate constant; a fast decomposition rate is expected at high temperature and slow decomposition rate constant with dry soil. When the soil experiences temperature changes to alter the SOC decomposition rate, the decomposition rate constants are modified by multiplying by one or more "rate modifiers".

The input data required to run the model include:

1. Monthly rainfall (mm)
2. Monthly open pan evaporation (mm)
3. Average monthly air temperature (°C)
4. Clay content of the top soil (as a percentage)
5. An estimate of the decomposability of the incoming plant material – DPM/RPM ratio

6. Soil cover – Is the soil bare or vegetated in a particular month?
7. Monthly input of plant residues (Mg-C ha^{-1})
8. Monthly input of farmyard manure (FYM, in Mg-C ha^{-1}), if any.

The PC software to run RothC can be downloaded free from the website http://www.rothamsted.ac.uk/aen/carbon/rothc.htm.

5.2.4. Modification of RothC for Andosols and Paddy Soils

RothC has been used to adequately simulate changes in the SOC content for a variety of upland soil types, including arable land and grassland in Europe and the USA. In Japan, Andosols and paddy soils are major arable soils in Japan, and they have slow organic matter decomposition rates. However, this model does not simulate SOC changes in Andosols and paddy soils in Japan with satisfactory results. Modifications of the RothC model have been performed in recent years for a better application to Japanese characteristic fields (Shirato et al., 2004; Shirato and Yokozawa, 2005).

The main reason that RothC does not work properly for Andosols is that active aluminum formed in the weathering process of the volcanic ash holds organic matter strongly. The presence of Al–humus complexes in Andosols makes the HUM pool extremely stable. This has not been considered in the original RothC model. In the modified RothC model for Andosols, a factor $H(f)$ is incorporated, and the value of $H(f)$ is calculated based on the quantity of pyrophosphate extractable aluminum (Al$_p$), i.e. $H(f) = 1.20 + 2.50 \times$ Al$_p$ (%). The decomposition rate of the HUM compartment is modified by dividing the original HUM by $H(f)$. The IOM pool is set to "zero" because the soil did not contain carbon when it was formed from fresh volcanic ash. The modified model has been evaluated with satisfactory results by using sets of long-term experimental data collected on Andosols under various climate conditions.

The main reason for the slow decomposition rate of organic matter during the rice-growing season in flooded paddy fields is that the submerged soils are waterlogged and subjected to anaerobic conditions to hamper the activities of soil microorganisms. On the other hand, the decomposition of organic matter is inhibited in paddy soils not only during the submergence period but also throughout the whole year because of the difference in the composition of microorganisms between upland and paddy soils. The decomposition rate in the modified RothC model for paddy fields is set to 0.2-fold in the submergence period (summer) and 0.6-fold in the period without submergence (winter) as compared with the original model. These modifications have been found to be satisfactory for the long-term experimental data sets collected from paddy soils.

5.2.5. Application of Modified RothC

Box 5.1 shows examples of utilizing the modified RothC. It becomes clear that cultivation of the forage rice for high crop yield will assist in alleviating global

BOX 5.1 Prediction of Soil Organic Carbon Sequestration Change using Modified Rothamsted Carbon Model

The method of predicting changes of soil organic carbon (SOC) using the modified Rothamsted Carbon model (RothC) is explained here.

Figure 5.1A shows a prediction of the changes of SOC in the field of Andosols, which is the upland located at 35°41′ N and 139°29′ E. Mean monthly temperatures at this location are 4.6°C in January and 27.0°C in August, with an annual mean temperature of 15.6°C. Mean precipitation is 38 mm in January and 315 mm in August, with an annual precipitation of 1639 mm. Dry density of the soil at this location is 0.68 Mg m^{-3}, and the clay content is 35.6%. This soil is Melanic Silandic Eutrosilic Andosols; the pyrophosphate extractable Al (Al$_p$) is 0.67%. Compost in an amount of 5.38 Mg-C ha^{-1} y^{-1} is applied to the compost application plot. When the plot is rotated with growing wheat and soybean crops, input of the plant residue becomes 4.44 Mg-C. The SOC of the chemical fertilizer plot was 60 Mg-C ha^{-1} in 2007, but is expected to decrease to 55 Mg-C ha^{-1} in 2050 when compost will not be applied. Compost application is known to be a necessary procedure for maintaining an appropriate SOC level in the soil; applying 5.38 Mg-C ha^{-1} of compost continuously will lead to a predicted SOC level of 105 Mg-C ha^{-1} in 2050. Hence, applying an adequate amount of compost is expected to either maintain or increase carbon sequestration in soil that will contribute significantly to alleviating the global warming problem.

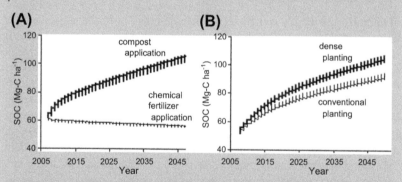

(A) **(B)**

FIGURE 5.1   Estimation of soil organic carbon sequestration changes from 2008 to 2050 with different fertilizer application in the upland field (A), and with different planting density of forage rice (Leaf Star) in the paddy field (B).

Figure 5.1B shows a prediction of changes of SOC in the soil of the paddy, located at 35°40′ N, 139°28′ E in the alluvial plain of the Tama River near the upland field. Both locations have almost the same climatic conditions. The soil at this location is Fluvic Hydragric Anthrosols; its dry density is 0.64 Mg m^{-3} and the clay content is 34.2%. The amount of compost application is 4.30 Mg-C ha^{-1} y^{-1}. In this field, forage rice of "Leaf Star" is cultivated for producing bioethanol. For the cultivation of "Leaf Star", input of the plant residue is 3.78 Mg-C ha^{-1} y^{-1} with dense planting and 2.30 Mg-C ha^{-1} y^{-1} with conventional

(Continued)

BOX 5.1 Prediction of Soil Organic Carbon Sequestration Change using Modified Rothamsted Carbon Model—Cont'd

planting. The cultivation of forage rice requires more plant residue to be plowed into the soil, and as a result SOC is predicted to increase to 92 Mg-C ha^{-1} for conventional planting and 106 Mg-C ha^{-1} for dense planting in 2050. Cultivation of the forage rice in high yield will contribute to alleviating the global warming problem through bioethanol production and by increasing soil carbon sequestration.

warming through not only bioethanol production but also increased soil carbon sequestration. Hence, a turnover of SOC varies according to the difference in cropping system so that changes of SOC sequestration can be predicted using an appropriate model such as RothC. Additionally, choosing an appropriate cropping system that does not alter the SOC sequestration process needs to be considered for future agricultural activities.

5.3. SOIL MICROORGANISMS

Akane Katsuta and Koki Toyota

Soil microbes are mainly composed of bacteria, fungi, protozoa, algae, archaea such as methanogens and halobacteria, and viruses. They play crucial roles in soil function, especially in nutrient cycling, and thus contribute to sustainable crop production. In agricultural systems, the contribution of synthetic nitrogen fertilizers to crop production is unequivocal. Surprisingly, the largest amount of nitrogen uptake by a crop is derived from the mineralization of soil organic matter in general. Soil microbes are essential to this mineralization process, and soils with higher nitrogen-supplying capacity are considered to have higher soil fertility. In this section, general aspects of soil microbes are presented by emphasizing the role of soil microbes involved in crop production. In the next section, microbial mediation of soil fertility is discussed.

5.3.1. Abundance and Biomass

In general, hundreds of millions to billions of bacteria and hundreds of thousands to millions of fungi exist in 1 g of soil, although the number of microorganisms differs depending on soil properties and environmental conditions (Table 5.14). The total amount of microbial biomass ranges from 0.1 to 1 mg g^{-1} of soil. Bacteria and fungi are the two predominant soil microbes in number and mass. The third most dominant category of microbes are protozoa consisting of ciliate, flagellate and ameba, and their numbers range from tens to hundreds of thousands per gram of soil, and their biomass varies from 1 to 10 μg g^{-1} of soil.

TABLE 5.14 Abundance and Biomass of the Major Soil Organisms

	Population		Biomass
	per m^2	per g-soil	wet kg ha^{-1}
Bacteria	$10^{13}-10^{14}$	10^8-10^9	300–3000
Actinomycetes	$10^{12}-10^{13}$	10^7-10^8	300–3000
Fungi	$10^{10}-10^{11}$	10^5-10^6	500–5000
Algae	10^9-10^{10}	10^3-10^6	10–1500
Protozoa	10^9-10^{10}	10^3-10^5	5–2000
Nematodes	10^6-10^7	10–100	1–100
Earthworms	30–300		10–1000
Arthropods	10^3-10^5		1–200

Data from Alexander (1977), Brady (1974), and Lynch (1983).

5.3.2. Habitats

Soil microbes live in the liquid and/or gas phases of soil and mostly by attaching themselves to soil particles. Microbes need to obtain energy and nutrients through their membranes in the presence of water, and thus they cannot grow without water. Therefore, pores filled with water or the surfaces of soil particles having water films are the primary microhabitats. Micropores with diameters of 3 μm or less, in which water is held by capillary force, are the major habitats for soil bacteria; macropores are habitats for fungi and protozoa.

Microbial number and activity are enhanced where organic matter such as compost and plant residues is abundant (see Box 5.2). The surrounds of crop residues and dead bodies of soil animals are the major sites colonized by abundant microbes. The rhizosphere, defined as the narrow zone of soil subject to the influence of living roots, is also a hot spot for microbes because organic matter such as the leakage or exudation from roots is plentiful.

BOX 5.2 Effects of Organic Supplements on Nematode Populations in Soil

Organic materials have some characteristics that cannot be replaced by chemical fertilizers. One such characteristic is to enhance the growth, activity, and diversity of soil microorganisms. An example relating to the impact on the populations of nematodes, which play an important role in nutrient cycling in soil, is presented here.

Twenty-one types of organic material including compost, crop or food residue were added into samples of Andosols, which is a typical soil covering 50% of total upland fields in Japan, at rates ranging from 1.2 to 100 g kg^{-1} soil on a fresh basis

(Continued)

BOX 5.2 Effects of Organic Supplements on Nematode Populations in Soil—Cont'd

and 0.9–45 g kg^{-1} soil on a dry basis. After incubation for 3 weeks at 25°C, nematodes were extracted from the soils by using the Baermann funnel method to be counted under a microscope.

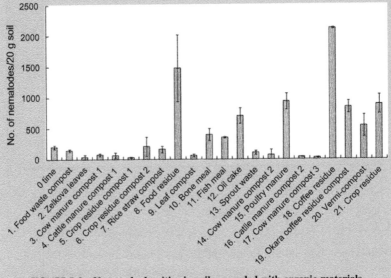

FIGURE 5.3 Nematode densities in soils amended with organic materials.

As shown in Figure 5.3, the nematode population varies greatly depending on the type of organic material applied. The nematode population increases in soils supplemented with nine materials such as food residue (No. 8; application rate 14 g (dry) kg^{-1} soil) and coffee residue (No. 18; 11 g (dry) kg^{-1} soil). Most nematodes in soil are either bacterial or fungal feeders, suggesting that the increasing nematode population reflects higher biomass of bacteria and fungi. Because there is no significant correlation observed between the organic matter application rates and the increase in nematode populations, the degree of increase in nematode populations is considered to vary depending on the quality rather than quality of organic materials added.

Since nematodes enhance the nitrogen-supplying capacity of soil through the mineralization of organic matter, the organic materials that enhance the nematode population are considered to be beneficial to soil fertility.

5.3.3. Taxonomy

With advances in molecular taxonomy, microorganisms are now divided into three domains, Bacteria, Archaea and Eucarya, according to the sequences of ribosomal RNA. Approximately 800 genera and 5000 species of bacteria have been described, and they are classified into more than 19 divisions or groups.

A bacterial group that used to be called Gram-negative bacteria mainly corresponds to Proteobacteria and Bacteroides/Chlorobi. Proteobacteria consist of five subdivisions, α, β, γ, δ, and ϵ. Symbiotic nitrogen-fixing bacteria *Rhizobium* and free-living nitrogen fixing bacteria *Azospirillum* are included in α-proteobacteria. *Burkholderia, Ralstonia*, and nitrifying bacteria (*Nitrosomonas*) are members of β-proteobacteria. *Pseudomonas* and Enterobacteriaceae such as *Escherichia coli* and *Salmonera* spp. are γ-proteobacteria, and sulfate-reducing bacteria (*Desulfovibrio*) are δ-proteobacteria. *Cytophagales*, as well as *Flavobacterium* and *Sphingobacterium*, are included in Bacteroides/Chlorobi. Bacteria that used to be called Gram-positive bacteria are divided into two divisions: Actinobacteria and Firmicutes; Actinobacteria contain filamentous bacteria such as *Streptomyces* and coryneform bacteria such as *Arthrobacter*, whereas Firmicutes include spore-forming bacteria such as *Bacillus* and *Clostridium*.

It is well known that the classical dilution plate method depicts only a part of soil bacteria. Recent advances in molecular techniques have revealed the presence of quite diverse bacteria, which are rarely deposited in culture collections. Strains belonging to *Arthrobacter, Streptomyces*, and *Bacillus* used to be considered as the major soil bacteria based on the dilution plate method, but it appears that they occupy a tiny portion of the total bacteria based on the molecular method (Table 5.15). A comparison between the dilution plate and molecular methods (Figure 5.2) shows that Actinobacteria, Firmicutes, and α-Proteobacteria are dominant in the dilution plate method whereas more diverse bacterial groups are observed in the clone library method.

Eighty thousand species of fungi have been described, and they are divided into the four classes Zygomycota, Ascomycotina, Basidiomycetes, and Chytridiomycetes. Mycorrhizal fungi, *Mucor* and *Rhizopus*, are classified as Zygomycota and ectomycorrhizal fungi as Basidiomycetes. Soil-borne plant pathogens belonging to imperfect fungi such as *Fusarium* and *Verticillium* are now classified as Ascomycotina. *Plasmodiophora brassicae*, clubroot fungi, which were previously classified as fungi but is now classified as Cercozoa, belong to a group of protists.

5.3.4. Microbial Functions

a. Decomposition

The principal role of soil microbes is the decomposition of organic materials. Once incorporated into soil, all biodegradable organic materials are readily decomposed by soil microorganisms into carbon dioxide eventually, although at different rates. The largest source of organic materials is the residues of plant that generally contain 15–60% cellulose, 10–30% hemicellulose, 5–30% lignin, 2–15% protein, and about 10% saccharides, amino sugars, organic acids, amino acids and nucleic acids. Soil microbes have the capability to decompose a variety of organic materials. In general, microbes that degrade hemicelluloses or cellulose are less diversified, and their degradation rates are slower than microbes

TABLE 5.15 Major Soil Bacteria Based on the Dilution Plate and Molecular Methods

Genus name	Ratio (%) by clone library method	Ratio* (%) by dilution plate method	Ratio (%) in the culture collection
Actinomadura	0	–	1.5
Actinoplanes	0.06	–	1.5
Agrobacterium	0	0–13	–
Alcaligenes	0.09	1–8	–
Arthrobacter	0.53	3–40	1.3
Bacillus	0.62	5–45	7.6
Clostoridium	0.09	–	1.6
Flavobacterium	0.38	1–7	–
Flexibacter	0	–	1.2
Hyphomicrobium	0.03	–	1.2
Micromonospora	0	0–5	2.1
Mycobacterium	0.50	–	2.6
Nocardia	0	3–10	–
Paenibacillus	0.18	–	1.4
Pseudomonas	1.60	2–10	6.0
Ralstonia	0	–	1.0
Rhodococcus	0	–	1.4
Streptomyces	0.06	23–30	25.2

*Calculated based on Alexander (1977).
Data from Janssen (2006).

degrading simple sugars. Lignin is known to be more resistant to microbial decomposition than hemicelluloses and cellulose. While their degradation rates are different, microbes degrading these organic materials are ubiquitous in soil.

b. Anaerobic Respiration

Although the degradation of organic materials by microorganisms is faster under aerobic conditions, degradation of organic materials will occur even under

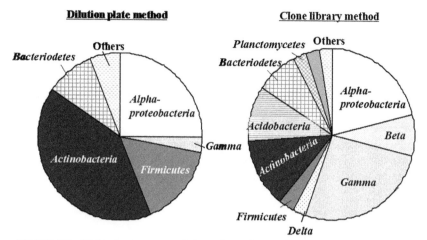

FIGURE 5.2   Molecular taxonomy of the major soil bacteria evaluated by the dilution plate and molecular methods. Thirty to forty isolates or clones were analyzed. *(Toyota and Kuninaga, 2006; Katayama, Yamakawa and Toyota, unpublished data)*

anaerobic conditions. In anaerobic respiration, nitrate, Mn^{4+}, Fe^{3+}, sulfate, acetate, and carbon dioxide are used as electron acceptors instead of oxygen. The degradation of organic materials is readily carried out under anaerobic conditions such as in a flooded paddy soil, if some of the electron acceptors are available. The anaerobic respiration is performed by diverse microbes. Nitrate respiration in which nitrate or nitrite is reduced to nitrous oxide or molecular nitrogen is performed by bacteria belonging to diverse genera, including α-, β- and γ-Proteobacteria and Firmicutes. Furthermore, some fungi, including the most typical genus *Fusarium*, obtain energy for their growth under anaerobic conditions by reducing nitrite to nitrous oxide.

Strictly anaerobic bacteria live even in upland soil, suggesting that anaerobic microsites are localized in soil, e.g. within macro-aggregates.

c. Nitrification

Autotrophic bacteria that gain energy by the oxidation of mineral materials and carbon from CO_2 in the air exist in soil. A typical example is ammonia-oxidizing bacteria that obtain energy by oxidizing ammonia to nitrite. Nitrite-oxidizing bacteria are also autotrophs that obtain energy by oxidizing nitrite to nitrate. Because nitrification is an energy-generating process, the reaction proceeds readily in soil if appropriate substrates are available. Nitrogen fertilizer applied in the form of urea or ammonium sulfate is transformed into nitrate through nitrification in soil. Because nitrifying bacteria are strictly aerobic, their growth is enhanced in upland soil. In general, most nitrogen applied as fertilizer is transformed into nitrate within weeks in agricultural soils. Nitrate is easily leached out of the soil layers along with water movement.

The leached nitrate that is a loss of the applied nitrogen fertilizer can lead to nitrate pollution of groundwater and surface water bodies such as lakes and rivers. Thus, the nitrification process is not always beneficial to agriculture. For this reason, nitrification inhibitors have been developed in order to mitigate the loss of nitrate in agricultural soils. Furthermore, it is known that a small but significant portion of nitrogen fertilizer applied to agricultural soil is transformed into nitrous oxide, a potent greenhouse gas, through the nitrification processes.

Paddy field soil is generally under anaerobic conditions but there are aerobic pockets in the soil, e.g. the oxidation layer (a few mm) at the upper part of the plow layer and the rhizosphere of rice plants, where aerobic nitrification occurs.

d. Nitrogen Fixation

The most beneficial microorganisms in agriculture are nitrogen-fixing bacteria, including prokaryotes, bacteria, and archaea of diverse groups. Some cyanobacteria have functions of both photosynthesis and nitrogen fixation, and therefore are expected to play a role in the restoration of infertile soil. Nitrogen-fixing bacteria are extremely diverse in terms not only of taxonomy but also physiology and ecology. Some nitrogen-fixing bacteria, such as *Rhizobium* and *Bradyrhizobium*, are symbiotic to leguminous plants, whereas other diverse bacteria have been reported as free-living nitrogen fixers. Filamentous bacteria such as *Frankia* are known to live in symbiosis with non-leguminous plants such as alder (*Alnus* sp.) and *Comptonia* sp. Among free-living nitrogen fixers, some are plant associated and others not. Nitrogen-fixing bacteria are also known in archaea, such as methanogens and *Halobacterium*. Some nitrogen fixers are strict aerobes, whereas some others are strict anaerobes such as *Clostridium,* or photosynthetic and facultative such as *Enterobacter* and *Klebisiella*. In any type of nitrogen-fixing bacteria, the major nitrogen-fixing enzyme nitrogenase functions only in the absence of oxygen and therefore anaerobic bacteria must have some mechanisms to avoid oxygen near the cells.

The total amount of nitrogen fixed biologically in terrestrial regions of 139 million ton N per year exceeds that fixed artificially (100 million ton N per year) (Brady and Weil, 2008). A plant's nitrogen-fixing potential differs depending on the type of land use (Table 5.16), with the highest in leguminous crops and relatively high nitrogen fixation in paddy fields. Cyanobacteria and photosynthetic bacteria play important roles in the nitrogen fixation of paddy soil. Many studies have revealed the presence of endophytic and/or plant-associated nitrogen-fixing bacteria, such as *Herbaspirillum*, *Azoarcus*, and *Burkholderia*, in non-leguminous plants.

Nitrogen-fixing activity is affected by different parameters. Firstly, inorganic nitrogen suppresses nitrogen-fixing activity significantly. Therefore, the

TABLE 5.16 Nitrogen-Fixing Potential under Different Land Uses

	kg-N ha^{-1} y^{-1}
Paddy rice	20
Legume	140
Non-legume	8
Pasture	15
Forest	10
Other vegetation	20

Data from Brady and Weil (2002) and Nishio (2005).

application rate of nitrogen fertilizer is generally low in leguminous crops. A future trend in research is the stimulation of biological nitrogen fixation in agricultural soils as an alternative to applying synthetic nitrogen fertilizers.

e. Plant Growth-Promoting Microorganisms

Fluorescent *Pseudomonas* is well known to suppress soil-borne plant diseases, and such types of bacteria are called plant growth-promoting rhizobacteria (PGPR). In addition to bacteria, some fungi like *Phoma* and *Trichoderma* also promote plant growth; they are called plant growth-promoting fungi (PGPF). The term PGPR is usually used for antagonists that colonize plant roots and suppress pathogens; however, PGPR also includes microorganisms that enhance plant growth by producing plant hormones such as auxin and supplying nutrients such as nitrogen or phosphorus (Table 5.17).

5.3.5. Soil Sickness due to Continuous Cropping

In general, crop yield decreases after continuous cropping. This is because some pathogenic soil microbes or nematodes accumulate in the soil and cause diseases in plants. Sugar cane, sugar beet, maize or corn, wheat, etc. have been evaluated as biofuel-producing plants, but many types of soil-borne disease are reported in these crops that may pose a threat to future crops. Crop rotation has been recommended and practiced to avoid such losses. Soil disinfection with synthetic chemicals or solar energy is also regularly used where crop rotation is difficult to implement. Such disinfection practice is feasible in relatively smaller land areas for growing vegetables but is not practical in large fields for growing cereals and energy

TABLE 5.17 Free-Living Nitrogen-Fixing Bacteria Isolated from Plants

Plants	Taxonomy
Sugar cane (*Saccharum* spp.)	*Azospirillum* sp., *Azotobacter* sp., *Beijerinckia indica, Beijerinckia fluminensis, Derxia* sp., Enterobacteriaceae, *Klebsiella* sp., *Paenibacillus azotofixans, Vibrio* sp.
Maize (*Zea mays*)	*Azospirillum lipoferum, Azotobacter vinelandii, P. azotofixans, K. terrigena,* Enterobacter sp., E. cloacae, Bacillus circulans, Burkholderia sp., B. vietnamiensis
Wheat (*Triticum* spp.)	*A. lipoferum, Azotobacter* sp., *Bacillus* sp., *E. cloacae, E. agglomerans, K. oxytoca, K. pnuemoniae*
Rice (*Oryza sativa*)	*A. lipoferum, Azotobacter* sp., *Clostridium* sp., *E. cloacae, E. agglomerans, K. oxytoca, K. pnuemoniae, P. azotofixans, Pseudomonas* sp., *H. seropedicae, Burkholderia brasilensis, Azospirillum irakense*
Paspalum notatum	*Acetobacter paspali, A. halopareferans, Enterobacter agglomerans*
Kallar grass (*Leptochloa fusca*)	*A. halopareferans, E. cloacae, E. agglomerans, Azoarcus* sp.
Spartina	*A. lipoferum, H. frisingense, A. brasilense, Gluoconacetobacter diazotrophicus, H. seropedicae, A. vinelandii, Azotobacter chroococcum, P. stutzeri*
Banana (*Musa* spp.)	*K. pnuemoniae, E. cloacae, H. seropedicae, H. rubrisubulbicans*
Sweet potato	*Pantoea agglomerans, Acetobacter diazotrophicus*
Oak	*P. putida, P. fluorescence, Xanthomonas oryzae, B. megaterium*
Cereals	*H. seropedicae, A. lipoferum, Azospirillum amazonense, Azospirillum irakense, A. brasilense, B. brasilense*

TABLE 5.17 Free-Living Nitrogen-Fixing Bacteria Isolated from Plants—Cont'd

Plants	Taxonomy
Sugar cane	H. seropedicae, B. tropicalis, G. diazotrophicus, G. azotocaptans, A. brasilense, A. lipoferum, A. amazonense, B. brasilensis, H. seropedicae, Pantoea sp.
Forage grasses	H. seropedicae, A. brasilense, A. lipoferum, B. brasilensis
Sago palm (Metroxylon sagu)	K. pnuemoniae, K. oxytoca, E. cloacae, P. agglomerans, B. megaterium
Coffee	Gluconacetobacter johannae, G. azotocaptans
Mangrove	A. chroococum, A. vinelandii, A. beijerinckii
Pineapple	Acetobacter diazotophicus, H. rubrisubulbicans
Dune grass	B. tropicalis, P. agglomerans, Sternotrophomonas maltophilia
Pine	B. gladioli, B. glathei, Bacillus sp.
Miscanthus spp.	A. doebereinerae, A. lipoferum, H. frisingense

crops. Most soil-borne plant pathogens are aerobic and have difficulty surviving in paddy fields, as seen by the continuous cropping of paddy rice for more than 1000 years. Hence, rice should be focused on as a sustainable energy crop in future studies.

5.4. MICROBIALLY MEDIATED SOIL FERTILITY

Kozue Sawada and Koki Toyota

5.4.1. Productivity and Environmental Impacts

Soil fertility has been considered as the characteristics of soils that support high crop yields in order to enhance crop production. With recent concerns about the environment, however, the developments of agriculture with less damage to the environment is desirable. Therefore, improving soil fertility should consider not only enhancing crop yields but also maintaining the balance between crop nutrient requirements and nutrient supplies because either excessive or deficient nutrients have major adverse impacts on the productivity and the environment.

Soil microbes govern the numerous reactions of nutrient cycles in soils. Transformations between organic and inorganic nutrients by microbes (i.e. biological mineralization and immobilization) contribute significantly to supplying nutrients to crops, especially nitrogen (N) and sulfur, and to some extent phosphorus. The availability of potassium, which is mostly fixed in minerals such as micas and feldspars, is not affected by microbial activities. In the agricultural field, nitrogen is one of the major limiting elements for crop yields. In addition, excessive nitrogen causes nitrate leaching, ammonia volatilization, and nitrous oxide emissions that often lead to adverse environmental effects, including air and water pollution, soil acidification, ozone destruction and global warming, as well as human health problems. This section focuses on the biological mineralization and immobilization of nitrogen, which contribute significantly to soil fertility in agricultural ecosystems.

5.4.2. Nitrogen Cycles Through Microbial Biomass in Soils

Major processes in the nitrogen cycle through microbial biomass in soil are illustrated in Figure 5.4. In agricultural ecosystems, both organic and inorganic nitrogen are supplied through the application of organic materials, such as crop

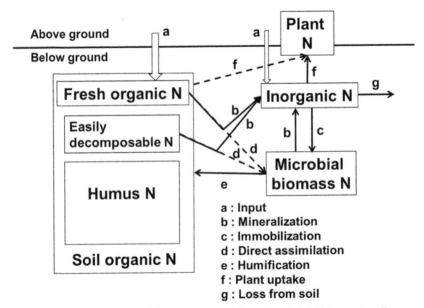

FIGURE 5.4   Diagram of the nitrogen cycle through microbial biomass in soils.

residues and manure, and mineral fertilizer (Figure 5.4a). The organic nitrogen, whether freshly added or already existing in soil, is continuously decomposed by soil microbes via two routes: (1) mineralization of organic nitrogen to inorganic forms (Figure 5.4b), and (2) immobilization of inorganic nitrogen to organic forms through microbial assimilations (Figure 5.4c). Although the organic nitrogen forms a large pool of nitrogen in soil, only a small fraction of soil organic nitrogen, i.e. easily decomposable nitrogen, can be decomposed biologically during the crop growing period. Recent studies show that microbes can assimilate significant amounts of organic molecules (e.g. amino acids) directly in their tissues (Figure 5.4d) (Geisseler et al., 2010). Turnover of microbial biomass nitrogen releases inorganic nitrogen (mineralization (Figure 5.4b)) and organic nitrogen (humification (Figure 5.4e)). Uptake of inorganic nitrogen and some organic nitrogen by plants occurs in the vegetated area (Figure 5.4f), and the excess inorganic nitrogen is lost from soil to groundwater (nitrate leaching), to surface runoff, or to the atmosphere as N_2, N_2O (dinitrification), and NH_3 (volatilization) (Figure 5.4g). Soil microbes play important roles in regulating the nitrogen cycles, especially the highly labile nitrogen pool.

The nitrogen cycles in soils and ecosystems are more complicated than those shown in Figure 5.4. For example, nitrification of ammonium to nitrate occurs continuously in the inorganic nitrogen pool. Ammonium in soil is strongly and almost irreversibly fixed by several 2:1-type clay minerals such as vermiculites, fine-grained micas and some smectites in soils. The conversion of atmospheric N_2 into soil inorganic nitrogen (nitrogen fixation) or the atmospheric input of inorganic nitrogen to soil also occur. In the soil environment, nitrification, denitrification, and nitrogen fixation are carried out respectively by some specific microbial groups (see reviews by Hayatsu et al., 2008). The decomposer system in soil is a diverse food web that involves numerous types of organisms; the most important groups are bacteria, archaea, fungi, and protozoa, collectively known as soil microbes, which account for about 80% of the decomposition activities (Brady and Weil, 2008). Soil fauna (e.g. nematodes, mites, earthworms) act as regulators of decomposition through feeding on microbes, and also alter the soil's physical environment through movement and burrowing of the soil.

5.4.3. Nitrogen Cycle Through Microbial Biomass in Japanese Agricultural Soils

a. Nitrogen-Supplying Capacity in Japanese Agricultural Soils

Mishima et al. (2006) evaluated recent trends in nitrogen application and crop production for Japanese agricultural land. Chemical nitrogen fertilizer application peaked in 1985 (near 130 kg-N ha^{-1}) and then decreased (less than 100 kg N ha^{-1}) whereas the total amount of livestock excreta applied to agricultural land peaked in 1990 to near 110 kg-N ha^{-1} and then decreased to less than 80 kg-N ha^{-1}. The nitrogen uptake by crop plants peaked in 1985 to about 65 kg-N ha^{-1} and then decreased to about 59 kg-N ha^{-1}. Total nitrogen surplus is the difference between the sum of nitrogen inputs to agricultural land including applications of chemical fertilizer, livestock excreta as manure as well as irrigation and N$_2$ fixation, and the sum of nitrogen outputs in crop yields from agricultural land such as the removal of by-products and natural denitrification. The nitrogen surplus peaked in 1985 at about 150 kg-N ha^{-1} and then declined to less than 100 kg-N ha^{-1} (Figure 5.5).

The mean content of total nitrogen in Japanese agricultural soils is ~2600 mg-N kg^{-1}, calculated based on 2272 sets of data collected from Japanese agricultural fields by the Soil Conservation Project (Oda and Miwa, 1987). Japanese agricultural lands vary widely in soil types, reflecting mainly the differences in parent materials, weathering degree and topography, as well as land uses, e.g. paddy, upland, pasture, and orchard. In general, the content of total nitrogen and distribution pattern of various nitrogen fractions are significantly affected by soil types and land use. Sano et al. (2004) grouped soil nitrogen into three fractions: inorganic, potentially mineralizable, and stable organic. The potentially mineralizable nitrogen is considered as the estimated net accumulation of inorganic nitrogen after long-term aerobic incubation of the soil sample for 22 weeks or until the net inorganic nitrogen accumulation reaches a plateau. This nitrogen pool that originates from the easily decomposable nitrogen and the microbial biomass nitrogen pools is shown in Figure 5.4. The sum of inorganic and potentially mineralizable nitrogen is thought to represent the amount of nitrogen available for crop uptake during the crop growing period. This nitrogen

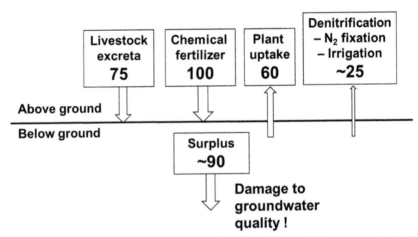

FIGURE 5.5 Average annual input, output and surplus of nitrogen in Japanese agricultural soils in 2002 (units: kg-N ha^{-1}).

available for plant growth amounts to 196 mg kg^{-1} (295 kg-N ha^{-1}) or 7.6% of total nitrogen assuming that the plow layer depth is 15 cm and the bulk density is 1.0 Mg m^{-3}. The potentially mineralizable nitrogen is higher in paddy soils (200 mg-N kg^{-1}) than in upland soils (108 mg-N kg^{-1}), probably because relatively abundant fresh organic matter is added through the application of rice straw, the natural depositions by algae, and N$_2$ fixation in paddy soils. Another possibility is that the organic matter in paddy soils is less degraded than that in uplands under anaerobic conditions. Stable organic nitrogen, which is unavailable for plant growth, is higher in volcanic soils than in non-volcanic soils because the presence of Al–humus complexes in volcanic soils gives humus strong stability. Volcanic soils cover more than 50% of the total upland fields in Japan, hence, Japan has higher total nitrogen in agricultural soil ($\sim$2600 mg-N kg^{-1}) than the average value in the world ($\sim$1600 mg kg^{-1}) (Batjes, 1996). However, the high total nitrogen in Japanese agricultural soil does not necessarily imply that the soil has high nitrogen-supplying capacity for crop production, although the nitrogen-supplying capacity in various soils is closely related to the total nitrogen content in the soil.

b. Microbial Biomass Nitrogen in Japanese Agricultural Soils

Soil microbial biomass nitrogen is usually determined by using the chloroform fumigation extraction method (Brookes et al., 1985). This method is based on the extraction of cell components immediately after chloroform fumigation. Although microbial biomass nitrogen occupies only 1–6% of the total nitrogen in soils, the biomass nitrogen pool has the potential to supply available nitrogen for crop growth because of rapid turnover rate of the biomass nitrogen. The microbial biomass nitrogen is reported to be highly correlated with the quantity of available nitrogen for crops to uptake in Japanese upland soils (e.g. Sakamoto and Oba, 1993). Inubushi and Watanabe (1986) also observed that microbial biomass nitrogen is highly correlated with nitrogen uptake by rice under conditions of nitrogen deficiency in flooded paddy fields, although biological processes in paddy soils are complicated (see review by Inubushi and Acquaye, 2004). Nira (2000) estimated both turnover rates of microbial biomass nitrogen and amounts of mineralized nitrogen for 1 year by applying ^{15}N-labeled crop residues to different types of upland soils. His results show that the flux of nitrogen through microbial biomass enhances the amounts of mineralized nitrogen by 39–138% (average 67%). Sakamoto et al. (1997) observed that the flux of nitrogen through microbial biomass is a major source of nitrogen uptake for upland rice in the vegetative growth stage (until 6 weeks). These results support that managing microbial biomass nitrogen is essential for the maintenance and improvement of soil fertility.

Guan et al. (1997) observed that fractions of microbial biomass nitrogen to total nitrogen in nine volcanic soils of 0.76% average are about one-fourth lower than those in seven non-volcanic soils (3.2% average) in Japan. This result further supports that total organic nitrogen is not a good indicator of the nitrogen-supplying capacity to crop uptake, especially in volcanic soils.

It is known that long-term applications of organic materials (e.g. crop residue or manure, etc.) to soils usually increase the total soil organic nitrogen, including microbial biomass nitrogen (e.g. Sakamoto and Oba, 1991), and improve the soil fertility profoundly. The present status of long-term field studies using chemical fertilizers and organic materials conducted at national and prefectural agricultural research stations in Japan is summarized by Kanamori (2000). Shirato et al. (2004) used a model approach and showed that soil organic matter can be more easily accumulated after long-term application of organic materials in Japanese volcanic soils than in non-volcanic soils due to the presence of active aluminum and iron in the latter. However, the information on the effect of long-term application of organic materials on microbial biomass nitrogen is relatively scarce, especially in volcanic soils (Sakamoto and Oba, 1991). This is because microbial biomass nitrogen must be measured using soil samples freshly collected from the field, whereas soils stored in air-dried conditions is acceptable for measuring total nitrogen.

5.4.4. Management of Microbial Biomass Nitrogen During a Crop Growth Period

a. The Effect of Drying–Rewetting on Microbial Biomass Nitrogen

(i) Basic Principle

Surface soils are sometimes subjected to rapid rewetting by rainfall following dry conditions, and the rewetting of dried soil is recognized to cause a significant and sudden increase in net nitrogen mineralization (Birch, 1958) through the enhanced availability of both microbial biomass and non-biomass soil organic nitrogen. Mineralized nitrogen is derived from lyses of microbial cells (Bottner, 1985) and/or release of microbial osmoregulatory compounds responding to the water potential shock (Halverson et al., 2000), as well as through the physical release of occluded soil organic matter from soil aggregates (Denef et al., 2001). Results of case studies conducted in Japanese paddy ecosystems and in dry tropical ecosystems on the turnover of microbial biomass nitrogen under drying–rewetting cycles are summarized in the following paragraphs.

(ii) Japanese Paddy Ecosystems

The amount of nitrogen mineralization after flooding following an air-drying period increases drastically in Japanese paddy fields in a phenomenon known as the "drying effect". Mineralized nitrogen associated with the drying effect in Japanese paddy fields is estimated to be 50–100 kg-N ha^{-1}, which corresponds approximately to the amount of chemical nitrogen fertilizer applied to paddy soils annually. Toriyama et al. (1988) evaluated the effect of soil water contents before flooding on the measured mineralized nitrogen after flooding, and found that the mineralized nitrogen increases linearly with decreasing soil water contents before flooding when the soil has been dried below pF 4. Ando et al. (1995) also observed that the amount of nitrogen taken up by rice plants increases with decreasing soil water contents before flooding. These results prove that the soil water content before flooding affects significantly the amount of nitrogen mineralized caused by the drying effect. According to Marumoto et al. (1997), the contribution of microbial biomass nitrogen to total nitrogen mineralized due to the drying effect is 20–48% in Japanese paddy soil. This result clearly shows that microbial biomass nitrogen in Japanese paddy soil acts as an available nitrogen pool for rice crops. Therefore, the management of microbial biomass nitrogen is essential for maintaining and improving rice production. Practicing long-term application of organic materials is usually recommended for enhancing the soil biomass nitrogen, especially for soils that have been subject to sufficient drying before flooding.

(iii) Dry Tropical Ecosystems

The availability of nitrogen limits crop production in dry tropical cropland with nutrient-poor soils. Soil microbes in this region act as an important nitrogen pool for crop nitrogen uptake. Singh et al. (1989) observed in dry tropical natural ecosystems in India that microbial biomass nitrogen is accumulated, probably due to accumulation of intracellular solutes during the dry seasons and the later depletion of these solutes during the early rainy seasons. Sugihara et al. (2010a) also observed that microbial biomass nitrogen in dry tropical cropland in Tanzania tends to decrease during rainy seasons but will increase and remain at high levels during dry seasons. These results suggest that the mineralization of biomass nitrogen stimulates plant growth during rainy seasons in nutrient-poor dry tropical ecosystems. However, if crops are unable to uptake the nitrogen mineralized during early rainy season because of their slow growth, the mineralized nitrogen would be potentially lost from soil through leaching.

b. Effect of Application of Organic Materials on Microbial Biomass Nitrogen

(i) Basic Principle

Application of organic materials (e.g. crop residue or manure, etc.) to soils immediately and significantly alters several fractions in nitrogen pools and enhances the turnover rates by increasing the soil microbial activities (Jensen and Magid, 2002). The rates of gross mineralization and immobilization differ depending on the type of organic materials applied to soil. When no organic material is applied to soil, gross mineralization is usually slightly higher than

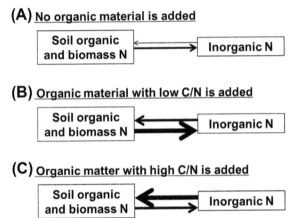

(A) No organic material is added

(B) Organic material with low C/N is added

(C) Organic matter with high C/N is added

FIGURE 5.6 The processes of gross mineralization and gross immobilization of nitrogen when (A) no organic material, (B) organic material with low C/N, and (C) organic material with high C/N are added. The thickness of the arrow signifies the magnitude of the flux rates.

gross immobilization, thus resulting in net mineralization (Figure 5.6A). When organic materials with low C/N ratio, e.g. animal manure, are applied to soil, the increased biological activity causes a greater increase of gross mineralization than gross immobilization, although gross immobilization increases. Therefore, the net mineralization of nitrogen is usually enhanced (Figure 5.6B). In contrast, when organic materials with high C/N ratio, e.g. crop straw, are applied to soil, the enhanced biological activities result in a greater increase of gross immobilization than gross mineralization. Therefore, the net immobilization of nitrogen (Figure 5.6C) causes a shortage of nitrogen for crops immediately after the applications of organic materials. This phenomenon is called "nitrogen starvation".

Nitrogen turnover is regulated by two microbial parameters, i.e. microbial growth efficiency and C/N ratio of newly formed biomass, in addition to the C/N ratio of organic materials applied. The microbial growth efficiency is defined as the ratio of substrate carbon and the carbon that is incorporated into microbial biomass. Substrate that is not incorporated is respired as CO_2. Substrate quality, defined as the ease with which a substrate is decomposed, also affects the temporal patterns of gross mineralization and immobilization of nitrogen.

(ii) Nitrogen Cycle Through Microbial Biomass after Application of Organic Materials

Katoh et al. (2009) evaluated the effects of combined applications of ^{15}N-labeled ammonium sulfate and cattle, swine and poultry manures (with C/N ratios of 6–26) on nitrogen cycles under upland soil conditions. They observed that ^{15}N-NH_4 derived from ammonium sulfate (5–60 mg-N kg^{-1} soil) is immobilized by microbes 7 days after the application of manure although the manure contained high concentrations of nitrogen. They also observed that the quantity of immobilized ^{15}N is significantly correlated with the amounts of easily decomposable carbon. This suggests that the quality of substrate carbon is an important factor to influence the gross immobilization of nitrogen immediately after the application of organic materials.

Herai et al. (2006) measured microbial biomass nitrogen, NO_3-N leaching and nitrogen uptake by corn over 4 months after application of sawdust compost (with C/N ratio of ~ 30) to a sandy soil in Japan. Their results clearly showed that the assimilation of nitrogen to microbial biomass decreases NO_3-N leaching without negative impact on nitrogen uptake by corn. This method is effective in reducing NO_3-N leaching.

Sugihara et al. (2010b) observed that a significant amount of inorganic nitrogen (~ 50 kg-N ha^{-1}) is lost from the surface soil through leaching during early rainy seasons, and that inorganic nitrogen remains low thereafter (~ 20 kg-N ha^{-1}) in dry tropical cropland in Tanzania. Therefore, Sugihara et al. (2012) evaluated the effect of the application of plant residue (with C/N ratio of ~ 70) on nitrogen leaching and crop productivity. They observed that microbial biomass nitrogen clearly increases (14.6–29.6 kg-N ha^{-1}) with the application of plant residue during the early crop growth period, and that potentially leachable nitrogen is also immobilized. They also reported that the crop nitrogen uptake is improved by applying the plant residue with the conclusion that the re-mineralization of immobilized nitrogen stimulates crop growth during the later period. Therefore, plant residue application is a promising option to improve crop production and reduce the leaching of nutrients.

5.4.5. Future Prospects

In developing countries, especially in low-input traditional systems, nitrogen availability limits crop production because of low soil fertility and the inability of farmers to afford chemical fertilizers. In these regions, soil microbes act as an important nitrogen pool for crop nitrogen uptake, and therefore the management of microbial biomass nitrogen has the potential to improve crop production (Sugihara et al., 2012). However, to date, little quantitative information concerning the effect of the management of microbial biomass nitrogen on crop production, especially the flux of nitrogen through microbial biomass, is available in literature. Therefore, further research needs to be initiated to carry out quantitative analyses on soil nitrogen, such as with the ^{15}N-labeled nitrogen.

In contrast, in developed countries such as Japan, intensive agricultural practice that has already increased crop production causes severe environmental problems. In Japanese agricultural ecosystems, surplus nitrogen adversely affects environment quality (Figure 5.5). To reduce the nitrogen surplus effectively, various types of organic materials need to be applied as an alternative to chemical fertilizers. When organic materials are applied to agricultural fields, synchronizing nitrogen supply with crop demand is needed in order to maintain crop production while minimizing the nitrogen surplus.

With recent concerns about environmental quality, further studies are needed to understand the biological mineralization and immobilization over different time scales, even though high agricultural productivity has already been achieved in Japan. Additional studies should be initiated to address solving problems for volcanic soils and paddy soils that are unique to Japan.

REFERENCES

Alexander, M. (1977). *Introduction to soil microbiology.* Wiley.

Ando, H., Marumoto, T., Wada, G., & Nakamura, T. (1995). Mineralization of soil organic nitrogen and nitrogen absorption by rice (*Oryza sativa*) plant as affected by drying duration. *Japanese Journal of Soil Science and Plant Nutrition, 66*, 499–505 [in Japanese].

Batjes, N. H. (1996). Total carbon and nitrogen in the soils of the world. *European Journal of Soil Science, 47*, 151–163.

Birch, H. (1958). The effect of soil drying on humus decomposition and nitrogen availability. *Plant and Soil, 10*, 9–31.

Bottner, P. (1985). Response of microbial biomass to alternate moist and dry conditions in a soil incubated with ^{14}C and ^{15}N labelled plant material. *Soil Biology and Biochemistry, 17*, 329–337.

Brady, N. C. (1974). *The nature and properties of soils* (8th ed.) MacMillan.

Brady, N. C., & Weil, R. R. (2002). *The nature and properties of soils* (13th ed.) Prentice Hall.

Brady, N. C., & Weil, R. R. (2008). Organisms and ecology of the soil. In *The nature and properties of soils* (14th ed.) (pp. 443–494). Upper Saddle River, NJ: Prentice Hall.

Brookes, P. C., Landman, A., Pruden, G., & Jenkinson, D. S. (1985). Chloroform fumigation and the release of soil nitrogen: A rapid direct extraction method to measure microbial biomass nitrogen in soil. *Soil Biology and Biochemistry, 17*, 837–842.

Coleman, K., & Jenkinson, D. S. (1996). RothC-26.3 – A model for the turnover of carbon in soil. In D. S. Powlson, P. Smith, & J. U. Smith (Eds.), *Evaluation of soil organic matter models: Using existing long-term datasets* (pp. 237–246). Berlin: Springer.

Denef, K., Six, J., Bossuyt, H., Frey, S. D., Elliott, E. T., Merckx, R., & Paustian, K. (2001). Influence of dry–wet cycles on the interrelationship between aggregate, particulate organic matter, and microbial community dynamics. *Soil Biology and Biochemistry, 33*, 1599–1611.

Geisseler, D., Horwath, W. R., Joergensen, R. G., & Ludwig, B. (2010). Pathways of nitrogen utilization by soil microorganisms – A review. *Soil Biology and Biochemistry, 42*, 2058–2067.

Guan, G., Marumoto, T., Shindo, H., & Nishiyama, M. (1997). Relationship between the amount of microbial biomass and physicochemical properties of soil – Comparison between volcanic and non-volcanic ash soils. *Japanese Journal of Soil Science and Plant Nutrition, 68*, 614–621 [in Japanese].

Hakamata, T., Hatano, R., Kimura, M., Takahashi, M., & Sakamoto, K. (2000). Interaction between greenhouse gases and soil ecosystem: 1. Carbon dioxide and terrestrial ecosystem. *Japanese Journal of Soil Science and Plant Nutrition, 71*, 263–274 [in Japanese].

Halverson, L. J., Jones, T. M., & Firestone, M. K. (2000). Release of intracellular solutes by four soil bacteria exposed to dilution stress. *Soil Science Society of America Journal, 64*, 1630–1637.

Hamazaki, T., & Micosa, A. G. (1991). Land capability classification in Japan – Productive capability classification of land based on soil survey. Collected papers on environmental planning/Division of Environmental Assessment. National Institute of Agro-Environmental Sciences, No. 7, 1–20.

Hayatsu, M., Tago, K., & Saito, M. (2008). Various players in the nitrogen cycle: Diversity and functions of the microorganisms involved in nitrification and denitrification. *Soil Science and Plant Nutrition, 54,* 33–45.

Herai, Y., Kouno, K., Hashimoto, M., & Nagaoka, T. (2006). Relationships between microbial biomass nitrogen, nitrate leaching and nitrogen uptake by corn in a compost and chemical fertilizer-amended regosol. *Soil Science and Plant Nutrition, 52,* 186–194.

Inubushi, K., & Acquaye, S. (2004). Role of microbial biomass in biogeochemical processes in paddy soil environments. *Soil Science and Plant Nutrition, 50,* 793–805.

Inubushi, K., & Watanabe, I. (1986). Dynamics of available nitrogen in paddy soils: II. Mineralized N of chloroform-fumigated soil as a nutrient source for rice. *Soil Science and Plant Nutrition, 32,* 561–577.

IPCC. (2001). *Climate change 2001: The scientific basis.* Cambridge: Cambridge University Press.

Janssen, P. H. (2006). Identifying the dominant soil bacterial taxa in libraries of 16S rRNA and 16S rRNA genes. *Applied and Environmental Microbiology, 72,* 1719–1728.

Jenkinson, D. S., & Rayner, J. H. (1977). The turnover of soil organic matter in some of the Rothamsted classical experiments. *Soil Science, 123,* 298–305.

Jensen, L. S., & Magid, J. (2002). Nutrient turnover in soil after the addition of organic matter. In L. S. Jensen (Ed.), *Plant nutrition, soil fertility, fertilizers and fertilization* (4th ed.) (pp. 15.1–15.20). Samfundslitteratur.

Kanamori, T. (2000). Present state of long-term field experiments on successive application of chemical fertilizers and composts as organic matters in national and prefectural agricultural research stations. *Japanese Journal of Soil Science and Plant Nutrition, 71,* 286–293 [in Japanese].

Katoh, M., Hayashi, Y., & Morikuni, H. (2009). Effect of combined application with ^{15}N-labeled ammonium sulfate and swine or poultry manure compost on mineralization of compost nitrogen. *Japanese Journal of Soil Science and Plant Nutrition, 80,* 152–156 [in Japanese].

Lynch, J. M. (1986). *Soil biotechnology.* Blackwell.

Marumoto, T., Andoh, H., & Wada, G. (1997). Air-drying treatment and mineralization of biomass nitrogen in paddy soils. *Japanese Journal of Soil Science and Plant Nutrition, 68,* 376–380 [in Japanese].

Mishima, S., Taniguchi, S., & Komada, M. (2006). Recent trends in nitrogen and phosphate use and balance on Japanese farmland. *Soil Science and Plant Nutrition, 52,* 556–563.

National Conference of Soil Conservation and Survey Project. (1979). *Actual condition and improvement of cultivated soils in Japan, National Conference of Soil Conservation and Survey Project.* Tokyo [in Japanese].

Nira, R. (2000). Comparison between flux of nitrogen through microbial biomass and estimated amount of nitrogen mineralization by kinetic method in upland soil. *Japanese Journal of Soil Science and Plant Nutrition, 71,* 388–390 [in Japanese].

Nishio, M. (2005). Agriculture and environmental pollution. *Rural Culture Association of Japan* [in Japanese].

Oda, K., & Miwa, E. (1987). Compact data base for soil analysis data in Japan. *Japanese Journal of Soil Science and Plant Nutrition, 58,* 112–131 [in Japanese].

Sakamoto, K., & Oba, Y. (1991). Relationship between the amount of organic material applied and soil biomass content. *Soil Science and Plant Nutrition, 37,* 387–397.

Sakamoto, K., & Oba, Y. (1993). Relationship between available N and soil biomass in upland field soils. *Japanese Journal of Soil Science and Plant Nutrition, 64,* 42–48 [in Japanese].

Sakamoto, K., Guan, G., & Yoshida, T. (1997). The contribution of nitrogen through microbial biomass to plant nitrogen uptake in upland soils with different physicochemical properties. *Japanese Journal of Soil Science and Plant Nutrition, 68*, 402–408 [in Japanese].

Sano, S., Yanai, J., & Kosaki, T. (2004). Evaluation of soil nitrogen status in Japanese agricultural lands with reference to land use and soil types. *Soil Science and Plant Nutrition, 50*, 501–510.

Shirato, Y., & Yokozawa, M. (2005). Applying the Rothamsted carbon model for long-term experiments on Japanese paddy soils and modifying it by simple turning of the decomposition rate. *Soil Science and Plant Nutrition, 51*, 405–415.

Shirato, Y., Hakamata, T., & Taniyama, I. (2004). Modified Rothamsted carbon model for Andosols and its validation: changing humus decomposition rate constant with pyrophosphate-extractable Al. *Soil Science and Plant Nutrition, 50*, 149–158.

Singh, J. S., Raghubanshi, A. S., Singh, R. S., & Srivastava, S. C. (1989). Microbial biomass acts as a source of plant nutrients in dry tropical forest and savanna. *Nature, 338*, 499–500.

Soil Science Glossary Terms Committee. (2008). *Glossary of soil science terms*. Soil Science Society of America.

Sugihara, S., Funakawa, S., Kilasara, M., & Kosaki, T. (2010a). Effect of land management and soil texture on seasonal variations in soil microbial biomass in dry tropical agroecosystems in Tanzania. *Applied Soil Ecology, 44*, 80–88.

Sugihara, S., Funakawa, S., Kilasara, M., & Kosaki, T. (2010b). Dynamics of microbial biomass nitrogen in relation to plant nitrogen uptake during the crop growth period in a dry tropical cropland in Tanzania. *Soil Science and Plant Nutrition, 56*, 105–114.

Sugihara, S., Funakawa, S., Kilasara, M., & Kosaki, T. (2012). Effect of land management on soil microbial N supply to crop N uptake in a dry tropical cropland in Tanzania. *Agriculture Ecosystems and Environment, 146*, 209–219.

Toriyama, K., Sekiya, S., & Miyamori, Y. (1988). Quantification of effect of air-drying of paddy soil before submergence on amount of nitrogen mineralization. *Japanese Journal of Soil Science and Plant Nutrition, 59*, 531–537 [in Japanese].

Toyota, K., & Kuninaga, S. (2006). Comparison of soil microbial community between soils amended with or without farmyard manure. *Applied Soil Ecology, 33*, 39–48.

Chapter 6

Machinery and Information Technology for Biomass Production

Tadashi Chosa, Takeshi Matsumoto and Masahiro Iwaoka

Chapter Outline

Research Approaches to Sustainable Biomass Systems. http://dx.doi.org/10.1016/B978-0-12-404609-2.00006-4

6.1. MACHINERY FOR PRODUCTION OF ENERGY CROP

Tadashi Chosa

6.1.1. Introduction

An energy crop refers to plants grown and harvested for energy by direct combustion to generate electricity or heat, or making biofuel. Energy crops are generally categorized as woody biomass or herbaceous biomass, and the machinery to be discussed in this chapter is used only for herbaceous biomass production.

A herbaceous biomass, e.g. rape, sunflower, sugar canes, and rice, is basically an agricultural product that can be produced and managed using agricultural machinery in the open field. The use of agricultural machinery to replace human labor for engagement in heavy and dangerous tasks has introduced evolutional profit for the farming industry by increasing the productivity of crops. This technical innovation improves farmers' life expectancy and farming management; it also contributes to a stable food supply for the modern society.

Conventional agricultural machinery can be used to produce energy crops. But commercial energy crops need to be planted more densely than food crops for high crop yields and better product quality. The most important point is that the costs of producing energy crops must be much lower than for producing conventional agricultural food and feed crops in order for the bio-energy to be competitive in price with low-cost conventional fossil energy.

The technical background for biomass production in Japan is completely different from other countries. Growing frequently mentioned energy crops such as copra, jatropha, palm, rapeseed or soybean is not feasible in Japan because of weather conditions. Additionally, the intensive agricultural production system currently practiced in Japan also poses a problem for producing energy crops. The sophisticated production system in Japan has the advantage of using narrow land effectively for stable production but the production costs become high. Hence, reducing the agricultural production cost is one of the most important concerns for agriculture in Japan to produce either food or energy crops. The intention of producing biomass for energy may provide an opportunity for Japan to address this concern by improving the current production system.

6.1.2. Farm Machinery for Low Land

Rice, which is suitable as a crop in the Asian monsoonal climate, is a major agricultural crop in Japan. More than half of arable land in Japan is paddy fields to grow rice for maintaining a high self-sufficiency rate of rice. The production system of rice has been studied as a unique technology because most farm work, including tillage, paddling, transplanting and harvesting, has been mechanized in Japan. The rice-transplanting machine developed in the 1960s is also one of Japan's original technologies. The establishment of "mat-type rice seedlings" and re-enactment of the sensitive planting operation were the breakthroughs in developing the rice-transplanting machine.

Transplanting rice in a traditional cropping system has many advantages. The nursery can be grown under controlled indoor conditions to avoid unstable changes of temperature in early spring and risk of damage caused by disease and insects. The grown nursery is more competitive than weed, and the root-cutting stress during transplantation is beneficial to the growth of rice seedlings. Recently, the technology of applying chemicals to nurseries before and during transplanting has been studied. Farmers will not have to care for the plants after transplanting until harvesting. On the other hand, the high production cost is the problem associated with the transplanting operation. High energy consumption and the resulting high costs are needed to manage the rice nursery. Additionally, the nursery will eliminate the labor involved in rice planting so that some farmers will become unemployed; hence, farmers in many regions of the world are choosing the direct seeding method without a rice nursery for producing rice. There are increasing numbers of reports on experimental and practical direct seeding in Japan published recently (e.g. Box 6.1).

Direct seeding in Japan can be roughly classified as direct seeding of rice on well-drained paddy field and submerged soil; it is also classified as broad casting, drill seeding, and hill seeding from the viewpoint of sowing method (Figure 6.1).

A seed-shooting seeder of rice (Togashi et al., 2001a) combined with a tractor-mounted paddy harrow is used in the submerged soil for hill seeding (Figure 6.2). Seeding is carried out simultaneously with the last paddling operation in a paddy field. The seeder employs a seed roller to feed the seeds, which are then coated with calcium dioxide (CaO_2) before being shot on to the

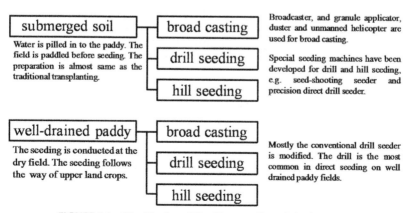

FIGURE 6.1   Classification of the direct seeding of rice in Japan.

FIGURE 6.2   A seed-shooting seeder of rice. *(Photo by Koichiro Fukami, National Agriculture and Food Research Organization, Japan.)*

soil surface of the paddy field with a saw-toothed disk. Both disk and roller are driven by a direct current motor powered with a tractor battery. Results of performance tests of the seeder (Togashi et al., 2001b) show that the depth of seeding is 5–20 mm if the seeds were shot into the submerged soil (viscosity 50–150 dPa•s) of the paddy field, with dimensions of the hill-seeded spot (ellipse in shape) in the paddy field of 6–10 cm (longer diameter) by 3–5 cm (shorter diameter). The hill-seeded rice has nearly the same lodging tolerance as the transplanted rice, and better lodging tolerance than the broadcasted rice. The yield of the hill-seeded rice is about 10%, inferior to that of the transplanted rice. A precision direct drill seeder (Nishimura et al., 2001) that is capable of controlling the seeding depth precisely is also used in the submerged soil for hill seeding (Figure 6.3). The seeder was mounted on a paddy vehicle consisting of a feeding unit and a furrow opener–closer unit, and the furrow opener–closer unit consisted of a float that follows the soil surface smoothly, a two-stage adjustable furrow opener, and a seeding depth controller with the soil surface hardness sensor. This device will enable stable germination by planting seeds at a precise depth so that inappropriate seed germination is reduced. According to a field trial in adaptability tests (Nishimura et al., 2003), the precision direct drill seeder is found to be capable of planting rice seeds with accuracy in various types of fields.

Direct seeding of rice on well-drained paddy fields does not need a paddling operation and the seeding vehicle can travel more smoothly. The seeders designed for seeding wheat, barley, and other upland crops can easily be modified for seeding rice. The rate of pulverization of soil is an important factor to consider to ensure satisfactory performance. Problems caused by water

FIGURE 6.3   Precision direct drill seeder.

leakage and uneven leveling may occur depending on the soil conditions, so that the soil properties in each region have to be considered. Using the winter-paddling seedbed preparation to prepare the land prior to seeding may help to solve these problems. The V-furrow no-till direct rice seeder is usually used to perform winter-paddling seedbed preparation. According to practical experiments by farmers in Anjo, Aichi (Hamada et al., 2007), seedling emergence is improved with increasing seedling stand number, substantially less seeds are eaten by birds, and control of weeds in winter-paddling seedbed preparation sites as compared with the complete no-till sites was easier. Furthermore, the soil of field surfaces of tested sites became as hard as that of fields with no till, but seeding was possible even after rainfall. The seedling emergence number and the grain yield of rice plants grown in interpaddling seeded preparation fields are as high as those of conventional transplanting rice cultures.

The management of cropping and harvesting for direct seeding plants are not very different from those for conventional transplanting plants. A head-feeding combine harvester can be used for any type of direct seeding plants. Recently, small conventional combine harvesters adaptable for both rice and other upland crops have been developed (Umeda et al., 2012); its adaptability to high-moisture paddy is studied (Kurihara et al., 2011). The achievements in flexible use of combine harvesters will also contribute to reducing the production cost.

6.1.3. Farm Machinery for Converted Fields and Uplands

Japan has abundant rainfall compared to other countries of moderate climate, and the groundwater level is thus quite high; this is one of the reasons why

FIGURE 6.4 Simultaneously tilling and ridging implement. *(Photo by Hisashi Hosokawa, National Agriculture and Food Research Organization, Japan.)*

paddy is the most popular land use for farming in Japan. However, these conditions sometimes become disadvantageous for upland crops. Recent oversupply of rice has led to conversion of paddy crops to upland crops.

The use of simultaneous tilling and ridging implements (Hosokawa, 2009) was originally developed to improve soybean cropping in converted fields with heavy clay that is very unstable for upland crops. The developed equipment (Figure 6.4), which simultaneously makes ridges, spreads fertilizers, and seeds when tilling the soil, reduces the adverse effects on wet injury. It features an up-cut rotary shaft and improved holder with excellent pulverization performance. Clods of soil are thrown into the center of the ridge by the curved tines. Then, by slightly raising the leveling plate, the soil can be tilled and piled to form ridges at the same time (Hosokawa, 2009). Various types of seeders for seeding soybean, corn, buckwheat, and rapeseed, among many others, have been developed.

6.2. INFORMATION SYSTEMS IN CROP PRODUCTION

6.2.1. Precision Agriculture

Information technology has developed rapidly in the last 20 years. GPS, which is a mobile phone with other useful terminals, has become widespread in our daily life. Additionally, the internet has become available almost anywhere. Nowadays, our daily living depends on these devices and technologies. Information technology is also useful for farming operations to produce crops. Some examples of the practical use of information technology are presented in the following section.

a. Precision Agriculture in Japan

Precision agriculture that began in Japan in the mid-1990s is about 15 years behind the USA and Europe; it may be implemented to solve problems of variability within and between fields that leads to economic and environmental benefits and stable production. Methods of precision agriculture are based on information technology; farmers make optimal decisions depending on site-specific information, about fertility, growth, and yield. Because the cropping system in Japan is completely different from those in the USA and Europe, the technology developed internationally must be modified to suit the conditions encountered in Japan. For example, because of small-scale Japanese farming operations, only undersized farming machinery with more sensitive monitoring devices can be used efficiently in Japan. Moreover, the main crop in Japan is rice rather than corn, barley and wheat, among many others, that are popularly grown in other nations; research on precision agriculture has been attracting attention with regard to the production of rice in Japan.

b. Yield Monitoring System

Many types of devices, techniques, systems, and technologies have been developed for implementing precision agriculture in Japan over the last decade. The system includes a soil sensor to monitor soil fertility, an image-mapping technique to monitor growth, a plant growth information-processing system using cellular phones, and a variable-rate application technology for paddies. The yield monitoring system is also a representative technology used in precision agriculture. An understanding of yield variations within a field or among fields has been used to evaluate growth and management profiles that provide important information to determine site-specific management strategies for the following year. Less fertilizer should be applied at sites where lodging has been observed, whereas more fertilizer needs to be applied at sites where yields have been low. Although general yield variation information can be obtained from experience with routine farming work, the results obtained with yield monitoring are expected to play an important role in establishing the technology for implementing site-specific crop management and spreading practices.

Although yield monitors are installed in combines as standard equipment, they have been marketed and installed as after-sale accessories throughout Europe and the USA and are not sufficiently accurate for head-feeding combine harvesters in Japan. The sensor and system have usually been designed for a conventional combine; their detecting signals (e.g. grain flow or change of mass in the grain tank) are inappropriate for a head-feeding combine. This is because the combine has narrower operating width than a conventional combine due to differences in feeding kernels to the grain tank. Kernels are continuously fed to the grain tank by a grain auger in the head-feeding combine, whereas kernels

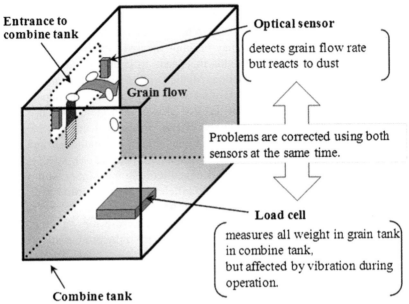

FIGURE 6.5 Principle behind hybrid yield monitoring system using optical sensor and load cell. *(From Chosa et al., 2005.)*

are intermittently fed to the grain tank by a bucket conveyer in a conventional machine. Also, the required accuracy of yield variations is different because of the difference in grid size. Consequently, developing an original yield monitoring technique for a head-feeding combine is necessary so that yield monitoring combines are generally adopted by Japanese farmers.

A hybrid yield monitoring system (Chosa et al., 2006) is one solution to the problems described above. The system applies one pair of optical sensors attached to the entrance of the grain tank coupled with a load-cell unit fixed at the bottom of the grain tank (Figure 6.5). The regression analysis results of signals from the optical sensor that reflect the grain particle flow rate at the top of the grain auger are not stable because of changes in coarse particulates or dusty conditions caused by variations in operating conditions, such as field and grain moisture and mixing of weeds and mud due to lodging. Additionally, the load-cell signal varies according to the weight of discharged grain; it is not adopted for the sequence yield monitor due to the effects of vibration. To solve these problems, a batch yield monitor is used to measure the grain weight at each stroke for calculating the accumulated variations in the particle flow sensor. Subsequently, the regression equation is defined by comparing the total grain weight and the accumulated variations in the particle flow sensor. Finally, sequence yield variations are calculated using the regression equation.

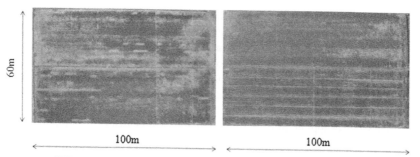

60m

100m   100m

FIGURE 6.6   Example of the effect of the site-specific management.

c. Comprehensive Trials

Several integrating systems, i.e. precision farming systems, using developed technologies have also been developed. One example is a site-specific nitrogen management system that applies fertilizer at a variable rate in accordance with fertility and growth information; results with reduced yield variability and higher quality of rice production have been reported (Toriyama et al., 2003). Some of the technologies developed for rice production are also useful in other types of field. The use of site-specific information is being applied to wheat production, resulting in higher quality (Sasaki et al., 2007). Morimoto et al. (2006) introduced a decision-making method for harmonizing objectively the sensed information with a farmer's knowledge. Chosa et al. (2009a) showed that the site specific management is effective to introduce newly developed technology to practical farm (Figure 6.6).

In addition to rice, other crops including wheat, soybean, vegetables, tea, forage crops, fruit, sugar cane, and many others are now being targeted, and a number of farmers are becoming interested in the information acquired. Unfortunately, the devices developed so far are too expensive or inconvenient for practical use by working farmers; hence, the first stage of developing new technologies for precision agriculture in Japan is coming to an end. The next stage must focus on improving and disseminating these technologies.

While "precision agriculture" is based on site-specific crop management for narrow field space, it is more practical and comprehensive because it includes not only the farming system but also a natural resource management model coupled with a business model. The concept of precision agriculture is being extended from production management marketing management. The new generation of precision agriculture is expected to be a tool of regional commutation between farmers, consumers, and the general public.

6.2.2. Use of a Personal Digital Assistant

Cell phones are common communication devices nowadays; farmers and consumers will have individual cell phones. Many studies on using cell phones to realize traceability systems or more effective production systems have been reported. SEICA (http://seica.info) that follows the virtually identified produce system (Sugiyama, 2004) is a brochure on the internet. Using a cell phone, farmers can input information not only on crop production but also crop shipment. Cell phones are available to input and browse the system using both text and images; QR code and IC tags can assist farmers in accessing the system in the field with ease.

Acquiring cropping information is also possible using a cell phone. A mobile phone-based field data logger system for mandarin orange production has been reported (Kamiya et al., 2011). The system focuses on collecting data that can be used to determine the watering schedule for cultivation using drip irrigation. Automated field monitoring was performed using Field Server, which is a server computer installed in the field, and a sensor network to collect weather information. The results can be visualized as a timeline using a web-based system; users can compare their results with the data provided by other users, and can then consult with experts in their area.

6.2.3. Optimization of Farm Operations

The farming operation is deeply related to regional culture and unpredictable climate, so that optimizing the farming operation depends on collections of information, knowledge, wisdom, and skill. However, the situation is changing dramatically in that the management size is increased and many operations are completed rapidly by using farming machinery. Hence, many types of weather information services are needed and have to be utilized properly to reduce production costs while maintaining high crop quality.

A database that consists of information on operation name, cost, time, and machines and materials can be used to develop proper farming operations including a technical approach such as the dependence of optimum cropping on farm machinery and climate, among many others, and a financial approach such as financial management. Both topics have been studied by researchers in other disciplines; how to implement the study results and concepts integrated with the knowledge and concepts accumulated on farming management has also been reported. FAPSDB (Nanseki and Matsushita, 2005) is designed as a web application with a unified concept and information; it generates 15 CSV files to describe the index of farming management using a prepared database.

Describing a farm operation as a mathematical model is another solution to achieve an optimum system. The model of operation is developed using linear programming, simulated annealing, probabilistic modeling, and other mathematical analysis. Optimization of the system for harvesting and shipping a potherb mustard crop including driving route within fields and other operations, has been introduced using mathematical modeling and analyses (Miyasaka and Ohdoi, 2005).

Integration of farming operations and a geographic information system (GIS) is valuable because the latter improves farming by covering a wider area. Records of the farming operation obtained with the GIS can be analyzed to determine the optimum operation schedule using a computer-assisted system. As reported by Daikoku (2005), based on practical experiments, tightly scheduled operations in early spring, paddling and transplanting can be properly scattered in hilly and mountainous areas.

6.3. HARVESTING THE FOREST BIOMASS

Takeshi Matsumoto

6.3.1. Forest Biomass

a. Introduction

The famous old Japanese tale "Momotaro" starts as follows: Long, long ago, there lived an old man and his old wife in a village. He went to the mountain to gather *Shiba* (*Shiba-kari*). She went to the river to wash clothes. The word "*Shiba*" refers to miscellaneous small trees grown in fields and mountains that can be used for fence or fire wood (*Maki*). The Momotaro, well known in Japan, starts the story by harvesting the biomass.

In Japan, biomass harvesting from fields and mountains traditionally constitutes a useful resource as fuel, life materials, and agricultural fertilizer (*Karishiki*). The area of fields and mountains needed to supply *Karishiki* is 10 times the size of cultivated land area (Mizumoto, 2003). Figure 6.8 shows the transportation of firewood by sledding in the Edo Era.

Today, the total quantity of usable wood biomass in Japan is estimated at 32 million tons per year from the following sources: logging residue (3 million tons), thinning timber (5 million tons), and low-level broad-leaved trees (9 million tons) (Yoshioka and Suzuki, 2012).

In Japan, forest biomass is produced by practicing forest management measures such as thinning or regeneration cutting, and also from fallen stands or broken limbs caused by snow or wind events. Straight logs are shipped as grade A timber for building materials; crooked or bronzing logs are shipped as grade B timber to be used mainly for plywood. The allowable level of crook is that the height of the curve is under 1% of the log length. Larger crook logs, short logs, heartwood rot logs, and distorted logs are shipped as grade C timber mainly for pulp or chip. Low-level logs, including small logs, logs from the basal part of the tree and logs from the top end of the tree, are shipped as grade D timber mainly for fuel. Traditionally, Japanese forest has produced mainly grade A timber, but recent demand for grade B timber has been increasing because of the growing number of plywood factories using domestic timber. However, most grade C and D timber is left in forest land.

BOX 6.1 Air-Assisted Strip Seeding

Direct seeding of rice instead of transplanting seedlings has been gradually developed in order to achieve a low-cost and labor-saving system in Japan. The most practical direct seeding methods are drill seeding and hill seeding using a seed-shooting seeder as well as a precision direct drill seeder in the flooded paddy. Direct seeding of rice in well-drained paddy fields is also becoming popular because the seeding machine can be used for both rice and other upland crops. The most efficient method, broad casting, is not very popular because the broad-cast seeds are not completely buried in soil; they are often eaten by birds in addition to a low rate of germination with and unstable initial germination.

The objective of the air-assisted strip seeder (Chosa et al., 2009b) is to achieve efficient operation as broad casting and stable initial growth as drill seeding. The machine for air-assisted seeding was developed using commercially available farm machinery. The original machine was a granule applicator with improved swath and fixing device for seeding rice in a paddy field (Goto, 1997). The tested applicator can be mounted on a conventional tractor using a three-point hitch and power take-off (PTO). The air-assisted seeding machine consists of a hopper, a boom-shaped application pipe, an application roll, DC motors, a ventilation fan, and 16 blow heads. The motor drives the application roll, which transports granular material from the hopper to the application pipe. The application rate of seeds can be controlled by the revolution rate of the DC motor, and the ventilation fan blows granular seeds through the pipe to the right and left, creating a 10-m wide swath (Chosa et al., 2009a) (Figure 6.7). A high-speed camera shows that the speed of rice seed falling is 4–7 m s^{-1}, and the direction of falling is 7.1° from the vertical. The rice seed falling speed is too slow to be buried properly under any conditions. The direction of falling should be lessened to narrow the strip width. Further experiments to improve the air-assisted seeder have continued (e.g. Loan et al., 2011); more practical machines and technology will be developed and implemented in new rice cropping systems in the future.

According to field trials using high-quality grain with good eating quality rice, *kosihikari* (Furuhata et al., 2011), air-assisted strip seeding methods lead to superior numbers of eras and crop yields than hill seeding using a seed-shooting seeding

FIGURE 6.7   Field operation by air-assisted seeding machine and effects of air-assisted seeding. *(From Chosa et al., 2009a.)*

BOX 6.1 Air-Assisted Strip Seeding–Cont'd

method. Mukouyama et al. (2012) applied the lodging tolerant rice cultivar, *Takanari*, and reported that the air-assisted drilled rice produces heavier dry matter and grain yield than the hill-seeded and transplanted rice because the former has a higher crop growth rate. Higher efficiency of the seeding operation using the air-assisted strip seeder has also been confirmed through field investigations (Chosa et al., 2009c).

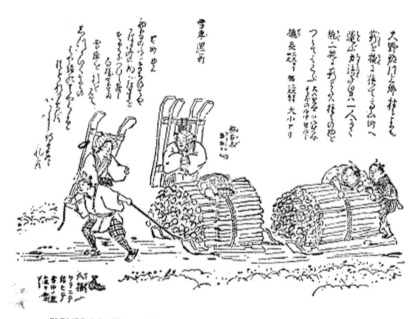

FIGURE 6.8   Firewood transportation by sledding (Yuzankaku, 1972).

b. Units for Measurement of Forest Biomass

SI units are used in the measurement of forest biomass. However, in the Japanese forestry sector, traditional units (*Shaku-Kan* system) are also frequently used.

Various methods such as the Huber formula, Rieke formula, Smalian formula, and square-a-shorter diameter method are used for calculating log volume. Japanese forestry and timber dealers use the square-a-shorter diameter method in which the log volume V (m^3) is calculated by substituting the small-end diameter D (m) and log length L (m) into the following formula for log length less than 6 m:

$$V = L \times D^2. \tag{6.1}$$

When the log length exceeds 6 m, the following modified formula is used:

$$V = \left(D + \frac{L' - 4}{2}\right) \times L, \qquad (6.2)$$

where L' is the integer of log length (m). The volume of a log 3 meters in length with a small-end diameter of 60 cm is almost 1 m^3.

The stem volume of standing trees can be estimated by using various methods, including the form factor method, Pressler method, log volume table method, and Denzin method. The Denzin method is very simple; it is used to calculate the stem volume V (m) from the diameter breast height D_B (cm) using the following formula:

$$V = \frac{D_B^2}{1000}. \qquad (6.3)$$

This method assumes a tree height of 25–30 m, and therefore the following corrections are needed: $+4\%$ per m for a tree height of over 30 m; and -4% per m for a tree height of under 25 m.

c. Current Status of Japanese Forests

Most of the Japanese region termed "forest" is synonymous with "mountain". Thus, in many cases, mountain means not only mountainous terrain but also forest.

The terrain index I (%) is calculated from the inclination I_i (%), the relief energy R (m), and the valley density V (valleys per km^2), using the following formula (Hori, 1965):

$$I = \frac{3I_i + I_r}{4} \qquad (6.4)$$

$$I_r = R(0.1 + 0.01 \times V). \qquad (6.5)$$

The following terrain classifications are used: gentle terrain ($I = 0$–19), middle terrain ($I = 20$–39), steep terrain ($I = 40$–69), and steeper terrain ($I > 70$). The Japanese forest land consists of 3.7 million ha of gentle terrain, 10.3 million ha of middle terrain, 11.4 million ha of steep terrain, and 1.1 million ha of steeper terrain (Sakai, 2000). The terrain index can be regarded as the inclination of the ground (Sakai, 2000), and therefore 50% of Japanese forest land is composed of steep and steeper terrain with inclinations greater than 40% (22°).

Seventy percent of the countryside land in Japan comprises forest land (25.1 million ha) and the growing stock is 4.4 billion m^3. Artificial forests constitute 10.4 million ha (41%), with a standing volume of 2.7 million m^3 (61%). Most of the species in artificial forests are coniferous, for example *Sugi* (*Cryptomeria japonica*; 4.5 million ha), *Hinoki* (*Chamaecyparis obtusa*; 2.6 million ha), *Karamatsu* (*Larix leptolepis*; 1.0 million ha), *Matsu* (*Pinus*; 0.9 million ha), and *Ezomatsu* and *Todomatsu* (*Picea jezoensis* and *Abies sachalinensis* respectively; 0.9 million ha) (Forestry Agency, 2012b). More than 50% of artificial forests, which are 36–50 years old, need thinning. In 2010, thinning was conducted across 556,000 ha, and the volume of wood produced by thinning was 6.7 million m^3. Based on a unit volume of forest thinning of 300 m^3 ha^{-1}, and a thinning ratio of 30%, the thinning volume will be 100 m^3 ha^{-1}. If the yield rate of felling and bucking trees is 50%, thinning will produce 50 m^3 of timber per ha. Accordingly, the total production potential in 2010 was estimated to be 27.8 million m^3 of wood resources; however, the actual volume produced was approximately 25% of the estimate.

Japanese forest resources (mainly artificial forests) are increasing at an annual rate of 80 million m^3; this volume exceeds the annual Japanese gross timber demand of 72 million m^3. In 1989, the annual gross timber demand was 116 million m^3, but has been decreasing in recent years. Today, Japanese forest resources are sufficient to meet domestic timber demand (i.e. 100% self-sufficiency) by harvesting only the annual growth volume. However, in 2010, the self-sufficiency rate was only 26% with a large gap between actual and potential self-sufficiency. The gross timber demand peaked at 113.7 million m^3 in 1995, and has been decreasing in recent years. In 2010, the gross timber demand was 72 million m^3, consisting of 35% sawn wood, 45% pulp and chip, 13% plywood, and 13% others. In 2009, the Japanese government developed the "Forest and Forestry Revitalization Plan", aimed at achieving a wood self-sufficiency rate of 50% or higher in 10 years.

Japanese forests are classified based on their ownership; national forest accounts for 31%, whereas non-national forest accounts for the remaining 69%. Non-national forest comprises 77% private forest, 19% public forest, and others. As for holding size of private forest, the number of owners with holding area smaller than 20 ha represents 81% of the total number of owners, and the total holding area represents 15% of the total private forest. Moreover, 24% of private forest owners are absentee owners who do not live in the same municipality where their forest is located.

d. Forest Planning System

The forest grows for a very long time. Therefore, the forest planning system in Japan is based on a long-term perspective with appropriate management practices aimed at stable supply of forest products and development of multiple functions for public benefits. According to the "Basic Plan for Forest and Forestry", the "Nationwide Forest Plan" that covers three 5-year periods is formulated by the Ministry of Agriculture, Forestry and Fisheries for managing 44 planning areas divided by major river basins. At prefecture level, the "Regional Forest Plan" for two 5-year periods for a total of 10 years is formulated by the prefectural governor for managing private forests that covers 144 forest planning blocks. For national forest, the "Regional Plan for National Forests" is formulated by the director general of the regional forest office.

At the municipality level, the "Municipality Forest Management Plan" is formulated by the head of municipality; this plan is adopted as the "master plan" for on-the-ground forest management.

At the individual forest owner level, a 5-year "Forest Management Plan" was until recently developed by the individual forest owners and others involved. However, in 2012, a new "Forest Management Plan" was introduced. This plan is formulated to cover forest operation and other practices to manage geographically continuous forest by forest owners or trustees of forest management. Under the new system, the costs of forest management (including thinning and construction of forestry road) are paid to the forest owners or trustees to meet the required conditions with the "Forest Management and Environmental Conservation Direct Support System".

e. Current Japanese Forestry Situation

The number of forestry workers in Japan was estimated to be 210,000 in 1970, fell to 40,000 in 2007, and subsequently increased to approximately 80,000 in 2010, which represents 0.13% of the total Japanese workforce of 62.6 million. In 1995, 70% of forestry workers were older than 50 years of age (Ministry of Health, Labor and Welfare, 2002). In recent years, this rate has decreased to 50%; nevertheless, forestry remains an aging industry in comparison with the overall average industry that has 40% of workers older than 50 years of age.

The total number of forestry worker accidents has decreased in recent years, but the frequency of accidents remains at a high level. The forestry industry has 15 times higher incidence ratio per 1000 workers than other industries, and six times higher than the construction industry.

The self-sufficiency rate of domestic timber peaked at 94.5% in 1955, fell to under 18% in 1995, and subsequently recovered to 26% in 2010. The timber price of middle-class *Sugi* log ($\varphi = 14$–22 cm, length $= 3.65$–4 m) was 8200 yen m^{-3} in 1960, increased to reach 38,700 yen m^{-3}, and subsequently fell to 10,900 yen m^{-3} in 2009 (lower than in 1960). The log price at roadside could pay the wages of 11.8 workers in 1960 but only 0.22 workers in 2010. In 2010, forestry production generated 157.6 billion yen, representing 0.03% of Japanese GDP (481.8 trillion yen).

f. Forest Management Prescription

The forest management prescription combines different practices aimed at developing sustainable forests.

(i) Afforestation

Planting seedlings or sowing seeds is practiced; the planting density depends on the species, region, and production objectives. For example, planting densities are 1500 ha^{-1} for Obi forest district in Miayzaki prefecture, 6000–8000 ha^{-1} for Owase forest district in Mie prefecture, 8000–10,000 ha^{-1} for Yoshino forest district in Nara prefecture, and 10,000–12,000 ha^{-1} for Kitayama forest district in Kyoto prefecture. Today, the standard planting density in Japan is 3000 ha^{-1}. Recently, the afforestation area has decreased because of the increasing non-clear-cut operations.

(ii) Tending

Tending is practiced to stimulate tree growth for developing healthy forest after planting until harvesting. The tending practice consists of weeding, snow-break preservation, pruning, cleaning cutting, climber cutting, thinning, etc.

Weeding Weeds and shrubs are removed to avoid suppression of planted seedlings. It is practiced during spring and summer for about 5 years after planting. In recent years, the weeding area has decreased because of reduction in the afforestation area. Weeding and planting that are labor intensive have previously contributed to spring and summer employment.

Snow-Break Preservation If planted seedlings felled by snow are left, the top of the seedlings will grow to above the fallen position with a crooked base. Therefore, the planted seedlings are raised for normal growth.

Pruning Unnecessary and lower branches are cut down to produce knot-free timber with high quality. Pruning is practiced during non-growing season in autumn and winter.

Climber Cutting This operation removes the climbers that wind around planted trees to prevent the latter growing healthily. Climbers often proliferate after weeding until crown closure; they are removed to maintain healthy tree growth and avoid debasement of the timber quality.

Cleaning Cutting This operation removes shrubs that prevent growth of the planted tree. Cleaning cutting is practiced several times after weeding to remove branches coming into contact with each other.

(iii) Thinning

Thinning is included in tending to remove some of the planted trees after crown closure in order to induce intraspecific competition. The adjustment of forest density is aimed at reducing competition, improving timber quality, promoting growth of floor vegetation, and developing the multiple functions of artificial forests. Thinning is indispensable in Japan for developing healthy artificial forests. There are various methods of thinning.

Thinning from Above With thinning from above, upper-story trees (dominant trees) are removed from stands, while lower-story trees (oppressed trees) are left in place. The main purpose of thinning from above is to generate large incomes.

Thinning from Below With thinning from below, only lower-story trees are removed from stands.

Thinning from All Stories With this method, trees of all stories are removed from stands. The same result is achieved by line thinning and zone thinning.

Qualitative Thinning With qualitative thinning, felling trees are determined beforehand based on their characteristics such as superiority or inferiority of crown, crook of stem, and flaw of stem among many others.

Quantitative Thinning With quantitative thinning, the quantity (number or volume) of felling trees is determined beforehand and some trees are selected for quantitative thinning based on similar tree characters.

Mechanical Thinning With mechanical thinning, felling trees are selected mechanically using line thinning or zone thinning. In Japan, line thinning was first practiced in the 1970s.

Traditionally, the thinning ratio was generally 20–25%. Today, however, thinning is often practiced with a ratio of 30–35%. Intense thinning (40–45%) is sometimes practiced; 10–15 years after thinning, the stand volume of most 36- to 50-year-old forests recovers to at least the same level as that prior to thinning.

(iv) Regeneration Cutting

Regeneration cutting involves removing trees that have reached their usable age to promote the growth of next-generation trees (regeneration). Clear cutting means cutting all or most stands. Although the technique and practice of cutting and afforestation is easy, public function must be secured. Selective cutting involves cutting mature trees every few years to several decades. This practice causes minimal change to the status of a forest, and can sustainably breed next-generation trees. There are two methods of selective cutting: single tree cutting and small group tree cutting.

Shelter wood cutting involves cutting and utilizing the forest where it divides several times (or dozens of times) and introducing next-generation trees.

g. Seasonality of Forestry Working

Forestry operations are outdoor occupations that are influenced by the weather; inclement weather leads to lost working days. This operation is also seasonal because it is practiced in accordance with the growth of trees. An example of an annual working schedule is shown as follows:

- **January–March:** Thinning or regeneration cutting
- **February–March:** Ground clearance
- **April:** Planting

- **May–August:** Weeding
- **September–October:** Cleaning cutting
- **October–December:** Pruning (thinning or regeneration cutting).

During spring and summer, trees are heavy, tarnishable, and rotable; therefore, they are not suitable for felling. Additionally, the timber price is lower during summer and winter than autumn and spring.

6.3.2. Harvesting Technology of Forest Biomass

a. Logging Operation

(i) Felling

This practice involves felling standing trees by using axes, saws, or chainsaws.

(ii) Yarding, Skidding

The felled trees or bucked logs are moved to the forest road or landing, where convenient transportation is available. *Kiyose* or *Yabudashi* are traditional terms for yarding; however, today, *Kiyose* refers to prehauling. Yarding methods include classified manpower yarding, animal skidding, and machine yarding. Today, machine yarding is mainly practiced. Depending on tree form, yarding consists of short-stem yarding, full-stem yarding, and whole-tree yarding.

Short-Stem Yarding Logs are felled and bucked in the forest and collected. Unnecessary parts of the felled tree (branches, tree tops, crooked stem parts, etc.) are removed on-site in the forest. In comparison with full-stem yarding and whole-tree yarding, short-stem yarding leaves the greatest amount of biomass in the forest.

Full-Stem Yarding Stems that are limed and not cut are collected. Branches are left in the forest.

Whole-Tree Yarding Felled trees that are not limed and cut are collected, along with the branches. All biomass is removed from the forest.

(iii) Limbing and Bucking, Processing

Branches of felled trees are limbed, and the stems are cut to logs. Sometimes barking is carried out at the same time. Traditionally, this operation was practiced by using chainsaws; however, recently, the use of processors has been increasing in Japan.

(iv) Log Transportation, Logging

This generally means log transportation from the landing or forest road, to the log market or manufacturer. Today, log transportation in Japan is practiced by using trucks.

b. *Present-Day Biomass Harvesting Technology*

(i) Chainsaws

The chainsaw is a hand-held machine, which mechanically turns a chain of saw blades at high speed. It is used to fell stands, limb branches and cut timber. The use of chainsaws increased rapidly in Japan after destruction of trees by the Toyamaru typhoon of 1954. Chainsaw operators are required to attend a course on tree-felling work. Chainsaw felling is the mainstream forest operation in Japan, where most forests are located on steep terrain. Hence, chainsaw accidents account for at least 50% of deaths within the forestry sector, with 20% of these deaths caused by hung-up trees.

(ii) Yarders

A yarder is a log-collecting machine that mechanically turns an internal drum to release and rewind a wire rope. According to the Industrial Safety and Health Law Enforcement Ordinance Article 6, the cable-yarding system with skyline is defined as equipment in which "timber or fuel wood is rolled up or carried in the air using power with skyline, carriage, prop and attachments." Employers must elect a person to be in charge of forestry cable-yarding with skyline. The use of cable-logging systems in Japan increased rapidly after World War II. Typical systems include the endless-Tyler system, downhill North Bend system, and double endless system. In addition, an H-type system that can collect logs from a wide area is in use. In recent years, cable logging has been decreasing and there are concerns that this will affect the technique of cable yarding. In 2009, the number of yarders accounted for slightly more than half of the total number of high-performance forestry machines (Forestry Agency, 2012a).

(iii) Processors

The processor takes limbed whole trees with branches and leaves, cuts them into logs, and bunches the logs. In Japan, many attachments are available for a processor head (e.g., backhoe, oil pressure shovel). With regard to movable equipment, crawler-type processors are common in Japan whereas wheel-based machines exist in Europe and America. The processor head consists of a grapple arm for catching timber, feeding equipment such as a feeding roller or crawlers, a cutter to remove branches while feeding timber, a head sensor to measure timber length while feeding, and a chainsaw for cutting timber into logs. The spread number of processors in Japan is increasing.

In recent years, a new type of processor has been developed to cut the top end of the tree and its branches by using scissors, and deposit the wood directly in a container. These features are added to its conventional logging and lumber-production function (Mozuna et al., 2011).

(iv) Harvesters

A harvester travels throughout the forest felling standing trees, limbing branches, cutting timber into logs, and bunching the logs. In addition to their processing role, harvesters operate as construction machines (e.g. backhoe, oil pressure shovel) with attached harvester heads. As far as movable equipment is concerned, crawler-type harvesters are common in Japan whereas wheel-based machines are popular in Europe and America. Harvesters are equipped with a tilt system that is capable of felling standing trees. In comparison with processors, the number of harvesters has increased in recent years because of their capability of handling timber on the narrow spur roads constructed on steep terrain. In addition to roller-type harvesters, stroke-type machines (Figure 6.9) and crawler-type machines (with a timber-feed system) have recently been introduced.

FIGURE 6.9   Stroke-type harvester.

(v) Tower Yarders

A tower yarder (classified as a cable-logging forestry machine) is a movable yarder equipped with an artificial prop. Traction types and self-run types (e.g. loaded on the truck loading) are available. In recent years, the number of tower yarders in Japan has remained relatively stable. Machines equipped with a processor and tower yarder are common in other countries. Typical yarding systems include the downhill North Bend system, running skyline system, and high-lead system. With regard to log carriage, various mechanisms are available, from a simple type equipped with only two guide brocks to a remote control type with a built-in drum for lifting logs.

(vi) Swing Yarders

A swing yarder (Figure 6.10) is a self-run movable yarder mounted on a construction machine (e.g. backhoe, oil pressure shovel) equipped with a winch, and using an arm and boom to substitute for a prop. Similar to a tower yarder, a swing yarder is classified as a cable-logging forestry machine. In contrast to tower yarders, swing yarders are becoming popular in Japan, but are uncommon in other countries.

(vii) Feller Bunchers

Feller bunchers travel throughout the forest to fell standing trees and bunch whole trees without limbing. The major type consists of a feller buncher head attached to a construction machine (e.g. backhoe, oil pressure shovel). Feller bunchers are uncommon in Japan.

FIGURE 6.10 Swing yarder.

FIGURE 6.11 Grapple and forwarder.

(viii) Grapples

This machine has an attached grapple head to catch logs and pass them to a construction machine. Grapples are used to collect, sort, load (Figure 6.11), and prehaul trees, and may therefore be considered as general high-purpose machinery. Although wheel-based machines exist, the most common type in Japan is the crawler-type grapple.

(ix) Skidders

A skidder travels throughout the forest and pulls whole trees that have been lifted at one end. They are increasingly used in Europe and America, but are uncommon in Japan.

(x) Forwarders

A forwarder (Figure 6.11) is a self-run machine that loads logs using a grapple crane, and transports them to the landing or forest road. In Japan, most movable forwarders use crawler-type equipment. The load capacity varies from 3 to 6 tons. A model equipped with a carrier structure to push the bulky biomass forward has recently been developed (Jinkawa et al., 2011).

(xi) Bundling Machines

This machine constructs bundles (Figure 6.12) by compressing the bulky biomass (mainly on the branch) obtained from the logging and timber production work of harvesters or processors. The resulting bundle can be handled (like a log) by a grapple. Transportation of bundles with trucks or trailers is much more efficient than transportation of raw branches. Domestic bundling

FIGURE 6.12 Bundling machine and bundle.

machines have recently been developed (Yogi et al., 2006, 2008; Murakami and Yamada, 2008).

(xii) Mobile Chippers

A mobile chipper (Figure 6.13) is a self-run machine that crushes raw wood on the forest road or landing to reduce the volume of harvested biomass so that it can be transported more efficiently. In some countries, combined machines, e.g. chipper truck (a chipper combined with a truck) and chipper forwarder (a forwarder combined with a truck), are available.

(xiii) Self-Propelled Carriages

A self-propelled carriage hanging from a wire rope suspended in the forest (without using a mechanical yarder) is used to collect timbers using an automatic winch. The number of mobile yarders in Japan had increased until 2000, but has recently been decreasing.

(xiv) Monorails

A monorail is a riding or baggage vehicle, which is connected to a powered vehicle equipped with engine, driving wheel, operation device, and braking system, and runs on a rail supported by a prop. Different types of running methods are available. In the chinning type the vehicle is hung under the rail, whereas in the mounted type the vehicle is mounted over the rail. With regard to drive type, in the pinion-rack type, the pinion pin of the driving wheel engages with a wave-shaped rack that is attached to the side and bottom of the rail. This allows the vehicle to climb steep slopes of around 45° so that it moves along the

FIGURE 6.13   Mobile chipper.

local topography to minimize disturbance to the forest floor or soil. The rail structure may differ among manufacturers but the versatility is low.

(xv) Helicopters

Helicopters can be loaded and unloaded while hovering (Figure 6.14) with a high transportation capability. However, helicopter transportation is limited to high-value or high-volume materials because of high operating costs. Several other factors including the requirement to obtain different licenses for specific models must be considered for using helicopters for forest-related transportation.

6.3.3. Forest Road Network as an Infrastructure for Biomass Harvesting

With regard to the forest management infrastructure, public road passing through or within 200 m from a forest and forest roads have been constructed for many years in Japan. However, the importance of the spur road network as an operating base for forestry machines or the base of warm forest management (detailed road network) has recently been emphasized. Successful examples and technical standards have been clarified, and the spread of spur or strip roads (Figure 6.15) is regulated. In 2006, the Forestry Agency designated the component of the road network for forestry, and the function of each category of forest road, spur road, and strip road. In addition, the "Forest and Forestry Revitalization Plan 2009" was developed to establish standards for the various roads serving forests, such as "forest road", "wood transportation road" for trucks, and "log salvage road" for forestry machines.

FIGURE 6.14 Helicopter yarding.

FIGURE 6.15 The spur road construction scene.

According to the current version of Forest Road Regulations Article 3, roads for forestry are defined based on the process of divergence with regard to forest management. A "main road" is defined as a core road of the exploitation area, a "branch road" diverges from "a main road", and a "lateral road" differs from a "branch road". In addition, the Forest Road Regulations Article 4 classifies the concrete standard of each road into two categories, "motor road" (grades 1–3) and "light motor road". For motor road, a first-grade road (width 4.0 m, single traffic lane) is a main road that connects public roads such as a national road, prefectural road, or municipal road, whereas a third-grade road (width 2.0 or 1.8 m) is a "branch road" or "lateral road" located in a small exploitation area. All other roads are classified as second grade, which has vague definitions. In Japan, second-grade roads account for 69.1% of all forest roads, with first-grade and third-grade roads representing 11.7% and 19.2% respectively (Forest Road Technology Meeting for the Study, 1999). Light motor roads (width between 1.8 and 3.0 m) are defined as roads that serve only light cars. Monorail was defined in the forest road regulations revised in 2001, and categorized as forest road. It can be constructed using subsidies for forest road projects. Matsumoto (2000) defined the following relationship in Aichi prefecture:

$$P = S^{0.66},\tag{6.6}$$

where S is the exploitation area of forest road (ha) and P is the number of forest owners of forest land in the exploitation area. The agreement of all owners (average 50 persons) is required for forest road construction (Matsumoto and Kitagawa, 1999b).

The standard for constructing a forest road network is related to the road network density D (m ha^{-1}), according to the following formula:

$$D = \frac{L}{S},\tag{6.7}$$

where L is the road length (m) and S is target forest area (ha).

In addition, the average yarding (arrival) distance from the road to forest land is used as an index to express a qualitative standard. When roads are placed in parallel in a rectangular area, the average yarding distance M_g (m) is expressed by the following formula:

$$M_g = \frac{10^4}{4D} \left(= \frac{2500}{D} \right),\tag{6.8}$$

where D is the road density (m ha^{-1}) (Segebaden, 1964).

If only one side of road is available, M_g is calculated using the following formula:

$$M_g = \frac{10^4}{2D} \left(= \frac{5000}{D} \right).\tag{6.9}$$

There is a difference between the calculated M_g (theoretical distance) and the actual average yarding distance (m). The parameter V_{corr} is the adjustment coefficient, and is expressed by following formula:

$$M_g = \frac{10^4}{4D} \times V_{corr} \left(= \frac{2500}{D} \times V_{corr} \right), \tag{6.10}$$

where V_{corr} is 1.0 in the above-mentioned rectangular model, but becomes large in actual road networks with many deviations, divergences, or detours. Examples of V_{corr} values in Japan are as follows:

- 1.18–1.68 for forest road networks in which road placement is as uniform as possible (Hori et al., 1971).
- 1.2 for forest roads in each exploitation area (Sakai, 1990).
- 1.42–2.41 for forest road networks of the national forest, and 1.8 for an outline price (Sawaguchi, 1996).
- 1.7 for a road network that includes public road and forest road in a municipality unit (Matsumoto and Kitagawa, 1999a).
- 1.91–2.15 for a road network that includes public roads, forest roads, and spur roads in a municipality unit (Nakazawa et al., 2005).

6.3.4. Harvesting System for Biomass

a. Productivity

Productivity refers to the amount of production from an individual machine or a single site; it is expressed as m^3/machine or m^3/day for timber, and t/machine or t/day for other wood biomass. Labor productivity, which is the productivity divided by the labor input, is expressed as the productivity per unit hour and unit labor input, i.e. m^3 per person per hour for timber, and t per person per hour for wood biomass, or m^3 per person per day for timber and t per person per day for wood biomass. Labor productivity is globally expressed as m^3 per person per hour (or t per person per hour), and it is often expressed as m^3 per person per day (or t per person per day).

Labor productivity in Japan is lower than that in Europe and America. However, the actual working hours of the day differ among companies. There are several types of unit processes related to forestry work, e.g. felling, prehauling, limbing, bucking, and yarding. On timber production sites, various machines operate in different unit processes collectively known as the operation system.

The labor productivities for each unit process within the operation system often differ. The total labor productivity (TLP) for the whole operation system is calculated using the following formula:

$$TLP = \frac{1}{\dfrac{1}{TLP \text{ of Process } 1} + \dfrac{1}{TLP \text{ of Process } 2} + \cdots + \dfrac{1}{TLP \text{ of Process } n}}. \tag{6.11}$$

In this formula, a smaller process number in whole system is preferable when the productivity is concerned. If the labor productivity of all processes is 10 m^3 per person per hour and if only one unit process is involved, the labor productivity of the whole operation system is thus 10 m^3 per person per hour. However, in case two or three unit processes are involved, the labor productivity for the whole system is 5 m^3 per person per hour for two unit processes, and 3.3 m^3 per person per hour for three unit processes. In addition, keeping the labor productivities at the same level as possible is preferred for the forestry operation.

The average Japanese labor productivity for the timber industry is 3.45 m^3 per person per day for thinning and 4.76 m^3 per person per day for regeneration cutting. In its "Basic Plan for Forest and Forestry", the Forestry Agency has set a goal for labor productivity to exceed 10 m^3 per person per day within 10 years (Forestry White Paper).

b. Harvesting System

The forest biomass is harvested using machines, staff, and infrastructure. The Japan Forestry Agency recommends a number of different operating systems according to terrain conditions as listed in Table 6.1 (Forestry Agency, 2010).

TABLE 6.1 Operating Systems According to Terrain Conditions

Terrain	Type	Felling	Prehauling Yarding	Bucking	Transportation
Gentle	Vehicle	Harvester	Grapple Forwarder	Processor	Truck
Middle	Vehicle	Harvester Chainsaw	Grapple Winch Forwarder	Processor	Truck
	Cable-logging	Chainsaw	Swing yarder Forwarder	Processor	Truck
Steep	Vehicle	Harvester	Grapple Forwarder	Processor	Truck
	Cable-logging	Chainsaw	Tower yarder Swing yarder Forwarder	Processor	Truck
Steeper	Cable-logging	Chainsaw	Tower yarder Swing yarder	Processor	Truck

(i) Gentle Terrain (0–15°)

- Vehicle machine system

- Felling: Harvester → Prehauling: Grapple → Bucking: Processor → Yarding: Forwarder → Transportation: Truck

- Forest road density: 100–250 m.

(ii) Middle Terrain (15–30°)

- Vehicle machine system

- Felling: Harvester or Chain-saw → Prehauling: Grapple or Winch → Bucking: Processor → Yarding: Forwarder → Transportation: Truck

- Forest road density: 75–200 m

- Cable-logging system

- Felling: Chainsaw → Prehauling: Swing yarder → Bucking: Processor → Yarding: Forwarder → Transportation: Truck

or

- Felling: Chainsaw → Prehauling: Swing yarder → Bucking: Processor → Transportation: Truck

- Forest road density: 25–75 m.

(iii) Steep Terrain (30–35°)

- Vehicle machine system

- Felling: Harvester → Prehauling: Grapple → Bucking: Processor → Yarding: Forwarder → Transportation: Truck

- Forest road density: 100–250 m

- Cable-logging system

- Felling: Chainsaw → Prehauling: Tower yarder or Swing yarder → Bucking: Processor → Yarding: Forwarder → Transportation: Truck

or

- Felling: Chainsaw → Prehauling: Tower yarder or Swing yarder → Bucking: Processor → Transportation: Truck

- Forest road density: 15–50 m

(iv) Steeper Terrain (>35°)

- Cable-logging system

- Felling: Chainsaw → Yarding: Tower yarder or Swing yarder → Bucking: Processor → Transportation: Truck

6.4. INFORMATION SYSTEM FOR FORESTRY PRODUCTION

Masahiro Iwaoka

6.4.1. Information for Forestry Production

a. Site Index

Site quality is a concept to classify the growth potential of forest trees for a particular site. It is a fixed constant with respect to tree species, and thus must be defined for each species. There are two methods for assessing site quality: (1) direct measurement of tree growth such as yield in volume or tree height growth, and (2) observation of environmental factors such as species or ground cover of vegetation under trees. The increment of wood volume during a cutting period indicates site quality directly; however, the result is influenced by stand density or duration of the cutting period so that it cannot be an objective indicator. Therefore, tree height, which is easy to measure, is usually used for indicating site quality because it is sensitive to site productivity but not sensitive to stand density. However, because tree height is the only parameter to index site productivity, it needs to be properly converted into yield using a specific yield conversion table. Site quality is generally classified into three or five classes designated as I, II, III or 1, 2, 3 (Shiraishi, 2001).

Tree height at specified stand age is thus used to indicate site quality. For example, the average height in feet of the top 150 dominant trees growing in one acre is used as the site index in the USA. The stand age is 50 years old on the east coast where the major species grow quickly, and 100 years on the west coast where the major species grow slowly. In Japan, the average height of dominant trees expressed by rounded-off integers is used as site index; the stand age is usually 40 or 50 years.

Site index can also be calculated based on environmental factors such as elevation, topography, geology, soil, meteorology, and others using score tables. The score tables for site indices are developed by using quantification theory I for calculating site indices from the environmental factors. The sum of all scores is site index (Ohta et al., 1996).

b. Evaluation of Potential for Forest Functions

The various forest functions have been classified by the Science Council of Japan classified into eight types: conservation of biodiversity, conservation of global environment, sediment disaster prevention and conservation of soil, water source cultivation, comfortable environment forming, health and recreation, culture, and material production. Prefecture governments in Japan try to evaluate the potential of each of the above functions based on official notices of the Forestry Agency. In this trial, score tables for evaluating four functions, i.e. flood prevention, water source cultivation, sediment disaster prevention, and health and culture, will be made based on quantification theory II, flood. Some prefectures use GIS to evaluate functions based on environmental information (Tanaka, 2001).

c. Yarding Distance of a Cable Logging System

A Scandinavian style cut-to-length mechanized logging system equipped with a harvester and a forwarder is spreading worldwide. It has also become popular in Japan for dense logging road networks. However, there are many sites not suitable for dense logging road networks because these sites have extremely steep topography. Cable yarding systems are needed at such sites because of the long yarding distances due to the low density of road networks. Usually, the cable yarding systems use aerial lines strung over the top of trees and uneven surfaces. Therefore, the area covered by a cable yarding system can be calculated using the buffering function available in GIS. This approach is similar to the basic concept of opening percentage as proposed by Backkmund (1966).

6.4.2. Using a Geographic Information System (GIS)

a. What is a GIS?

A GIS is a technique to manage and process spatial data, and display the results visually to enable advanced data analyses and timely judgment using geographic position as a clue (Geospatial Information Authority Japan, 2012). The company ESRI that developed the popular ArcGIS states: "a geographic information system (GIS) integrates hardware, software, and data for capturing, managing, analyzing, and displaying all forms of geographically referenced information."

A GIS can display a map and information, extract information, patchily paint map, overlay figures, measure distance, measure area, judge osculation or overlap of features, and search neighborhoods, among many other things. Its functions are based on Boolean operations such as logical AND, OR and XOR in figures, relations within neighborhoods such as buffering, analysis on aging, simulation on virtual reality, analysis of three-dimensional space, and many others. GIS manipulates data in either vector or raster mode. Any shape of figures can be handled in vector mode whereas only fixed shape and sized figures can be handled in raster mode. The positioning accuracy depends on the scale of the based map in a vector model and minimum units such as meshes. Objects are represented by nodes, arcs or polygons in a vector model and unit area in a raster model. Attributes are linked to node, arc or polygon in a vector model and unit area in a raster model. A vector model has functions to process graphics, and a raster model does not have functions. On the other hand, data structures are more complex in a vector model than in a raster model.

b. GIS in Forestry Industries

GIS is used in many fields to manage geographic data. In forestry industries, GIS is a useful tool to manage forest data because forest land spreads over wide areas with drastically different conditions. The forest data include lot number, stand compartment number, stand area, stand type (natural forest and plantation), tree species, blending ratio of species, stand age, soil type, and many others. In Japan, many prefectures use GIS to manage these forest data. For example, Gifu prefecture began to publish web GIS in 2005 and then switched the web GIS system to Gifu prefecture unified GIS. Data on forest type, vegetation type, forest road network, and others are open to the public via the internet (Gifu prefecture, 2012).

GIS can be used for calculating site index and yarding distance, as well as analyzing potential of forest functions and alignment of road network, etc. It is also used as part of the unified information system. For example, Komatsu Maxi is the umbrella term for Komatsu Forest's comprehensive control and information system for effective and profitable logging. It has all the necessary components to give sawmills, forestry companies, and contractors full control over the logistics chain from order to roadside delivery. The GIS is also used to control machines for optimum productivity and flexibility. GIS results and records are plotted on the map and site information stored in map layers is added by the operator.

c. Positioning Method with GPS

Global positioning system (GPS) is now a popular and useful tool for positioning. At first, methods that give high accuracy of positioning were studied by many researchers because the GPS positioning data includes selective availability (SA) with the signal intentionally degraded for civilian use. High-quality positioning data became available after 2000 because the SA had been turned off. After this, any user can get highly accurate positioning data with accuracy of less than 20 meters in addition to many additional available applications. GPS has been used with GIS to position specific points, to measure forest road networks, and sometimes to fix maps. Nowadays, new applications are developed to analyze forestry operations. For example, GPS is mounted on forwarders for positioning in order to control their operations. Moreover, GPS is also mounted on the carriage of a cable yarder to position the carriage and to control the carriage operation.

6.4.3. Applications of GIS for Forest Management

a. Site Index

The author has calculated site indices of Japanese cedar (*Cryptomeria japonica*) and Japanese cypress (*Chamaecyparis obtusa*) growing in the forest area of Tokyo using the score table of Tokyo and ArcGIS. The environmental factors used were elevation, inclination, soil types, geology, local topography, deposition pattern, effective depth, soil texture, and aspect. The elevation, inclination, and aspect were calculated from a digital elevation map (DEM) with a horizontal grid spacing of 50 meters; other elements were collected from forest registers. The site indices were calculated regarding all stands as Japanese cedar plantations or as Japanese cypress plantations.

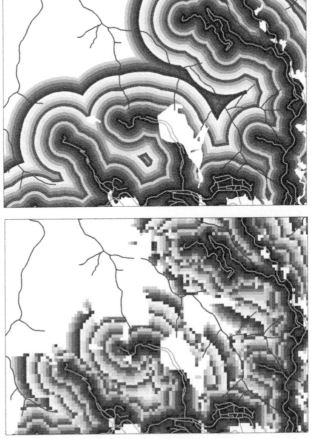

FIGURE 6.16   Difference of calculating methods of yarding distances. Top: usual buffering; bottom: considering the possibility of setting up a yarding cable.

A comparison of site indices for Japanese cedar and Japanese cypress reveals that the Japanese cypress has much higher site indices than Japanese cedar in this area. This result is contrary to the present situation of forest cover type. Japanese cedar, however, grows faster than Japanese cypress, and small-diameter logs of Japanese cedar have been produced historically for scaffolding in this area. As shown in this example, site indices can be easily calculated and recalculated for a wide area using GIS.

b. Yarding Distance of a Cable Logging System

The area covered by a cable yarding system is usually calculated with the GIS buffering function as described above. However, the author advocates that the local topography must be considered for evaluating mountainous diversified terrains. Cable yarding systems must also be influenced by local topography; for example, the usual cable yarding system is difficult to string over high ridges. The possibility of setting up a yarding cable is evaluated by investigating the production area; only the stands with the possibility of setting up a yarding cable can be used for timber production. The author has calculated yarding distances and area covered by cable yarding systems at each stand in the forest area of Tokyo to consider the possibility of setting up a yarding cable by using the Spatial Analyst extension of ArcGIS 8. The viewshed analysis is carried out on the forest terrain from points 10 meters above forest roads. The distance of 10 meters indicates the height of head spar for cable yarding. The local topography of forest terrain is collected from the DEM, and the visible points are selected as sites with the possibility of setting up a yarding cable. This method was applied on each actual stand. For stands with the potential of setting up a cable yarding system, the cable span is calculated as the inclined distance between the visible point of the nearest visible forest road and the farthest point of the stand from the visible point. As a result, the distribution of covered area for each cable span is lower than the usual results obtained using the buffering method that does not consider terrain elevation. This result means that yarding distances become longer when the possibility of setting up a cable is considered; this causes a reduction in productivity with higher logging costs. This type of analysis could not be done without GIS. The analytical results are shown in Figure 6.16.

REFERENCES

Backmund, F. (1966). Kennzahlen für den grad der erschließ ung von forstbetrieben durch auto-fahrbare wege. *Forstwissenschaftliches Centralblatt, 85*(12), 342–354.

Chosa, T., Shibata, Y., Kobayashi, K., Daikoku, Omine, M., Toriyama, K., Araki, K., & Hosokawa, H. (2006). Yield monitoring system for a head-feeding combine. *Japan Agricultural Research Quarterly, 40*(1), 37–43.

Chosa, T., Furuhata, M., Omine, M., & Sugiura, R. (2009a). Development and improvement of air-assisted seeding for paddy field following map information. In E. J. van Henten,

D. Goense, & C. Lokhorst (Eds.), *Precision agriculture '09* (pp. 127–132). Wageningen: Wageningen Academic Publishers.

Chosa, T., Furuhata, M., Omine, M., & Matsumura, O. (2009b). Development of air-assisted strip seeding for direct seeding in flooded paddy fields. *Japanese Journal of Farm Work Research, 44*(4), 211–218 [in Japanese with English summary].

Chosa, T., Furuhata, M., Motobayashi, K., Hosokawa, H., Aoki, M., Tsuchiya, M., Morita, K., & Sugimori, S. (2009c). Operation analysis of an air-assisted strip seeding using GPS. *Japanese Journal of Farm Work Research, 44*(Extra 1), 15–16 [in Japanese].

Daikoku, M. (2005). Development of computer-assisted system for planning work schedule of paddling and transplanting in distributed fields. *Japanese Journal of Farm Work Research, 40*(4), 210–214 [in Japanese with English summary].

Forestry Agency. (2010). *Final report of forest road network and operation system committee.* <http://www.rinya.maff.go.jp/j/seibi/saisei/pdf/romousaisyuu.pdf> Accessed 23.08.2012 [in Japanese].

Forestry Agency. (2012a). *Annual report on forest and forestry in Japan 2012.* Tokyo: Association of Agriculture and Forestry Statistic [in Japanese].

Forestry Agency. (2012b). *The forest and forestry statistics handbook.* <http://www.rinya.maff.go.jp/j/kikaku/toukei/pdf> Accessed 22.08.2012 [in Japanese].

Furuhata, M., Chosa, T., Shioya, Y., Tsukamoto, T., Seki, M., & Hosokawa, H. (2011). Current status, issue and perspective of direct seeding culture using an air-assisted row seeder. *Hokuriku Crop Science, 46*, 45–48 [in Japanese].

Geospatial Information Authority Japan. (2012). *What is GIS?.* <http://www.gsi.go.jp/GIS/whatisgis.html> Accessed September 2012.

Gifu prefecture, Gifu Forenavi. (2012). <http://www.gis2.pref.gifu.jp/MyMap2_0/GifuAdvanceMap/GifuAdvanceMap.jsp> Accessed September 2012.

Goto, T. (1997). Paddy management works by riding machine. *Journal of the Japanese Society of Agricultural Machinery, 59*(4), 131–135 [in Japanese].

Hamada, Y., Nakajima, Y., Hayashi, M., & Shaku, I. (2007). Development of the V-furrow no-till direct seeding of rice: Stabilization of culture by adopting the water-paddling seedbed preparation. *Japanese Journal of Crop Science, 76*(4), 508–518 [in Japanese with English summary].

Hori, T. (1965). A classification of ground configuration for logging planning. *Journal of Japanese Forestry Society, 47*, 168–170 [in Japanese].

Hori, T., Kitagawa, K., & Hasegawa, Y. (1971). Study on the area distribution of the terrain distances in a forest. *Journal of Japanese Forestry Society, 53*, 355–358 [in Japanese].

Hosokawa, H. (2009). Rotary tilling and ridge-making implement of soybean for avoiding wet injury. *Proceedings of the NARO international symposium on agricultural machinery*, 189–192.

Inoue, Y. (1974). *Forest management.* Tokyo: Chikyusha [in Japanese].

Jinkawa, M., Yoshida, C., Mozuna, M., Nakazawa, M., Ikami, Y., Furukawa, K., Usuda, H., Iwaoka, M. M., & Morooka, N. (2011). Development of a forwarder for biomass. *Journal of the Japan Forest Engineering Society, 26*, 227–231 [in Japanese].

Kamiya, T., Numano, N., Yagyu, H., & Shimazu, H. (2011). Field datalogger system using mobile phones and a web-based interface to allow the local community to determine the watering schedule for mandarin orange. *Agricultural Information Research, 20*(3), 95–101.

Komatsu Forest, & Maxi. (2012). <http://www.komatsuforest.com/default.aspx?id=2138&productId=&rootID=1475> Accessed September 2012.

Kurihara, E., Umeda, N., Hidaka, Y., Sugiyama, T., & Nonami, K. (2011). Study on increased high-moisture-paddy adaptability in combine (Part 3) – Schema of practical model and result of proof test. *Journal of the Japanese Society of Agricultural Machinery, 73*(6), 387–396 [in Japanese with English summary].

Loan, N. T. T., Chosa, T., & Tojo, S. (2011). Effects of auxiliary airflow in external side of seeding tube in air-assisted seeding for rice paddy. *Proceedings of the 2011 CIGR international symposium on "Sustainable bioproduction – Water, energy, and food", No. 22D07, on CD-ROM.*

Matsumoto, T. (2000). Evaluation of forest road networks to forestry and social infrastructure in a mountainous area. *Nagoya University Forest Science, 19*, 55–100 [in Japanese with English summary].

Matsumoto, T., & Kitagawa, K. (1999a). Relationship between current forest operations and provision of roads in private forests. *Journal of Forest Planning, 5*, 19–27.

Matsumoto, T., & Kitagawa, K. (1999b). Current problems of consensus building between forest owners in forest road planning. *Journal of the Japan Forest Engineering Society, 14*, 85–94.

Ministry of Health, Labor and Welfare. (2002). *Handbook of forestry employment management 2002.* Tokyo: Koyo-Mondai-Kenkyukai [in Japanese].

Miyasaka, J., & Ohdoi, K. (2005). Optimization methods and case studies for farm work scheduling. *Japanese Journal of Farm Work Research, 40*(4), 205–209.

Mizumoto, K. (2003). *Early modern Japan recounted by the grass mountain.* Tokyo: Ymakawa-Shuppansha [in Japanese].

Morimoto, E., Umemoto, H., Shimada, Y., Sugano, H., Nakade, D., Hirako, S., Osaki, H., Chosa, T., Omine, M., Iida, M., Shibusawa, S., & Kaho, T. (2006). Feasibility study of precision agriculture in Ishikawa rice production. *Conference abstracts of the 8th international conference on precision agriculture, 21.*

Mozuna, M., Jinkawa, M., Yoshida, C., Nakazawa, M., Ikami, Y., Iwaoka, M., Furukawa, K., Usuda, H., Kusano, Y., & Tanaka, S. (2011). Trial manufacturing of processor that can be used also for biomass harvesting. *Journal of the Japan Forest Engineering Society, 26*, 221–225 [in Japanese].

Mukouyama, T., Motobayashi, T., Chosa, T., Ookawa, T., Furuhata, M., Tojo, S., & Hirasawa, T. (2012). Dry matter production and grain yield of high yielding rice cultivar, Takanari, directly seeded in the paddy field with an "air-assisted drill": Effects of planting pattern on ecophysiology of direct seeded rice. *Japanese Journal of Crop Science, 81*(4), 414–423 [in Japanese with English summary].

Murakami, M., & Yamada, T. (2008). Volume reduction of forest biomass with large scale bundling machine. *Journal of the Japan Forest Engineering Society, 23*, 175–178 [in Japanese].

Nakazawa, M., Matsumoto, T., Yamada, Y., & Kondo, M. (2005). Route planning and road construction priority based on forest operations in Aichi prefecture, Japan. *Journal of the Japan Forest Engineering Society, 20*, 71–82.

Nanseki, T., & Matsushita, S. (2005). Farm planning support system with FAPSDB. *Japanese Journal of Farm Work Research, 40*(4), 199–204 [in Japanese].

Nishimura, Y., Hayashi, K., Gotoh, Horio, M., Ichikawa, T., Asano, S., Ueda, Y., Fukuma, H., & Nakao, T. (2001). Development of a precision direct drill seeder for direct sowing of rice on paddy field (Part 2) – Schematic of a precision direct drill seeder and seed depth control. *Journal of Japanese Society of Agricultural Machinery, 63*(6), 114–121 [in Japanese with English summary].

Nishimura, Y., Hayashi, K., Gotoh, H. M., & Ichikawa, T. (2003). Development of a precision direct drill seeder for direct sowing of rice on paddy field (Part 4) – Regional adaptability of developed seeder. *Journal of Japanese Society of Agricultural Machinery, 65*(1), 143–151 [in Japanese with English summary].

Ohta, T., Kitamura, M., Kumazaki, M., Suzuki, K., Sutou, S., Tadaki, Y., & Fujimori, T. (Eds.). (1996). *Encyclopedia of forests*. Tokyo: Maruzen.

Sakai, T. (2000). Geographical features of Japanese forest. *Journal of the Japan Forest Engineering Society, 15*, 221–224 [in Japanese].

Sasaki, Y., Tasaka, K., Tsuchiya, S., Ishii, K., Horio, M., Nishimura, H., Hamada, Y., Matsuo, Y., & Chosa, T. (2007). Protein content of wheat for bread by variable fertilization in southwestern Japan. *Proceedings of the 2nd Asian conference on precision agriculture (on CD-ROM)*.

Sawaguchi, I. (1996). Studies on forest-road evaluation and forest-road standards in mountain forests (I) - Characteristics of parameters for forest-road evaluation. *Bulletin of the Forestry and Forest Products Research Institute, 372*, 1–110.

Segebaden, G. V. (1964). Studies of cross-country transport distances and road net extension. *Studia Forestalia Suecica, 18*, 5–70.

Shiraishi, N. (2001). Site quality. In *The Japan Forestry Association Encyclopedia of forests and forestry* (pp. 653). Tokyo: Maruzen.

Sugiyama, J. (2004). Information technology and accountability system for produce. *Journal of Japanese Society of Agricultural Machinery, 66*(4), 16–20.

Tanaka, K. (2001). Quantification theory. In *The Japan Forestry Association Encyclopedia of forests and forestry* (pp. 533). Tokyo: Maruzen.

Togashi, T., Shimotsubo, K., & Yoshinaga, S. (2001a). Development of seed-shooting seeder of rice combined with a paddy harrow and characteristics of the sowing depth. *Japanese Journal of Farm Work Research, 36*(4), 179–186 [in Japanese with English summary].

Togashi, T., Shimotsubo, K., & Yoshinaga, S. (2001b). Performance of seed-shooting seeder of rice combined with a paddy harrow and the growth characteristics of rice. *Japanese Journal of Farm Work Research, 36*(4), 195–203 [in Japanese with English summary].

Toriyama, K., Sasaki, R., Shibata, Y., Sugimoto, M., Chosa, T., Omine, M., & Saito, J. (2003). Development of a site-specific nitrogen management system of rice in a large paddy field. *Japan Agricultural Research Quarterly, 37*(4), 213–218.

Umeda, N., Kurihara, E., Shimazu, M., Miyahara, Y., Sugiyama, T., Omoto, K., Nonami, K., Kimura, A., & Morihiro, T. (2012). Development of the small conventional combine harvester. *Proceedings of the joint conference on environmental engineering in agriculture, No. D11, on CD-ROM* [in Japanese].

Yogi, K., Okabe, S., & Takuwa, M. (2008). A study on the low cost harvesting and transporting of forest biomass – Development of No. 2 bundling machine for branches and treetrops. *Journal of the Japan Forest Engineering Society, 22*, 285–288 [in Japanese].

Yogi, K., Okabe, S., Tsuge, Y., & Takuwa, M. (2006). A study on the low cost harvesting and transporting of forest biomass – Development of a new bundling machine for branches and treetops. *Journal of the Japan Forest Engineering Society, 20*, 229–232 [in Japanese].

Yoshioka, T., & Suzuki, Y. (2012). Unutilized resource amount of forest biomass. *Journal of the Japan Forest Engineering Society, 27*, 111–117 [in Japanese].

Chapter 7

Pretreatment and Saccharification of Lignocellulosic Biomass

Eika W. Qian

7.1. COMPOSITION AND STRUCTURE OF LIGNOCELLULOSIC BIOMASS

Biomass is the only known sustainable bioresource that can be used to produce liquid transportation fuels (Qian, 2011). It is produced via photosynthesis to convert light energy and atmospheric carbon, i.e. CO_2, into chemical energy to be stored as carbohydrates in plant biomass or living carbon. This is a natural process to capture and store atmospheric CO_2. Lignocellulosic biomass is the most economical and highly renewable natural resource in the world. It is estimated that the annual production of lignocellulosic biomass is over 200 billion tons on Earth, and only 3% of this has been used in the pulp industry. In this chapter, the availability, potential, and recalcitrant structure of lignocelluloses are described. Further, saccharification, which is a very important process in the production of bioethanol, is introduced in order to develop a novel process to produce bioethanol for transportation fuels from lignocellulosic biomass.

7.1.1. Renewable, Low-Cost, and Abundant Lignocellulosic Biomass

Plants use solar energy to combine carbon dioxide and water, forming a sugar building block $(CH_2O)_n$ to produce oxygen, as shown in equation (7.1). The sugar is stored in a polymeric form such as cellulose, starch, and/or hemicellulose. Most biomass is approximately 75% sugar polymer.

$$nCO_2 + nH_2O + light \rightarrow (CH_2O)_n + nO_2 \tag{7.1}$$

The first step for biofuels production is obtaining an inexpensive and abundant biomass feedstock available from the following sources: waste materials (agriculture wasters, crop residues, wood wastes, urban wastes), forest products (wood, logging residues, trees, shrubs), energy crops (starch crops such as corn, wheat, barley; sugar crops; grasses; woody crops; vegetable oils; hydrocarbon plants), or aquatic biomass (algae, water weed, water hyacinth). Table 7.1 shows the growth rate or productivity, the lower heating value, the total production energy, and the chemical composition of different types of biomass (Towler et al., 2004). The plant growth rate varies with a typical range from 6 to 90 metric tons per ha-year that is equivalent to 19–280 barrels of oil equivalent per ha-year (Klass, 1998). Plants typically capture 0.1–1.0% of solar energy, with the percentage of solar energy capture proportional to the plant growth rate. The energy inputs reported in Table 7.1 include the energy required to make fertilizer as well as the energy needed to transport the harvested crop. The rates and the energy requirements for plant growth are dependent on plant species (Table 7.1). The development of more efficient plant materials with faster growth rates while requiring less energy input will depend on progress in plant breeding, biotechnology, and genetic engineering.

Although photosynthesis has much lower sunlight-to-chemical energy efficiency than photovoltaic, it is regarded as "the natural solar cell" that collects low-energy-density solar radiation from large areas, fixes CO_2, and generates a chemical energy carrier (i.e. biomass) at nearly zero cost. On average, the net primary productivity of terrestrial biomass equals only 0.3% of the average energy density of sunlight. With improvements in plant characteristics by using modern biotechnology and cultivation technologies, a 10% increase in global photosynthesis efficiency for territorial plants would

TABLE 7.1 Chemical Composition, Energy Content, and Yield of Various Terrestrial Biomass Species[*]

Biomass component	Corn grain	Corn stover	Switchgrass	Sugar cane	Sweet sorghum	Eucalyptus	Pine
Productivity (dry metric tons per ha-year)[†]	7	13–24	8–20	73–87	43.8	40.0	11.6
Lower heating value (MJ per dry kg)[‡]	17	17.5	17	16.8	17.3	18.1	18.6
Energy inputs (MJ per dry kg)[§]	1.35	1.20		0.346	2.82	5.57	7.43
Energy content (MJ per ha-year)	120	228–420	134–340	1230–1460	760	720	210
Energy content (boe per ha-year)	20	40–70	23–58	210–250	128	123	37
Representative components (dry wt%)							
Celluloses	3	36	40–45	22	35	48	46–50
Hemicelluloses	6	23	31–35	15	17	14	19–22
Extractives (starches, terpenes)	72	6	0	43	23	2	3
Lignins	2	17	6–12	11	17	29	21–29
Uronic acid			0	0	1	4	3
Proteins	10		5–11				
Ash	10	10	5–6	9	5	1	0.3

Adapted from: [*]Towler et al. (2004); Lynd et al. (1999); and Klass (1998). [†]Klass (1998). [‡]Towler et al. (2004). [§]O'Sullivan (1997). [§]Towler et al. (2004).

assimilate more than 8.8 billion tons of carbon from the atmosphere (not accounting for carbon re-emission), which is equal to the total global current net carbon emission.

7.1.2. Composition and Structure of Lignocellulosic Biomass

The major components of plant cell walls are cellulose, hemicelluloses, and lignin, which together form a complex and rigid structure (Figure 7.1). Plant cell walls of different plant types vary greatly in appearance and property. The complicated structure of lignocellulosic biomass contributes to its resistance to biological and chemical degradation. For example, in nature, natural biodegradation of lignocellulosic biomass is slow because it requires the collective actions of many hydrolytic enzymes, including cellulase (endoglucanase, cellobiohydrolase, and beta-glucosidase), hemicellulase, and lignin-degrading enzymes. Two main root causes of the recalcitrance of lignocellulosic biomass to cellulose enzymatic hydrolysis are believed to be: (1) low accessibility of (micro)crystalline cellulose fibers, which significantly reduces the efficiency of cellulose, and (2) the presence of lignin (mainly) and hemicellulose on the surface of cellulose, which prevents cellulase from accessing the substrate efficiently.

The structured portion of lignocellulosic biomass is composed of cellulose, hemicellulose, and lignin. Cellulose (a crystalline glucose polymer) and hemicellulose (a complex amorphous polymer, whose major component is a xylose monomer unit) make up 60–90 wt% of terrestrial biomass, as shown in Table 7.1. Lignin, which is a large polyaromatic compound, is another major component of biomass. The term "extractives" in Table 7.1 refers to those compounds that are not an integral part of the biomass structure (Huber et al., 2006), and are soluble in solvents such as hot and cold water, ethers, or

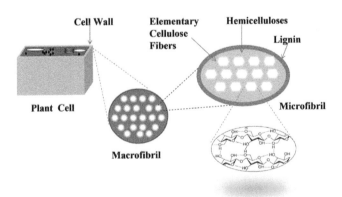

FIGURE 7.1 **Schematic structure of a lignocellulosic biomass containing cellulose, hemicellulose, and lignin.**

methanol. Extractives include different types of carbohydrates such as sucrose from sugar cane and amylose from corn grains. Ash listed in Table 7.1 is biomass material that cannot be blazed.

Extractives are mainly starch in crops and its wastes. Starch is a glucose polysaccharide that has α-1,4-glycoside linkages (Huber et al., 2006); they also have a large quantity of α-1,6-glycoside linkages that make the polymer amorphous. Human and animal enzyme systems can easily digest starches due to the presence of α-linkages. Starches are commonly found in the vegetable kingdom, e.g. corn, rice, wheat, beans, and potatoes. When treated in hot water, the starch forms two principal components: water-soluble amylose (10–20 wt%) and water-insoluble amylopectin (80–90%). Amylose contains only α-1,4-glycoside linkages, whereas amylopectin contains both α-1,4- and α-1,6-glycoside linkages with an approximate α-1,4 to α-1,6 linkage ratio of 20:1.

Cellulose is a linear biopolymer of anhydroglucopyranose connected by β-1,4-glycosidic bonds. Coupling of adjacent cellulose chains by orderly hydrogen bonds and van der Waal's forces lead to parallel alignment and a crystalline structure, resulting in low accessibility to enzymes (Figure 7.1). Recently, quantitative data of cellulose accessibility to cellulase clearly suggest that only a small fraction of β-glucosidic bonds of cellulose, ranging from 0.0023 to 0.041, are accessible by cellulase. The evidence from cell wall biophysics, biosynthesis, genomics, and atomic force microscopy (AFM) images suggests that elementary cellulose fibrils are synthesized by a cellulose synthase complex locus that contains 36-glucan chains with both crystalline and subcrystalline structures (Ding and Himmel, 2006). A number of elementary fibrils coalesce into much larger microfibrils and then into macrofibrils.

Unlike starch, cellulose is a crystalline material with an extended flat twofold helical conformation (Wyman et al., 2005). Hydrogen bonds maintain and reinforce the flat linear conformation of the chain. The top and bottom of the cellulose chains are essentially completely hydrophobic. The sides of the cellulose chains are hydrophilic and capable of bonding hydrogen because all the aliphatic hydrogen atoms are in axial positions, whereas the polar hydroxyl groups are in equatorial positions. The degree of polymerization of cellulose is approximately 10,000 glucopyranose monomer units in wood and up to 15,000 glucopyranose monomer units in cotton. Partial acid hydrolysis breaks down cellulose into cellobiose (glucose dimer), cellotriose (glucose trimmer), cellotetrose (glucose tetramer), and ultimately into glucose when the acid hydrolysis is complete.

Hemicellulose is situated between lignin and a collection of cellulose fibers underneath; it is a sugar polymer that typically consists of 20–40 wt% of biomass (Huber et al., 2006). In contrast to cellulose, which is a polymer of only glucose, hemicellulose is a copolymer of five different sugars. This complex polysaccharide occurs in association with cellulose in cell walls. Five-carbon sugars (usually xylose and arabinose) and six-carbon sugars (galactose, glucose, and mannose), which are highly substituted with acetic acid, are contained in hemicellulose. The most abundant building block of hemicellulose is xylan (a xylose polymer linked at the 1 and 4 positions). Hemicellulose is amorphous because of its branched nature that is relatively easy to hydrolyze into its monomer sugars as compared with cellulose.

Furthermore, consistent with their structural chemistry and side-group substitutions, xylans are interspersed, interweaved, and ester-linked at various points with the overlying "sheath" of lignin, as well as to produce a coat around underlying strands of cellulose via hydrogen bonds. The xylan layer with its covalent linkage to lignin and its non-covalent interaction with cellulose may be important in maintaining the integrity of the plant cell wall in situ, and in helping to protect the fibers against degradation by enzymes. At least two types of covalent crosslinks have been identified between hemicellulose and lignin: (1) diferulic acid bridges and (2) ester linkages between lignin and glucuronic acid attached to xylans.

Ten to twenty-five weight percent of lignocellulosic biomass is typically composed of lignin, which is a highly branched, substituted, mononuclear aromatic polymer found in the cell walls of certain biomass, particularly woody biomass. Lignin is often associated with the cellulose and hemicellulose materials making up lignocellulosic compounds. It is an irregular polymer, which is formed by an enzyme-initiated free-radical polymerization of the alcohol precursors. The bonding in the polymer can occur at many different sites in the phenylpropane monomer due to electron delocalization in the aromatic ring, the double bond-containing side chain, and the oxygen functionalities. The manner in which lignin is produced from lignocellulosic biomass affects its structure and reactivity in hydrolysis. Softwood lignins are formed from coniferyl alcohol as monomer units (Qian, 2011), whereas grass lignin contains coniferyl, sinapyl, and coumaryl alcohol.

7.2. PRETREATMENT OF LIGNOCELLULOSIC BIOMASS

7.2.1. Importance of Pretreatment

Lignocellulosic biomass can be converted into liquid fuels via three primary routes, including syngas production by gasification, bio-oil production by pyrolysis or liquefaction, and hydrolysis of lignocellulosic biomass to produce sugar monomer units (Huber et al., 2006). The hydrolysis route to produce liquid fuels includes two subprocesses: (1) hydrolysis of cellulose in the lignocellulosic materials to fermentable reducing sugars, and (2) fermentation of reducing sugars to target products (especially cellulosic ethanol or bioethanol). The hydrolysis is usually catalyzed by using enzymes such as cellulase, and the fermentation is carried out with suitable microorganisms. Bioconversion of lignocellulosic biomass to liquid fuels and commodity products has many potential benefits, with enzymatic hydrolysis of cellulose to glucose, followed by fermentation to ethanol, a particularly attractive route to produce low-cost sustainable liquid transportation fuels (Wyman et al., 2005).

However, enzymatic hydrolysis may be the most complex step in the bioconversion process because of the combination of substrate-related and enzyme-related effects. Although the hydrolysis mechanism and the relationship between the structure and function of various cellulases have been studied extensively (Converse, 1993), the complex biomass structure confounds understanding of the relative importance of these features and their roles; reducing one barrier to digestion can enhance or mask the importance of others.

Lignocellulosic biomass has a complicated structure because it contains many components such as cellulose, hemicelluloses, lignin, extractives and ash, as mentioned above (Table 7.1). Thus, these components bring about a barrier to the enzymatic hydrolysis of lignocellulosic biomass. Therefore, removing hemicelluloses and/or lignin before the bioconversion process is a prerequisite to increasing the accessible surface area of cellulose by enzymes to enhance the subsequent enzymatic hydrolysis reactions.

The biological conversion of lignocellulose to fuel generally has three main steps: (1) lignocellulosic biomass pretreatment to convert the recalcitrant lignocellulosic structure into reactive cellulosic intermediates, (2) enzymatic cellulose hydrolysis by cellulase to hydrolyze and convert reactive intermediates to fermentable sugars, e.g. glucose, and (3) fermentation to produce cellulosic ethanol (bioethanol) or other bio-based chemicals, e.g. lactic acid and succinic acid. Pretreating the recalcitrant structure of lignocellulosic biomass efficiently to release the locked polysaccharides is one of the most important factors concerning the success of R&D for the emerging biofuel and other biomass-based industries because lignocellulosic biomass pretreatment is one of the most expensive unit operations and/or processes, and has a major influence on the costs of both prior operations, e.g. reduction of lignocellulosic biomass particle size, and subsequent operations, e.g. enzymatic hydrolysis or fermentation (Wyman et al., 2005).

The purposes of pretreatment are to decrease the crystallinity of cellulose, increase biomass surface area, remove hemicellulose, and break the lignin seal (Mosier et al., 2005). The removal of lignin and/or hemicellulose can substantially reduce the recalcitrance of lignocellulosic biomass to enzymatic hydrolysis (Wyman et al., 2005). This pretreatment changes the biomass structure and improves downstream processing. On the other hand, pretreatment is also one of the least understood processing options. Generally, the following is expected for a desirable attribution of pretreatment (Mosier et al., 2005):

1. Low cost of chemicals for pretreatment, neutralization, and subsequent conditioning.
2. Minimal waste emission.
3. Limited size reduction to avoid energy-intensive and expensive milling operations.
4. Fast reactions and non-corrosive chemicals to minimize pretreatment capital investment.

5. Sugar concentrations higher than 10% derived from hydrolysis of hemicellulose to keep fermentation reactor sizes at a reasonable level and facilitate downstream recovery.
6. Promoting high product yields in subsequent enzymatic hydrolysis or fermentation operations with minimal conditioning costs.
7. By-products not posing additional processing or disposal challenges.
8. Low enzyme loading to realize greater than 90% digestibility of pretreated cellulose in less than 5 days and preferably 3 days.
9. Easy recovery of lignin and other constituents for conversion to valuable co-products and simplification of downstream processing.

7.2.2. Various Pretreatments

A variety of physical and chemical pretreatment methods and their combinations have been developed to improve the accessibility of enzymes to cellulosic fibers (Mosier et al., 2005). Physical pretreatment methods include ball milling, comminution (mechanical reduction of biomass particulate size), and compression milling; chemical pretreatment methods include hydrothermolysis with acid, alkali, and solvents (e.g. H_2O_2, glycerol, dioxane, phenol and ethylene glycol, ozone). Physicochemical pretreatment methods including uncatalyzed steam explosion, hot compressed water treatment, ammonia fiber explosion, and biological pretreatment techniques have also been used.

Table 7.2 shows the effect of various pretreatment methods on the chemical and physical structure of lignocellulosic biomass. Uncatalyzed steam explosion is used commercially to remove hemicellulose for manufacturing fiberboard and other related products in the Masonite process (Mosier et al., 2005). High-pressure steam is applied to wood chips for a few minutes without adding chemicals. This process increases the wood surface area and improves the cellulose digestibility significantly without decrystallizing the cellulose.

Hot water treatment using subcritical water (200–250°C) and supercritical water (higher than 374°C and above 22 MPa) treatment at elevated temperatures and pressures can increase the biomass surface area and remove hemicellulose (Mosier et al., 2005; Wyman et al., 2005). Three types of reactors used for hot water pretreatment include co-current (lignocellulose and water flow in the same direction), counter-current (water and lignocellulose move in opposite directions), and flow-through (hot water passes over a stationary bed of lignocellulose). The advantage of hot water treatment is the elimination of chemical usage to achieve size reduction in the pretreatment of biomass, but a major disadvantage is the formation of sugar degradation products (e.g. furfural from xylose and 5-hydroxymethylfurfural (5-HMF) from glucose) in addition to requiring expensive devices because of the high pressure (10–15 MPa).

Pretreatment with dilute sulfuric acid-based chemical pretreatment (Mosier et al., 2005) is the most popular method; it hydrolyzes hemicellulose to sugars with high yields, changes the structure of the lignin, and increases the cellulosic

TABLE 7.2 Effect of Promising Pretreatment Methods on the Structure and Composition of Lignocelluloses Biomass (Mosier et al., 2005)

Pretreatment method	Increase in surface area	Decrystalization of cellulose	Removal of hemicellulose	Removal of lignin	Change in lignin structure
Uncatalyzed steam explosion	+		+		−
Liquid hot water	+	N.D.	+		−
PH controlled hot water	+	N.D.	+		N.D.
Flow-through liquid hot water	+	N.D.	+	−	−
Dilute acid	+		+		+
Flow-through acid	+		+	−	+
Ammonia fiber explosion (AFEX)	+	+	−	+	+
Ammonia recycled percolation (ARP)	+	+	−	+	+
Lime	+	N.D.	−	+	+

+: major effect; −: minor effect; N.D.: not determined.

surface area (Mosier et al., 2005; Wyman et al., 2005). This pretreatment process pioneered by Grethlein and others can break the lignin–hemicellulose shield in agricultural residues and woody biomass usually accompanied by degradation of hemicellulose. The acid-catalyzed treatment has been widely assessed from the viewpoint of efficiency improvement, kinetic analysis, and cost-effectiveness, among many others. This mainstream process, however, might have some undesirable effects. For example, the formation of aldehydes such as furfural due to excessive degradation of the produced monosaccharide is essentially inevitable; this results in lowering the conversion yield of poly-saccharides and inhibiting the subsequent ethanol fermentation process. An additional problem caused by using this process is that it uses a corrosive acid such as sulfuric acid that needs downstream neutralization in addition to requiring special materials for reactor construction. The recovery of the spent acid also complicates the downstream processing steps.

Ammonia fiber/freeze explosion (AFEX) in which anhydrous ammonia is used to react with lignocellulose can increase the surface area of the biomass, decrease crystallinity of cellulose, dissolve part of the hemicellulose, and remove lignin. Pretreatment of the lignocellulose with a less concentrated ammonia solution is known as ammonia recycled percolation (ARP). The AFEX pretreatment is a novel alkaline pretreatment process that effects a physicochemical alteration in the lignocellulosic ultra and macro structure. Studies have shown that the AFEX pretreatment increases the biomass enzymatic digestibility several folds over the untreated lignocellulosic (Teymouri et al., 2005). The AFEX pretreatment results in the decrystallization of cellulose (Gollapalli et al., 2002), partial depolymerization of hemicellulose, deacetylation of acetyl groups, cleavage of lignin–carbohydrate complex (LCC) linkages and lignin C–O–C bonds, increase in the accessible surface area due to structural disruption (Turner et al., 1990), and increase in wettability of the treated biomass (Sulbaran de Ferrer et al., 1997). The AFEX process has proved to be attractive economically compared with several other leading pretreatment technologies based on a recent study using an economic model (Eggeman and Elander, 2005) for evaluating bioethanol conversion of corn stover.

Wet oxidation (WO) has been shown to be an efficient pretreatment method of wood and wheat straw; it dissolves the hemicellulosic fraction and makes the solid cellulose fraction susceptible for enzymatic hydrolysis and fermentation (Bjerre et al., 1996). The WO process is carried out using hot water under pressure with the addition of oxygen. Once oxygen is not applied, the process is essentially a hydrothermal pretreatment that is comparable to the well-known steam pretreatment. Addition of carbonate during the hydrolysis (alkaline WO) results in low sugar degradation, and reduced formation of 2-furfural (Bjerre et al., 1996) and phenol aldehydes (Klinke et al., 2002). In steaming and dilute acid hydrolysis pretreatments, sugar monomers are produced whereas the alkaline WO treatment of wheat straw only produces soluble polymeric hemicellulose sugars (arabinoxylan).

Silverstein et al. (2007) compared the pretreatment effect for using acid, alkali and wet oxidation (H_2O_2 and ozone) to pretreat cotton stalks. Cellulose conversion of the pretreated samples after 72 h of enzymatic hydrolysis is shown in Table 7.3. Pretreatment using sodium hydroxide shows the highest cellulose conversion of 60.8%, followed by hydrogen peroxide (49.8%) and sulfuric acid (23.8%). The hydrolysis pretreatment using sodium hydroxide has the highest (62.6%) conversion of xylan to xylose, whereas the hydrogen peroxide pretreatment averages 7.8% conversion. For the acid pretreatment, no xylan is detected in the treated solids during initial carbohydrate, but an average of 14.3 mg xylose per g dry biomass is detected in the supernatant after

TABLE 7.3 Effect of Various Pretreatment Methods on Glucan and Xylan Conversion after Enzymatic Hydrolysis of Cotton Stalks (Eggeman and Elander, 2005)

Pretreatment agent	Compositon of hydrolysis supernatant and pretreated solid*†‡			Glucan conversion (%)	Xylan conversion (%)
	Lignin	Xylose	Glucose		
Sulfuric acid	–	1.43 (0.16)	11.03 (0.66)	23.85 (1.21)	0.00 (0.00)[§]
	40.68 (1.44)	0.00 (0.00)	46.3 (2.89)		
Sodium hydroxide	–	8.34 (0.15)	30.57 (0.56)	60.79 (2.75)	62.57 (2.57)
	18.40 (0.16)	12.13 (0.40)	50.33 (1.84)		
Hydrogen peroxide	–	0.90 (0.14)	17.21 (0.84)	49.82 (1.40)	7.78 (1.13)
	25.59 (2.30)	10.00 (0.26)	34.53 (0.86)		

*Composition percentages calculated from values on a dry weight basis.
†Data are averages of three replicates. Numbers in parentheses standard deviations.
‡Compositions of xylose and glucose in the hydrolysis supernatant are given in the upper rows while compositions of pretreated solids are given in the bottom rows.
§See text for explanation.

enzymatic hydrolyses. The difference in cellulose conversion during enzymatic hydrolyses is hence largely dependent on the difference in lignin composition. On the other hand, the ozone pretreatment does not cause any significant changes in lignin, xylan, or glucan contents during the treatment period.

Wyman et al. (2005) reported the results of using different pretreatment methods followed by enzymatic hydrolyses to produce sugars from corn stover feedstock; over 90% yields of sugar are obtained with the various pretreatments. Hot water treatment using a flow-through reactor has the highest overall soluble product yield; however, the xylose monomer yield is only 2.4%, indicating that this method does not produce xylose monomers. The dilute acid pretreatment method produces the highest quantity of sugar monomers, with 92% yields. Results are expected to be different with other feedstocks (Yang, and Wyman, 2004).

The order of bioethanol production costs is: dilute acid < AFEX < lime < ARP < hot water. Hot water pretreatment is the most expensive method because it requires more enzymes to break down the xylose oligomers. If the oligomers could be successfully converted into ethanol (or other products), the unit cost of producing ethanol would decrease for the hot water, ARP, and lime methods, because these pretreatment methods yield a significant quantity of oligomers.

Recently, novel pretreatment methods such as microwave treatment (Nakayama and Okamura, 1989), and combined ethanol cooking and ball milling (Teramoto et al., 2008) have also been developed. The latter is similar to organosolv pulping (Aziz and Sarkanen, 1989; Chum et al., 1990; Grethlein and Converse, 1991), which is generally implemented using a strong acid catalyst in order to remove the lignin component completely. This pretreatment alleviates the problems associated with the use of strong acid catalysts so that the production of furfural due to excessive degradation of polysaccharide components is extremely low with insignificant delignification (Teramoto et al., 2008). Therefore, the cooking process is regarded not as a delignification process but a process to activate the original wood. Subsequently, the activated solid products are pulverized by using ball-milling in order to improve their enzymatic digestibility. The enzymatic hydrolysis experiments demonstrate that the conversion of cellulosic components into glucose can attain 100% under optimal conditions.

7.3. PRETREATMENT AND SACCHARIFICATION OF LIGNOCELLULOSIC BIOMASS USING SOLID ACID CATALYSTS

7.3.1. Solid Acid Catalysts

Various methods to pretreat lignocellulose with advantages and disadvantages have been developed as mentioned above. On the other hand, there is no single best pretreatment method for all types of lignocellulose, including soft biomass such as rice straw, wheat straw, and corn stover, etc. as well as hard biomass such as wood, etc. A new way to efficiently exploit lignocellulose by avoiding the drawbacks associated with the aforementioned technologies is the use of solid acid catalysts. This new method is capable of overcoming the above-mentioned problems such as the use of environment-unfriendly chemicals and the need to purify final products. The solid acid catalyst has an exceptional catalytic performance in pretreating biomass with high selectivity because both the quantity and strength of acid applied can be controlled. Furthermore, the use of solid acid catalyst facilitates the subsequent separation of final products and recovery of the catalyst after pretreatment in addition to eliminating the neutralization step.

Using heterogeneous solid acid catalysts to replace conventional homogeneous acids has been applied across many fields, including petroleum refining, organic synthesis, and fine chemicals. Solid acid catalysts offer the great advantages of efficient recovery, low or even no corrosion, and mild reaction conditions. As shown in Figure 7.2, the solid catalysts include silica–alumina, zeolites, niobic acid, sulfonated carbon-based catalyst, mesoporous silica-based catalyst and strong ion-exchangeable resins, as well as sulfated transition metal oxides including zirconia, titania, etc., which have equivalent acidity or stronger acidity than liquid acid. In fact, using solid acid catalysts to

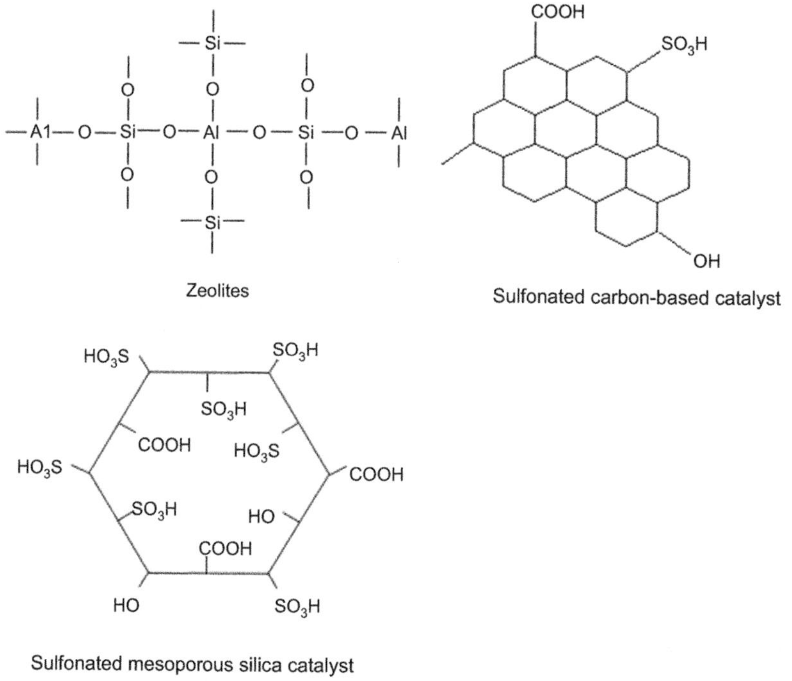

Zeolites

Sulfonated carbon-based catalyst

Sulfonated mesoporous silica catalyst

FIGURE 7.2   Several solid acid catalysts.

hydrolyze lignocellulosic biomass has been studied recently. Patents awarded to methods for lignocellulosic biomass saccharification and liquefaction have been published by Qian and Hosomi. (2009a, b). The hydrolysis of cellulose using amorphous carbon-bearing sulfonic acid groups (Suganuma et al., 2008) and acidic ionic liquid-modified silica catalysts (Amarasekara and Owereh, 2010; Goderis et al., 2010) have been carried out.

In this section, the utilization of solid acid catalysts in pretreatment and direct saccharification of lignocellulosic biomass is presented.

7.3.2. Pretreatment of Rice Straw

a. Pretreatment of Rice Straw using Solid Acid Catalysts

A novel pretreatment process that is mild and cost-effective for enzymatic hydrolyses of lignocellulosic biomass needs to be developed. Most hemicelluloses and some lignin and ash are removed after catalytic hydrothermal saccharification of rice straw using solid acid catalysts, and the major solid residues are cellulose (Li and Qian, 2011). Thus, solid acid catalysts are feasible in the hydrothermal pretreatment of soft biomass such as rice straw, grass, etc. Novel processes to pretreat lignocellulosic biomass using solid acid catalysts have thus been investigated, and the influence of pretreatment conditions such as pretreated temperature, pretreated time, and types of catalyst on the efficiency of subsequent enzymatic saccharification was evaluated.

The hydrothermal pretreatment of rice straw using solid acid catalysts was carried out in an 80-cm³ autoclave using a commercial solid acid catalyst (Amberlyst 35 Dry), sulfated zirconia (SA-J1), and a novel sulfonated mesoporous silica catalyst (MPS-1). After pretreatment, the solid acid catalyst was recovered, and the mixed slurry of rice straw and water was put in an L-type glass reactor to be hydrolyzed using cellulase (Onozuka R-10). Enzymatic saccharification of the pretreated rice straw was carried out at 45°C for 1–3 days to investigate the efficiency of solid acid catalysts.

The various chemical species before and after enzymatic saccharification of the pretreated rice straw at 180°C for different reaction times using SA-J1 catalyst were investigated. Before enzymatic hydrolyses, the yield of monosaccharide increased as pretreatment time increased from 10 to 60 min (Li et al., 2012a). At the same time, low levels of by-products such as furfural, 5-HMF and organic acids are formed, and their quantities increase with longer pretreatment time. On the other hand, there is no significant change in the yield of oligosaccharide. However, in the subsequent enzymatic saccharification of pretreated rice straw, the yield of monosaccharide increases as pretreatment time is extended from 10 to 30 min, but then remains constant up to 60 min. In comparison with the pretreatment process, the yield of oligosaccharide decreases significantly, indicating that oligosaccharide is converted into monosaccharide during the process of enzymatic saccharification.

Various by-products such as organic acids, furfural etc. are formed after pretreatment. Also, no significant variations in the yields of various by-products after enzymatic saccharification are observed for all samples pretreated for different durations. This means that no additional by-products are formed during the process of enzymatic saccharification.

b. Effect of Pretreatment Conditions and Solid Acid Catalysts

In order to investigate the effects of pretreatment temperature on the enzymatic saccharification of pretreated rice straw, studies on pretreating rice straw were carried out at 110–180°C for 3 h. With the pretreatment temperature elevated from 110 to 150°C, the yields of all products except glucose increased. The yields of oligosaccharide and monosaccharide reached 250 and 105 g kg^{-1} respectively when the pretreatment was maintained at 150°C for 3 h. Further, enzymatic saccharification of the pretreated rice straw was carried out at 45°C for 3 days. As shown in Figure 7.3, in the case of the enzymatic saccharification of pretreated rice straw, the yield of monosaccharide rises with increasing pretreatment duration up to 3 h at 150°C, and the maximum yield of 450 g kg^{-1} monosaccharide is obtained. Based on a composition analysis of raw rice straw, approximately 76% holocellulose in rice straw is converted into monosaccharide. In addition, the higher pretreatment temperature accelerates the

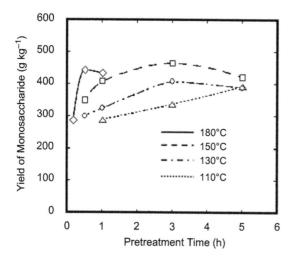

FIGURE 7.3 Effects of pretreatment temperature and time on yields of monosaccharides after enzymatic hydrolysis of pretreated rice straw using SA-J1 catalyst.

enzymatic hydrolysis rate of the rice straw. As listed in Table 7.4, the decreasing rates of xylan and glucan are accelerated at higher pretreatment temperatures for the same pretreatment duration. Lignin has the same tendency to decrease this rate as xylan and glucan.

Scanning electron microscope (SEM) images show that the raw rice straw sample exhibits rigid and highly ordered fibrils, whereas pretreated samples

TABLE 7.4 Effects of Pretreated Time and Pretreated Temperature on the Change in Composition of Pretreated Rice Straw using SA-J1 Catalyst

Content (%, w/w)	Raw rice straw	110°C/3 h	130°C/3 h	150°C/3 h	150°C/0.5 h	150°C/5 h
		Reaction conditions				
Xylan	11.2	7.32	6.73	3.53	9.62	1.89
Glucan	49.7	24.1	21.3	21.7	29.1	18.4
Acid-insoluble lignin	15.8	14.4	13.9	10.0	10.8	12.5
Acid-soluble lignin	1.70	0.67	0.54	0.24	0.87	0.27
Ash	13.9	9.16	8.48	9.20	9.49	10.0
Extractives	3.46	ND	ND	ND	ND	ND

have loose and fluty fibers with a more disordered orientation for higher pre-treatment temperature. The microfibers are also separated from the initial connected structure and fully exposed, thus increasing the external surface area and internal porosity, which enhances ability of the cellulase to attack cellulose contained in the pretreated rice straw. In addition, with increasing pretreatment temperature, the CrI index of cellulose in rice straw before and after enzymatic hydrolysis increases from 13.6% to 44.9%. Both the surface area and pore volume are higher after pretreatment than those of the untreated raw rice straw. At the same time, the molecular weight of cellulose in the pretreated rice straw decreases from 622,490 to 579,802, which indicates the reduced degree of polymerization of cellulose after pretreatment.

The effect of using four different catalysts (Lewis-type solid acids: SA-J1 and SA-J2; Brönsted-type solid acids: MPS-1 and Amberlyst 35 Dry) on the enzymatic hydrolysis of rice straw was elucidated. After pretreatment, there is no significant difference among the liquefaction rates of the pretreated rice straw using these four catalysts. However, SA-J1 and SA-J2 show higher yields of oligosaccharide after pretreatment than MPS-1 and Amberlyst 35 Dry cat-alysts. In contrast, MPS-1 and Amberlyst 35 Dry have higher yields of monosaccharide (xylose and glucose) than SA-J1 and SA-J2 catalysts. These results indicate the possibility of a direct saccharification of lignocellulosic biomass using a solid acid catalyst with stronger acidity such as a Brönsted-type solid acid without subsequent enzymatic hydrolysis. After enzymatic hydrolysis of the pretreated rice straw, SA-J1 and SA-J2 catalysts have higher yields of oligosaccharide than MPS-1 and Amberlyst 35 Dry catalysts. In addition, the oligosaccharide yield obviously decreases after enzymatic saccharification of the pretreated rice straw using SA-J1 and SA-J2 catalysts; it remains relatively constant if MPS-1 and Amberlyst 35 Dry are used. Most monosaccharides obtained using MPS-1 and Amberlyst 35 Dry catalyst treat-ment come from pretreated rice straw, whereas most monosaccharaides using SA-J1 and SA-J2 catalyst pretreatment come from the enzymatic hydrolysis of pretreated rice straw. As shown in Table 7.5, the Amberlyst 35 Dry catalyst has a quicker rate of decrease of xylan in the pretreated rice straw than SA-J1 catalyst. However, the rates of decrease of glucan, lignin, and ash in rice straw pretreated using SA-J1 catalyst are faster than those using Amberlyst 35 Dry catalyst.

c. Hydrothermal Pretreatment of Rice Straw using Solid Acid Catalysts

The effects of hydrothermal pretreatment of rice straw using solid acid catalysts are shown in Scheme 7.1. In hydrothermal pretreatment of rice straw in the presence of a solid acid catalyst, part of the hemicellulose in rice straw is first attacked by protons bonded to the functional groups of the solid acid catalyst. At the same time, part of the lignin soluble in water decomposes via

TABLE 7.5 Change in Composition of Pretreated Rice Straw in the Presence and Absence of a Solid Acid Catalyst at 150°C for 3 h

Content	Raw rice	Catalysts		
(%, w/w)	straw	Blank	SA-J1	Amberlyst 35Dry
Xylan	11.2	6.01	3.53	0.43
Glucan	49.7	25.6	21.7	25.4
Acid-insoluble lignin	15.8	12.0	10.0	10.0
Acid-soluble lignin	1.70	0.43	0.24	0.32
Ash	13.9	9.93	9.20	10.6
Extractives	3.46	ND	ND	ND

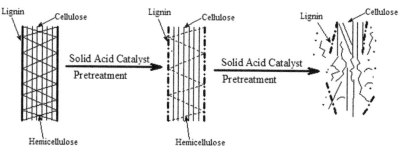

SCHEME 7.1 Mechanism of pretreatment of rice straw using solid acid catalysts.

hydrothermal hydrolyses. This leads to the structure of rice straw being broken up because lignin is soluble in water so that part of the hemicelluloses is removed. Hemicellulose is mainly converted into oligosaccharide via catalytic hydrolyses and/or hydrothermal hydrolyses. As the pretreatment process proceeds, hemicelluloses are removed completely, and the oligosaccharide is converted into monosaccharide. Simultaneously, part of the cellulose is converted into oligosaccharide, and the structure of rice straw becomes more irregular and disordered, resulting in low crystallinity of cellulose or an amorphous structure. Thus, the resulting celluloses become more susceptible to enzymatic hydrolysis in the subsequent enzymatic treatment.

7.3.3. Saccharification of Lignocellulosic Biomass using Solid Acid Catalysts

As described in the above section, when a stronger Brönsted acid catalyst such as MPS-1 is used in the pretreatment of rice straw, the formation of monosaccharide is observed. This suggests that direct saccharification of lignocellulosic biomass using a solid acid catalyst without subsequent enzymatic hydrolysis is possible. In this section, direct saccharification of lignocellulosic biomass using novel acid catalysts is introduced.

Selective and efficient hydrolyses of cellulose to glucose is a challenging task; the application of solid catalysts has been studied in recent years. Hara and co-workers demonstrated that a sulfonated carbon prepared from cellulose is capable of hydrolyzing cellulose to glucose and oligosaccharides at a low temperature of 373 K (Suganuma et al., 2008). Likewise, sulfonated carbon materials (e.g. activated carbon, silica–carbon composites, and a mesoporous carbon, CMK-3) convert cellulose into glucose with 40–75% yields for 24 h reaction time (Onda et al., 2008). Meanwhile, Rinaldi et al. (2008) showed that as a solid acid catalyst, a polystyrene sulfonated resin (Amberlyst 15DRY) in [BMIM]Cl can hydrolyze cellulose, although this observation deviates from the fact that the resin merely works as an acidifier under the aqueous conditions without the ionic liquid. On the other hand, lignocellulosic biomass is a solid and is known to be composed of cellulose, hemicelluloses, lignin, extractives, and other components. Hence, increasing the porous size and surface area of the solid catalyst is expected to improve the contact between the solid biomass and solid acid catalysts. On the other hand, mesoporous silicas such as SBA-15, MCM-41, etc. with a large surface area and mesoporous pore size of 2–50 nm have been developed recently. Further, the capability of sulfonated groups to be incorporated on to the surface of mesoporous silica SBA-15 has been reported (Margolese et al., 2000). Thus, novel solid acid catalysts with mesopores and large surface area to hydrolyze lignocellulosic biomass directly can be developed (Li et al., 2012a, b). In this section, the catalytic activities of several sulfonated group-bearing mesoporous silica SBA-15 catalysts in the saccharification of cellulose and rice straw, which is selected as a resource for bioethanol because of its abundance in Japan and Asia, have been assessed.

a. Novel Solid Acid Catalysts

X-ray photoelectron spectra (XPS) of the novel sulfonated mesoporous solid acid catalysts developed recently are shown in Figure 7.4 (Li et al., 2012a). The chemical groups that influence the acidity of the catalyst include OH, COOH, and SO_3H groups, and their distributions depend upon the conditions used to prepare the catalysts. The S_{2p} XPS peak at 168 eV is assigned to the sulfo group, which is consistent with the study of Nakajima et al. (2006). IR spectra confirm that the sulfo groups and –OH groups are bonded on the prepared catalyst. The X-ray diffraction (XRD) patterns show that the SBA-15 material structure has formed successfully after the preparation of the silica-based catalyst. When aged for 24 h (MPS-A and MPS-C), the surface area and pore volume increase with lower aging temperature. If aged at 90°C (MPS-B, MPS-C, and MPS-D), the surface area and pore volume increase with longer

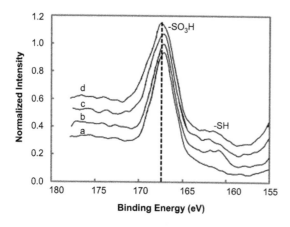

FIGURE 7.4 S_{2p} XPS spectra of mesoporous silica-based solid acid catalysts: (a) MPS-A; (b) MPS-B; (c) MPS-C; (d) MPS-D.

aging time. Furthermore, decreasing the molar ratio of MPTMS/H_2O_2 (MPS-A, MPS-E, and MPS-F) from 11.1% to 5.6% under the same aging conditions, the surface area increases initially and then decreases but the pore volume keeps increasing. For cases using 24-h aging time (MPS-A and MPS-C), the total quantity of acid increases with decreasing aging temperature. With 90°C aging temperature (MPS-B, MPS-C, and MPS-D), the total amount of acid increases with increasing aging time. With decreasing molar ratio of MPTMS/H_2O_2 (MPS-A, MPS-E, and MPS-F) under the same aging conditions, the total quantity of acid increases. The reason why the aging conditions enhance the quantity of acid contained in the prepared catalysts so significantly is not clear. Also, for these catalysts, the total quantity of acid determined by using a titration method is higher than that calculated from sulfo groups determined using XPS. There is no doubt that the sulfo groups originate from MPTMS; however, it has been reported that the total amount of acid from OH or COOH groups exist on the catalyst. Therefore, it was considered that the amount of OH or COOH groups is also affected by aging conditions.

b. Saccharification of Rice Straw

The catalysts mentioned above were tested in the saccharification of rice straw to investigate their catalytic characteristics (Li et al., 2012b). The mixture of 3 g rice straw and 30 g distilled water was treated with 0.5 g catalyst (weight ratio of 0.5 g/30 g/3 g for catalyst/distilled water/rice straw) at 180°C for 1 h. Results of post-reaction analyses of the liquid products show low content of mono-saccharide and low levels of by-products in a blank sample without catalyst (NSC). In contrast, samples treated with various solid acid catalysts show the presence of monosaccharide as the main product along with various organic acids, furfural, and 5-HMF.

In addition, both the liquefaction rate of rice straw and yields of by-products increase with more surface area, higher pore volume, and a greater amount of solid acid catalyst applied. The yield of total monosaccharide initially increases and then decreases. For the three solid acid catalysts studied, MPS-D catalyst has the highest activity for the saccharification of rice straw.

The effect of reaction temperature on saccharification of rice straw using MPS-D catalyst was investigated. Increasing reaction temperature leads to a higher liquefaction rate of rice straw and yields of furfural, as well as more 5-HMF and total organic acids produced. At 220°C for 1 h, the liquefaction yield of rice straw reaches 40%. The yield of total monosaccharide initially increases with higher reaction temperature up to 180°C, but decreases thereafter at temperatures higher than 180°C. The maximum yield of total monosaccharide is 38% at 180°C for 1 h. When the reaction temperature increases from 180 to 220°C, the yields of glucose and xylose decrease while the yields of furfural, 5-HMF, and total organic acids increase.

The effect of reaction time on saccharification of rice straw using MPS-D catalyst at 180°C was investigated. With the reaction time increased from 0.5 to 2 h, the liquefaction rate of rice straw and the yields of furfural and 5-HMF as well as total organic acids increase. The liquefaction yield of rice straw is 38.9% at 180°C for 2 h. The yield of total monosaccharide initially increased with longer reaction times up to 1 h, but then decreased thereafter. The maximum yield of total monosaccharide is 38% at 180°C for 1 h. The yield of monosaccharides decreases while the yields of furfural, 5-HMF, and total organic acids increase when the reaction time increases from 1 to 2 h.

Changes in the property of rice straw before and after saccharification at different reaction temperatures for 1 h were determined by following the procedure described in the NREL Chemical Analysis and Testing procedure and XRD respectively. The XRD patterns and CrI index of α-cellulose in rice straw before and after saccharification are shown in Table 7.6 and Figure 7.5. With the reaction temperature raised from 150 to 220°C, the content of xylan in rice straw decreases from 10.7% to 0%, the content of glucan decreases from 46.4% to 10.0%, and the content of acid-soluble lignin decreases from 1.9% to 0.7%. Quantities of acid-insoluble lignin and ash also decrease slightly. In Figure 7.5, (a) is the reference of the XRD pattern for cellulose, and (b)–(f) are the XRD patterns of rice straw before and after saccharification. The CrI index of α-cellulose in rice straw increases from 13.6% (before saccharification) to 56.6% (after saccharification).

c. Hydrolysis Mechanism of Rice Straw on Solid Acid Catalysts

The mechanism for saccharification of rice straw proposed in Scheme 7.2 includes three steps (Li et al., 2012a, b). In Step I, part of the hemicellulose and cellulose contained in rice straw are attacked by protons in the functional groups bonded to a catalyst and converted into various water-soluble oligosaccharides with different molecular weights via catalytic hydrolyses and/or partial hydrothermal hydrolyses. In Step II, water-soluble oligosaccharides

TABLE 7.6 Change in Property of Treated Rice Straw at Different Reaction Temperatures using MPS-D Catalyst

Content (%)	Feed	Run A	Run B	Run C	Run D
Xylan	10.7	6.3	2.3	0.2	0.0
Glucan	46.4	27.6	23.7	19.2	10.0
Acid-insoluble lignin	14.4	11.0	11.2	10.4	10.0
Acid-soluble lignin	1.9	1.1	0.9	0.7	0.6
Ash	13.7	13.6	13.2	12.2	10.6
Extractives	4.4	ND	ND	ND	ND
CrI	13.6	27.4	23.9	44.9	56.6

Run A: 150°C/1 h; Run B: 180°C/1 h; Run C: 200°C/1 h; Run D: 220°C/1 h.

with small molecular weight are adsorbed into the pores of the catalyst and then mainly converted into xylose and glucose via catalytic hydrolyses. In Step III, part of the monosaccharide produced inside the pore of the catalyst was further converted into by-products because of overreaction via catalytic hydrolyses. Xylose will be converted further into furfural, which will subsequently be converted into furan (not detected in this study because of its high volatility), formic acid, and/or lactic acid and acetic acid. Similarly,

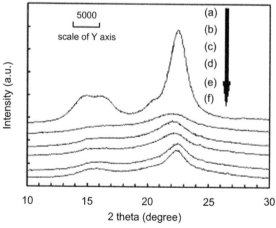

FIGURE 7.5   XRD patterns of: (a) cellulose; (b) rice straw; (c) reacted rice straw (150°C/1 h); (d) reacted rice straw (180°C/1 h); (e) reacted rice straw (200°C/1 h); (f) reacted rice straw (220°C/1 h).

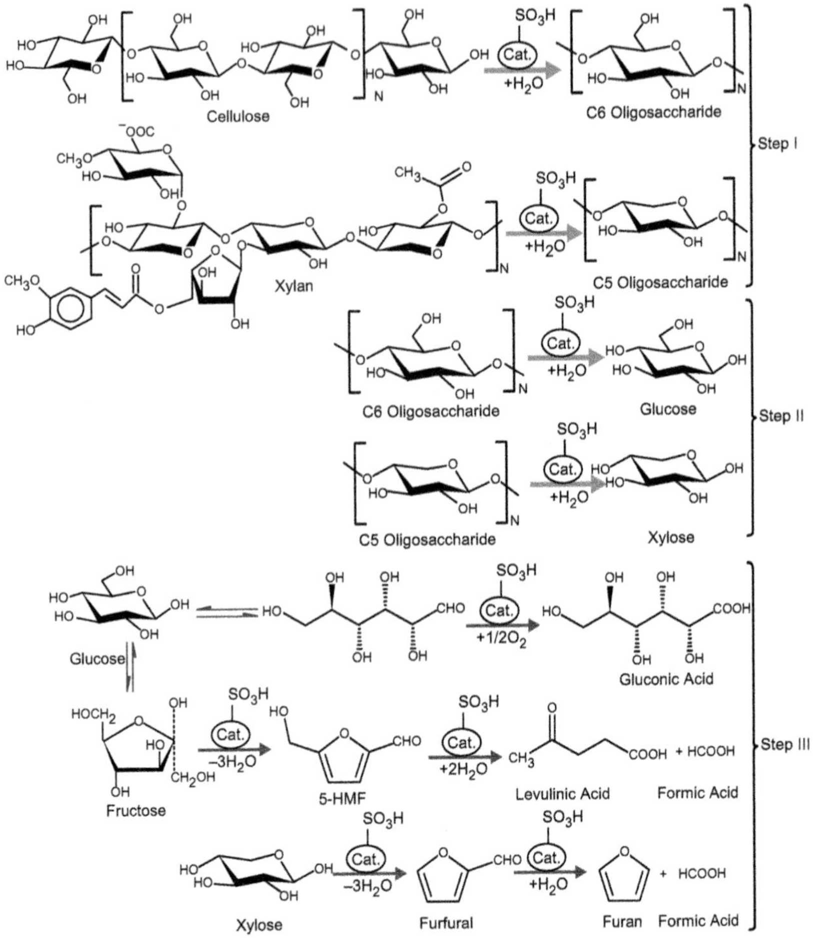

SCHEME 7.2 Mechanisms of rice straw saccharification using sulfonated mesoporous silica-based solid acid catalyst.

isomerization of glucose into fructose will also occur. Allotropes of glucose and fructose will be converted into 5-HMF, which will further be converted into levulinic acid and formic acid. The prepared catalyst plays a major role in all hydrolysis steps.

REFERENCES

Amarasekara, A. S., & Owereh, O. S. (2010). Synthesis of a sulfonic acid functionalized acidic ionic liquid modified silica catalyst and applications in the hydrolysis of cellulose. *Catalysis Communications, 11*, 1072–1075.

Aziz, S., & Sarkanen, K. V. (1989). Organosolv pulping—A review. *Tappi J. 72*, 169–175.

Bjerre, A. B., Olesen, A. B., Fernqvist, T., Plöger, A., & Schmidt, A. S. (1996). Pretreatment of wheat straw using alkaline wet oxidation and alkaline hydrolysis resulting in convertible cellulose and hemicellulose. *Biotechnology and Bioengineering, 49*, 568–577.

Chum, H. L., Johnson, D. K., & Black, S. K. (1990). Organosolv pretreatment for enzymatic-hydrolysis of poplars. 2. Catalyst effects and the combined severity parameter. *Industrial and Engineering Chemistry Research, 29*, 156–162.

Converse, A. O. (1993). Substrate factors limiting enzymatic hydrolysis. In J. N. Saddler (Ed.), *Bioconversion of forest and agricultural plant residues* (pp. 93–106). Wallingford: CAB International.

Ding, S. Y., & Himmel, M. E. (2006). The maize primary cell wall micro-fibril: A new model derived from direct visualization. *Journal of Agricultural and Food Chemistry, 54*, 597–606.

Eggeman, T., & Elander, R. T. (2005). Process and economic analysis of pretreatment technologies. *Bioresource Technology, 96*(18), 2019–2025.

Goderis, B., Jacobs, P. A., & Sels, B. F. (2010). Sulfonated silica/carbon nanocomposites as novel catalysts for hydrolysis of cellulose to glucose. *Green Chemistry, 12*, 1560–1563.

Gollapalli, L. E., Dale, B. E., & Rivers, D. M. (2002). Predicting digestibility of ammonia fiber explosion (AFEX) treated rice straw. *Applied Biochemistry and Biotechnology, 98–100*, 23–35.

Grethlein, H. E., & Converse, A. O. (1991). Common aspects of acid prehydrolysis and steam explosion for pretreating wood. *Bioresource Technology, 36*, 77–82.

Huber, G. W., Iborra, S., & Corma, A. (2006). Synthesis of transportation fuels from biomass: Chemistry, catalysts, and engineering. *Chemical Reviews, 106*, 4044–4098.

Klass, D. L. (1998). *Biomass for renewable energy, fuels and chemicals.* San Diego: Academic Press.

Klinke, H. B., Ahring, B. K., Schmidt, A. S., & Thomsen, A. B. (2002). Characterization of degradation products from alkaline wet oxidation of wheat straw. *Bioresource Technology, 82*, 15–26.

Li, S., & Qian, E. W. (2011). Direct saccharification of rice straw using a solid acid catalyst. *Journal of the Japan Institute of Energy, 90*, 1065–1071.

Li, S., Qian, E. W., Hosomi, M., & Fukunaga, T. (2012a). Preparation of sulfo group bearing silica-based solid acid catalysts and its application to direct saccharification. *Journal of Chemical Engineering of Japan, 45*, 1–9.

Li, S., Qian, E. W., Shibata, T., & Hosomi, M. (2012b). Catalytic hydrothermal saccharification of rice straw using sulfonated mesoporous silica-based solid acid catalysts. *Journal of the Japan Petroleum Institute, 55*, 250–260.

Lynd, L. R., Wyman, C. E., & Gerngross, T. U. (1999). Biocommodity engineering. *Biotechnology Progress, 15*, 777–793.

Margolese, D., Melero, J. A., Christiansen, S. C., Chmelka, B. F., & Stucky, G. D. (2000). Direct syntheses of ordered SBA-15 mesoporous silica containing sulfonic acid groups. *Chemistry of Materials, 12*, 2448–2459.

Mosier, N., Wyman, C., Dale, B., Elander, R., Lee, Y. Y., Holtzapple, M., & Ladisch, M. (2005). Features of promising technologies for pretreatment of lignocellulosic biomass. *Bioresource Technology, 96*, 673–686.

Nakajima, K., Tomita, I., Hara, M., Hayashi, S., Domen, K., & Kondo, J. N. (2006). Development of highly active SO_3H-modified hybrid mesoporous catalyst. *Catalysis Today, 116*, 151–156.

Nakayama, E., & Okamura, K. (1989). Influence of a steam explosion and microwave irradiation on the enzymatic hydrolysis of a coniferous wood. *Mokuzai Gakkaishi, 35*, 251–260.

Onda, A., Ochi, T., & Yanagisawa, K. (2008). Selective hydrolysis of cellulose over solid acid catalysts. *Green Chemistry, 10*, 1033–1037.

O'Sullivan, A. C. (1997). Cellulose: the structure slowly unravels. *Cellulose, 4,* 173–207.

Qian, E. W. (2011). Development of production method and catalysts for biomaterials. *Journal of the Japan Institute of Energy, 90,* 518–525.

Qian, E. W., & Hosomi, M. (2009a). Saccharification method of cellulosic biomass. *J. P. Patent,* 254283A.

Qian, E. W., & Hosomi, M. (2009b). Method of cellulosic biomass liquefaction. *J. P. Patent,* 296919A.

Rinaldi, R., Palkovits, R., & Schüth, F. (2008). Depolymerization of cellulose using solid catalysts in ionic liquids. *Angewandte Chemie International Edition, 47,* 8047–8050.

Silverstein, R. A., Chen, Y., Sharma-Shivappa, R. R., Boyette, M. D., & Osborne, J. (2007). A comparison of chemical pretreatment methods for improving saccharification of cotton stalks. *Bioresource Technology, 98,* 3000–3011.

Suganuma, S., Nakajima, K., Kitano, M., Yamaguchi, D., Kato, H., Hayashi, S., & Hara, M. (2008). Hydrolysis of cellulose by amorphous carbon bearing SO_3H, COOH, and OH groups. *Journal of the American Chemical Society, 130,* 12787–12793.

Sulbaran de Ferrer, B., Ferrer, A., Byers, F. M., Dale, B. E., & Aristiguieta, M. (1997). Sugar production fron rice straw. *Archivos Latinoamericanos de Production Animal, 5*(Suppl. 1), 112–114.

Teramoto, Y., Tanaka, N., Lee, S.-H., & Endo, T. (2008). Pretreatment of eucalyptus wood chips for enzymatic saccharification using combined sulfuric acid-free ethanol cooking and ball milling. *Biotechnology and Bioengineering, 99*(1), 75–85.

Teymouri, F., Laureano-Perez, L., Alizadeh, H., & Dale, B. E. (2005). Optimization of the ammonia fiber explosion (AFEX) treatment parameters for enzymatic hydrolysis of corn stover. *Bioresource Technology, 96*(18), 2014–2018.

Towler, G. P., Oroskar, A. R., & Smith, S. E. (2004). Development of a sustainable liquid fuels infrastructure based on biomass. *Environmental Progress, 23,* 334–341.

Turner, N. D., McDonough, C. M., Byers, F. M., Holtzapple, M. T., Dale, B. E., Jun, J. H., & Greene, L. W. (1990). Disruption of forage structure with an ammonia fiber explosion process. *Proceedings, Western Section, American Society of Animal Science, 41.*

Wyman, C. E., Decker, S. R., Himmel, M. E., Brady, J. W., Skopec, C. E., & Vikari, L. (2005). *Polysaccharides* (2nd ed.). New York: Marcel Dekker.

Yang, B., & Wyman, C. E. (2004). Effect of xylan and lignin removal by batch and flow through pretreatment on the enzymatic digestibility of corn stover cellulose. *Biotechnology and Bioengineering, 86,* 88–95.

Chapter 8

Energy-Saving Biomass Processing with Polar Ionic Liquids

Yukinobu Fukaya and Hiroyuki Ohno

Chapter Outline

Research Approaches to Sustainable Biomass Systems. http://dx.doi.org/10.1016/B978-0-12-404609-2.00008-8

8.1. CELLULOSE DISSOLUTION AND IONIC LIQUIDS

8.1.1. Cellulose Dissolution

Cellulosic biomass is the most abundant bioresource produced on earth. In view of a perpetual supply of food, converting edible food into fuel is out of the question in spite of the easiness of producing bioethanol from starch. The process of producing biofuel from inedible biomass has been developed for the purpose of preserving edible food. Use of ubiquitously available cellulosic biomass should be a key strategy for supplying sustainable energy. The polysaccharides contained in biomass are insoluble in water owing to their intra- and intermolecular hydrogen bonding. Moreover, polysaccharides exist as supramolecular complexes in nature that are almost impossible to dissolve in water at the molecular level; thus, they are not easily extracted from biomass under mild conditions. In order to realize the bioenergy conversion with minimum consumption of energy, implementing the following steps is especially important: (1) dissolution and extraction of cellulose from biomass; (2) depolymerization of cellulose to glucose or oligosaccharides; and (3) energy conversion of these saccharides to generate electrical energy. In any process to generate energy from the biomass, the energy consumed should be kept as low as possible. At present, the energy needed to generate energy from polysaccharides is greater than the total energy yielded.

8.1.2. Initial Stage of Ionic Liquid Design as Solvents for Cellulose

Since it is not easy to produce energy from biomass with molecular liquids, there is increasing attention on using ionic liquids (ILs) as a new class of solvents for enhancing the production of energy from biomass (Rogers and Seddon, 2003). ILs are organic salts that exist as liquid at normal room temperature or ambient temperature below 100°C (Welton, 1999). They have unique characteristics that include extremely low volatility, low flammability, and high thermal stability. Because of these promising features, ILs have been studied extensively to examine their physicochemical properties as well as their potential applications as solvents and electrolytes. There have been two major waves of studying ILs as a new class of solvents (Wassersheid and Welton, 2008) and electrolyte materials used in energy-related devices (Ohno, 2011). The application of ILs in bioscience is set to become the third wave; processing of biomass with ILs to enhance energy conversion from biomass is a big issue (Ohno and Fukaya, 2009).

ILs are composed wholly of ions; they are often mistakenly regarded as highly polar liquids. However, not all ILs show high polarity (Anderson et al., 2002); the physicochemical property essential for using the solvent is its polarity in addition to its hydrogen-bonding ability, affinity with compounds, viscosity, and usable temperature range as liquid. All these properties can be tuned by designing the component ions. The diversity of organic ions leads to the development of ILs with desired characteristics and functions. If designed appropriately, the component ions contained in ILs give rise to polar ILs that exhibit much higher polarity than conventional molecular liquids. Hence, ILs with unusual high polarity should be the next generation of liquids used to dissolve molecular insoluble materials such as components in biomass in an efficient and environmental-friendly closed system for producing bioenergy.

The challenge of dissolving cellulose with ILs has been considered over the past decade. Rogers et al. (2006) first demonstrated that some ILs are capable of solubilizing cellulose (Swatloski et al., 2002). They studied a series of 1-n-butyl-3-methylimidazolium ([C$_4$mim]) salts as solvents for solubilizing cellulose and found that 1-n-butyl-3-methylimidazolium chloride ([C$_4$mim]Cl) solubilizes cellulose under conventional heating conditions. For example, clear solutions containing 3 wt% and 10 wt% pulp cellulose (DP $\approx$ 1000) can be obtained by simply mixing the cellulose with [C$_4$mim]Cl at 70 and 100°C respectively. After this pioneering work, numerous papers have been published on cellulose dissolution and cellulose product fabrication, among many other related topics.

In spite of the potential of using chloride-based salts as solvents for dissolving cellulose and even other biomass components, most chloride salts have relatively high melting temperature. In order to improve the physicochemical properties of [C$_4$mim]Cl, several ILs with low melting point and relatively low viscosity have been synthesized to overcome the drawbacks of the chloride salts. For instance, introducing an allyl group instead of a butyl group onto the imidazolium cation was found to give relatively low viscosity and low melting temperature to the resulting chloride salts (Mizumo et al., 2004). While these chloride salts are obtained as liquid at room temperature through the modification of cations, they are still highly viscous and accordingly inadequate to be used as solvents under mild conditions.

We have focused on the flexibility of ion selection, and developed a new class of polar ILs suitable for carrying out biomass processing under mild conditions to achieve energy-saving biomass processing. Figure 8.1 shows the strategy to obtain energy from biomass using the new class of polar ILs. Three steps are involved in this strategy: (1) extraction of cellulose from biomass; (2) hydrolysis of cellulose into glucose; and (3) electrochemical oxidation of glucose into electric energy. In this chapter, the first step, namely the extraction of cellulose from biomass, is described.

8.2. REQUIRED FACTORS OF IONIC LIQUIDS FOR CELLULOSE DISSOLUTION

8.2.1. Interaction of Cellulose and Ionic Liquids

The conditions for ILs to dissolve cellulose are first explored in order to develop a new class of polar ILs that is capable of dissolving cellulose under mild conditions.

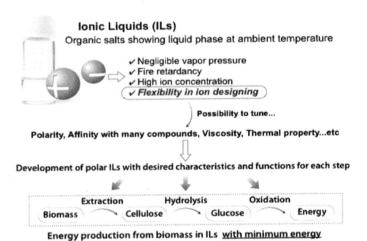

FIGURE 8.1 Strategy for energy production from biomass with ionic liquids.

Chloride salts have been used by Rogers and colleagues for studying the solubilizing mechanism of cellulose by ILs using NMR spectroscopy (Moulthrop et al., 2005). The ^{13}C-NMR spectra of cellulose in chloride salts show that cellulose is totally disordered in ILs. Moreover, Monya and colleagues determined the NMR longitudinal (T_1) and transverse (T_2) relaxation times of [C$_4$mim]Cl, both for the cation (^{13}C) and the chloride anion ($^{35/37}$Cl) in order to examine the interaction between IL and cellulose at the molecular level (Remsing et al., 2006). These relaxation times were analyzed as functions of concentrations of cellobiose, glucose, and glucose pentaacetate because, in general, variations in relaxation times yield useful information about the molecular dynamics of the compound. Investigation of the relaxation time as a function of increasing cellobiose concentration reveals that the interaction between chloride anion and the sugar unit is very strong; glucose interacts strongly with the anion whereas glucose pentaacetate has almost no effect on the relaxation rate of ^{35}Cl. The dissolving of cellobiose or glucose in [C$_4$mim]Cl involves the formation of hydrogen bonds between the Cl anions of ILs and the hydroxyl groups of sugar with a stoichiometric ratio. Youngs et al. (2006) investigated the interaction between D-glucose (model compound) and [C$_4$mim]Cl by means of molecular dynamic simulation. Their findings reveal that the cations interact only weakly with the sugar; this agrees with the results of previous studies that the most dominant interaction is clearly between chloride anions and the cellulose hydroxyl groups.

8.2.2. Hydrogen-Bonding Characteristics of ILs

The hydrogen-bonding characteristics of ILs have been empirically determined based on their solvent-dependent absorptions or emissions measured with a variety of solvatochromic probes (Crowhurst et al., 2003; Reichardt, 2005).

While $E_T(30)$ (in kcal mol^{-1}) values proposed by Crowhurst et al. have been widely used as an empirical polarity scale, the $E_T(30)$ values are generally affected by the basic structure of cations as well as by anion and alkyl chain length to some extent. On the other hand, three Kamlet–Taft parameters, i.e. the hydrogen-bonding acidity (α), hydrogen bonding basicity (β), and dipolarity/polarizability (π^*), are quite useful for discussing the polarity of ILs (Kamlet and Taft, 1976). These parameters are determined by using visible spectral measurements of prove dyes as shown in Chart 8.1.

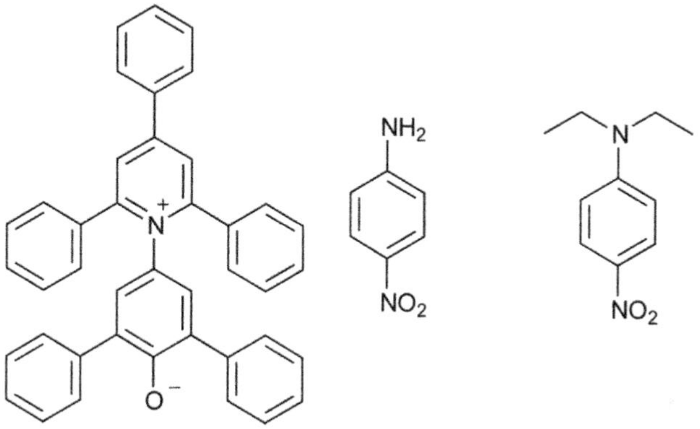

CHART 8.1   Chemical structure of probe dyes.

8.2.3. Polar ILs Containing Carboxylate Anions

The basic properties of a series of ILs have been analyzed especially for their hydrogen-bonging basicity, i.e. Kamlet–Taft parameter β value. Previous studies on the Kamlet–Taft parameters have shown that the β value strongly depends on the anion structure. Results of Kamlet–Taft parameter measurements reveal that chloride anions show strong hydrogen-bonding basicity. On the other hand, salts containing Br$^-$, PF6$^-$, or BF4$^-$ show lower hydrogen-bonding basicity so that they are considered inappropriate for dissolving cellulose.

A series of alkylimidazolium salts were prepared by coupling with various anions, and the carboxylate-based ILs show higher β values than conventional ILs (Fukaya et al., 2006). Alkylimidazolium acetate dissolves cellulose at lower temperature than chloride salts (Figure 8.2). The authors examined a series of carboxylate salts with different side chains, i.e. formate, propionate, and butylate. These carboxylate series show higher hydrogen-bonding characteristics (Table 8.1); the results clearly indicate that carboxylate anions are effective for preparing highly polar ILs with strong hydrogen-bonding basicity.

Compared with chloride salts, carboxylate-based ILs are effective for cellulose dissolution (Table 8.2). For example, to dissolve 10 wt% cellulose,

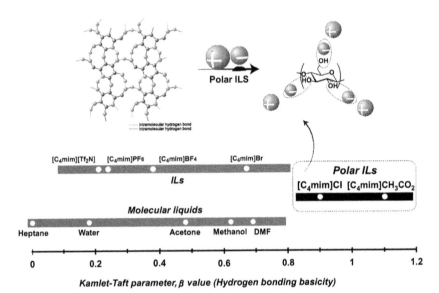

FIGURE 8.2 Polarity scale of ILs and molecular liquids.

[C$_4$mim]Cl needs to be heated to 85°C whereas [C$_4$mim][HCOO] will dissolve 10 wt% cellulose at 35°C. As mentioned in previous paragraphs, the hydrogen bond basicity is necessary to dissolve cellulose by weakening the inter- and intramolecular hydrogen bonds of the cellulose chains. Because [t-C$_4$COO] salt exhibits higher hydrogen bond basicity (β value) than any other carboxylate

TABLE 8.1 Transition Temperature and Kamlet–Taft parameters of ILs

Anions	$T_g^{\dagger}$	$T_m^{\dagger}$	Kamlet–Taft parameters at 25°C		
			α	β	π^*
[HCOO]	−73	‡	0.56	1.01	1.03
[C$_1$COO]	−67	‡	0.55	1.09	0.99
[C$_2$COO]	−64	‡	0.57	1.10	0.96
[C$_3$COO]	−63	‡	0.56	1.10	0.94
[t-C$_4$COO]	−49	‡	0.54	1.19	0.91
Cl	‡	66	0.47§	0.87§	0.10§

†Temperature (°C) at signal peak.
‡Not detected.
§Measured at 75°C.

TABLE 8.2 Solubilization Temperature of Cellulose in [C$_4$mim]-Type ILs

Anion	Solubilization temperature (°C) of cellulose		
	2 wt%[*]	6 wt%[*]	10 wt%[*]
[HCOO]	30	30	35
[C$_1$COO]	40	45	50
[C$_2$COO]	45	50	55
[C$_3$COO]	50	55	55
[t-C$_4$COO]	80	85	90
Cl	75	80	85

*Final concentration of cellulose.

salts, it is expected to dissolve cellulose more efficiently than the latter. However, the temperature for [t-C$_4$COO] salt to solubilize cellulose is higher than that for other ILs evaluated in this study. The alkyl side chain of carboxylate anions may have an effect on the solubilization of cellulose. The cellulose solubilization temperature increases with larger alkyl side chains for the imidazolium cation but the temperature tends to decrease with decreasing viscosity (Figure 8.3). The viscosity of ILs appears to be a further important factor in dissolving cellulose under mild conditions because cellulose powder is dispersed more efficiently in ILs with lower viscosity.

ILs composed of imidazolium salts with a short alkyl side chain tend to have low viscosity. Moreover, allylimidazolium-based salts have been reported to result in ILs with low viscosity as well. Hence, novel formate salts can be prepared with both 1-ethyl-3-methylimidazolium ([C$_2$mim]) and 1-allyl-3-methylimidazolium ([Amim]) cations to render lower viscosity to the resulting salts (Chart 8.2). Because [Amim][HCOO] has a low viscosity of 66 cP at 25°C, and high hydrogen bond basicity, cellulose can be dissolved in concentrated [Amim][HCOO] under mild conditions (Tables 8.3 and 8.4).

In spite of their strong hydrogen-bonding basicity and potential to solubilize cellulose, carboxylate salts have relatively poor thermal stability. Furthermore, carboxylate-type ILs need to be prepared by a three-step reaction: (1) imidazole quternizes with alkyl halide; (2) it is converted to hydroxide; and (3) it is subsequently coupled with the desired carboxylate anion. Designing a new class of ILs that are easy to prepare with sufficient polarity to dissolve cellulose is of great importance. The drawbacks associated with current ILs have motivated us to develop further new polar ILs with satisfactory thermal stability.

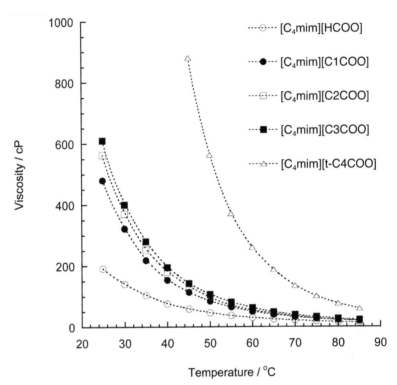

FIGURE 8.3   Viscosity of a series of [C$_4$mim] carboxylates.

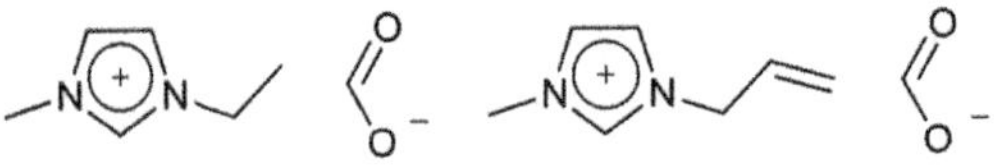

CHART 8.2   Structure of formate salts.

TABLE 8.3 Physicochemical Properties of Formate Salts

ILs	$T_g^\dagger$	$T_m^\dagger$	η (cP)‡	Kamlet–Taft parameters‡		
				α	β	π^*
[C$_2$mim][HCOO]	§	52	**	**	**	**
[Amim][HCOO]	−83	c	66	0.50	0.99	1.08
[C$_4$mim][HCOO]	−73	c	190	0.56	1.01	1.03

†*Temperature (°C) at signal peak.*
‡*Measured at 25°C.*
§*Not detected.*
**Not measured.

TABLE 8.4 Solubilization Temperature of Cellulose in ILs

ILs	Solubilization temperature (°C) of cellulose		
	2 wt%[*]	6 wt%[*]	10 wt%[*]
[Amim][HCOO]	25	25	25
[C$_4$mim][HCOO]	30	30	35

*Final concentration of cellulose.

8.3. DESIGN OF A NEW CLASS OF POLAR IONIC LIQUIDS

8.3.1. Requirement of New Polar ILs

As previously discussed, ILs such as 1,3-dialkylimidazolium chloride salts are capable of dissolving cellulose without any pretreatment under conventional heating conditions. 1,3-Dialkylimidazolium chloride salts can be readily prepared easily by using the one-pot procedure. However, this procedure has two major disadvantages: (i) it is a time-consuming process and (ii) an excess of alkyl halide is required to achieve good yield. The latter renders an undesired polluting nature to the quaternization process, especially when long chain derivatives are produced. Alkyl halides with long alkyl chains are difficult to remove from the reaction mixture due to their high boiling point. This disadvantage makes the procedure costly and inefficient for synthesizing 1,3-dialkylimidazolium chloride. Moreover, most 1,3-dialkylimidazolium chloride ILs are obtained as a solid or sticky paste at room temperature; they must be handled under heating. This makes the processing of cellulose in 1,3-dialkyl-imidazolium chloride even more costly.

For efficient processing of cellulose, heating should be avoided to reduce the energy cost. In Section 8.2, we proposed a series of novel ILs, namely 1,3-dialkylimidazolium carboxylate, as potential candidates for solubilization of cellulose. These 1,3-dialkylimidazolium carboxylates were obtained as low-viscosity liquids at room temperature. Moreover, they are more effective in solubilizing cellulose at lower temperature than chloride salts. However, 1,3-dialkylimidazolium carboxylates show relatively poor thermal stability. Furthermore, 1,3-dialkylimidazolium carboxylates are prepared by following the aforementioned three-step reactions in which the halide counter anion of the imidazolium cation is first converted to hydroxide, then coupled with the desired carboxylate anion (Scheme 8.1).

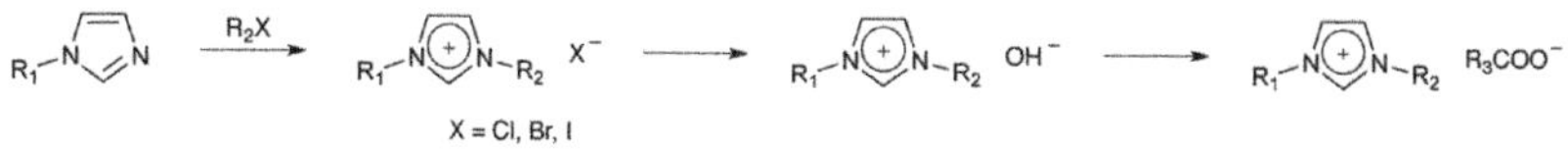

X = Cl, Br, I

SCHEME 8.1 Synthetic procedure for a series of halide salts and carboxylate salts.

8.3.2. Facile Preparation of Polar ILs

To overcome the above drawbacks, a new class of easily preparable ILs having sufficient polarity to dissolve cellulose is necessary. Preparing ILs based on the reaction between tertiary amines and alkyl esters of organic acids to form quaternized salts with desired anions derived from acid esters can be carried out using a straightforward one-pot procedure (Scheme 8.2).

SCHEME 8.2 Synthetic procedure for a series of polar ionic liquids.

Because many types of acid esters are commercially available, several alkylimidazolium salts having alkylsulfate, alkylsulfonate, and alkylphosphate anion derivatives can now be prepared by using the one-step quaternization procedure. However, there has been no evaluation of the ILs thus prepared as solvents for cellulose. Therefore, the preparation of a series of ILs with fixed cations, i.e. 1-n-butyl-3-methylimidazolium ([C$_4$mim]) cation, and a wide variety of anions using the one-pot procedure is described in this chapter.

Here the ILs that are readily prepared are emphasized. Among numerous methods for preparing ILs, the straightforward one-pot reaction between tertiary amines and alkyl esters to form quaternized salts with anions derived from acid esters (Fukaya et al., 2008) is presented. The 1-butyl-3-methylimidazolium ([C$_4$mim]) salts, with anions such as methanesulfonate ([MeSO$_3$]), methylsulfate ([MeOSO$_3$]), and dimethylphosphate ([(MeO)$_2$PO$_2$]), were synthesized and evaluated as solvents for dissolving cellulose. The polarity and hydrogen-bonding characteristics of these ILs were evaluated as the Kamlet–Taft parameters at room temperature. Among these prepared ILs, [C$_4$mim] [(MeO)$_2$PO$_2$] exhibits higher hydrogen bond basicity ($\beta = 1.00$) than [MeOSO$_3$] ($\beta = 0.61$) or [C$_4$mim][MeSO$_3$] ($\beta = 0.70$). As expected based on the β value, only [C$_4$mim][(MeO)$_2$PO$_2$] is confirmed to dissolve cellulose (Table 8.5).

8.3.3. Further Design of Polar ILs

To improve the physicochemical properties and cellulose solubility of the prepared ILs, further investigation was undertaken in order to reveal better anions. As discussed in previous sections, alkyl chains attached to the anion strongly affect solvent properties of the ILs. The reaction between 1-ethylimidazole and the corresponding dimethyl phosphite or dimethyl phosphonate among the many anions examined including alkyl phosphonate and alkyl phosphite can be used for easy preparation of both [C$_4$mim] [(MeO)(H)PO$_2$] and [C$_4$mim][(MeO)(Me)PO$_2$].

TABLE 8.5 Transition Temperature and Kamlet–Taft Parameters of [C$_4$mim]-Type Salts

Anions	$T_g{}^\dagger$	$T_m{}^\dagger$	Kamlet–Taft parameters at 25°C		
			α	β	π^*
[MeOSO$_3$]	−85	§	0.59	0.62	1.10
[CF$_3$COO]	−84	§	0.62	0.81	0.97
[MeSO$_3$]	§	76	**	**	**
[(MeO)$_2$PO$_2$]	−66	§	0.52	1.00	1.02
[HCOO]	−73	§	0.56	1.01	1.03
Cl	§	66	0.47††	0.87††	1.10††

†Temperature (°C) at signal peak.
§Not detected.
**Not measured.
††Measured at 75°C.

These ILs have almost similar π^* and α values. On the other hand, the ILs prepared here have high β value that exceeds 1.0, e.g. 1.07 for the β value of [C$_4$mim][(MeO)(Me)PO$_2$]. The phosphate derivative ILs as prepared here generally has better thermal stability than previously reported polar ILs; the prepared [C$_4$mim][(MeO)(R)PO$_2$] salts are thermally stable at temperatures up to 260–290°C. Their viscosity at 25°C is in the range of 100–500 cP, which is much lower than that for chloride salts (Tables 8.6 and 8.7).

These ILs are capable of dissolving cellulose owing to their strong hydrogen-bonding basicity (Table 8.8). The temperature needed to solubilize the same amount of cellulose powder depends strongly on the anionic structure of the ILs. Cellulose powder of 2.0 wt% concentration is dissolved completely in these ILs at room temperature (25°C) within 3 hours, and the 4.0 wt% cellulose powder is completely dissolved within 5 hours. These ILs also act as

TABLE 8.6 Thermal Properties of [C$_4$mim]-Type Salts

Anion	$T_g{}^*$	$T_m{}^*$
[(MeO)(H)PO$_2$]	−77	−†
[(MeO)(Me)PO$_2$]	−65	−†
[(MeO)$_2$PO$_2$]	−66	−†

*Temperature (°C) at signal peak.
†Not detected.

TABLE 8.7 Kamlet–Taft Parameters of [C$_4$mim]-Type ILs

| Anions | Kamlet–Taft parameters[†] | | |
	α	β	π^*
[(MeO)(H)PO$_2$]	0.52	1.02	1.01
[(MeO)(Me)PO$_2$]	0.51	1.09	1.01
[(MeO)$_2$PO$_2$]	0.52	1.00	1.02

[†]At 25°C.

TABLE 8.8 Relationship between Temperature and Solubility of Cellulose in [C$_4$mim]-Type Salts

| Anions | Solubilization temperature (°C) | | |
	2 wt%[*]	6 wt%[*]	10 wt%[*]
[(MeO)(H)PO$_2$]	40	45	50
[(MeO)(Me)PO$_2$]	50	55	65
[(MeO)$_2$PO$_2$]	55	60	70
[HCOO]	30	30	35

*Final concentration of cellulose.

solvents for a series of naturally obtained cellulose samples of high molecular weights, including cotton. These results strongly suggest that using phosphate and phosphonate derivative anions is effective in designing ILs with high polarity, low viscosity, and high thermal stability.

8.4. POLAR IONIC LIQUIDS FOR BIOMASS PROCESSING WITHOUT HEATING

8.4.1. Physicochemical Properties of New Polar ILs

Recent efforts have been concentrated on dissolving naturally obtained cellulosic biomass, and several previous reports have indicated that certain ILs can dissolve cellulosic biomass. Previously, the extraction of polysaccharides from plant biomass using ILs required continuous stirring at temperatures above 100°C for at least 24 hours. Much energy can be saved if the extraction is carried out rapidly at 50°C or below. As mentioned above, 1-ethyl-3-methylimidazolium methylphosphonate is effective for processing cellulosic biomass that consists of three main fractions: cellulose, hemicellulose, and lignin. The authors focused on the factors relevant to using ILs to dissolve and extract component cellulose from biomass under mild conditions (Abe et al., 2010). Because there is a wide variety of commercially available phosphonate-derived acid esters and alkyl halides, it is a good idea to combine diverse dialkylimidazolium cations with phosphonate-derived materials as anions in the search for polar ILs. The extraction of polysaccharides from biomass using a series of alkylimidazolium salts coupled with phosphonate-derived anions was investigated with respect to the requirements for efficient biomass processing under mild conditions.

To determine the principal factor governing the degree of extraction and find the ILs that are better suited for extracting cellulose, the authors have prepared a series of new ILs through modification of both cations and anions. The effect of alkyl side chain attached to the imidazolium cation ring on the extraction of polysaccharides from bran was analyzed by studying the effect of four cations with different alkyl side chain lengths on the amount of polysaccharide extracted (Chart 8.3). All the ILs prepared in this study were in liquid form at room temperature. They have one glass transition temperature at around $-80°C$, and no crystallization was detected when these ILs were kept at $-20°C$ for more than 2 weeks. Results of thermogravimetric analysis (TGA) show that all these ILs are stable under temperature of up to $250°C$ (Table 8.9).

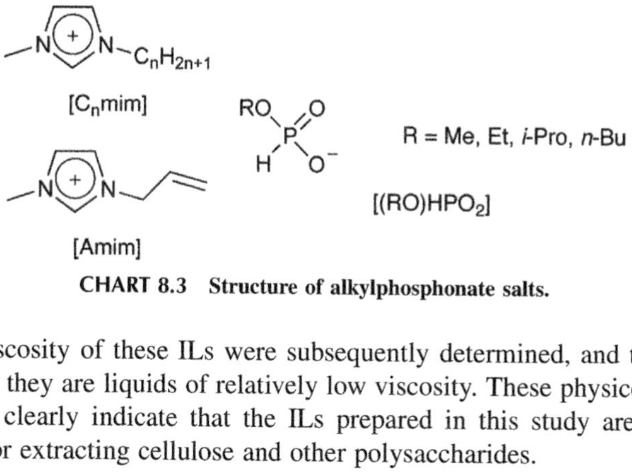

CHART 8.3 Structure of alkylphosphonate salts.

The viscosity of these ILs were subsequently determined, and the results reveal that they are liquids of relatively low viscosity. These physicochemical properties clearly indicate that the ILs prepared in this study are potential solvents for extracting cellulose and other polysaccharides.

8.4.2. Extraction of Cellulose from Bran with Phosphonate-Type ILs

The extraction of polysaccharides from bran is described herein as an example of cellulose extraction from biomass. Bran is the hard shell of wheat and comprises complexes of polysaccharides and lignin, thus making it a sparingly soluble material. The extraction and dissolution of polysaccharides from bran

TABLE 8.9 Physicochemical Properties of ILs Examined in this Study

ILs	T_g (°C)[†]	T_m (°C)	T_{dec} (°C)[‡]	Kamlet–Taft parameters at 25°C		
				α	β	π^*
[C_2mim][(MeO)HPO$_2$]	−86	–	278	0.52	1.00	1.06
[Amim][(MeO)HPO$_2$]	−82	–	256	0.51	0.99	1.06
[C_3mim][(MeO)HPO$_2$]	−79	–	277	0.54	1.00	1.02
[C_4mim][(MeO)HPO$_2$]	−77	–	277	0.52	1.02	1.01

[†]*Temperature at the signal peak.*
[‡]*Temperature at 10% weight loss.*

(Figure 8.4) was carried out by mixing bran (0.15 g) with ILs (2.85 g), and gently stirring the mixtures at 50°C to disperse bran in the ILs. Some polysaccharides were extracted within 10 min under gentle stirring at 50°C. The ability of these ILs to solubilize major component polysaccharides of bran such as cellulose, hemicellulose, and residual starch is thus confirmed. The remaining insoluble portion that consists of lignin and its complexes was filtered out. Excess ethanol was then added to the filtrate under continuous stirring. The resulting precipitate was collected by a second filtration and washed repeatedly with ethanol to remove residual ILs. The extracted materials were analyzed using FT-IR, NMR, and TGA measurements, and the results confirm that the extracted materials are a mixture of cellulose, hemicellulose, and residual starch.

The efficiency of extracting polysaccharides from bran using the four ILs depends on the cation structure. Moreover, these ILs are capable of dissolving polysaccharides at 50°C within 10 min. As mentioned in the previous section, dissolution of cellulose within a short mixing period depends to a great extent on the viscosity and hydrogen bond basicity of ILs. These ILs show almost the

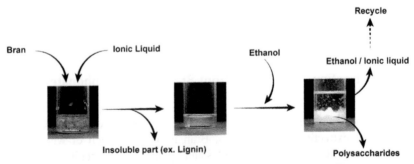

FIGURE 8.4 Extraction of polysaccharides from bran with ILs.

same hydrogen-bonding basicity. The degree of extraction using these four ILs is, however, independent of their β value. Furthermore, there is no relation between the hydrophobicity of the cations and the degree of extraction. Figures 8.1 and 8.3 show that ILs with lower viscosity are capable of extracting more polysaccharides because of better bran dispersion in ILs with less viscosity.

As described above, 1-ethyl-3-methylimidazolium salts are efficient in extracting polysaccharides. Four 1-ethyl-3-methylimidazolium salts coupled with phosphonate derivative anions having different alkyl side chains, i.e. methylphosphonate, ethylphosphonate, isopropylphosphonate, and n-butyl-phosphonate anions, were prepared. As shown in Table 8.10, these ILs are stable supercool liquids showing only glass transition at around $-80°C$. Results of TGA analyses reveal that these ILs remain stable at temperatures of up to 250°C.

These ILs have relatively strong hydrogen-bonding characteristics, especially the β value, which is generally affected mainly by the nature of the anionic species. Additionally, they have similar but slightly different β values in the following order of magnitude: $[(MeO)(H)PO_2]$ ($\beta = 1.00$) < $[(EtO)(H)PO_2]$ ($\beta = 1.02$) < $[(i\text{-}PrO)(H)PO_2]$ ($\beta = 1.03$) < $[(n\text{-}BuO)(H)PO_2]$ ($\beta = 1.06$). Because the series of anions studied here have the same basic unit structure ($[(R)(H)PO_2]$), the disparity between β values of these ILs can be explained by the difference in the electron-releasing capability of the alkyl side chain. As mentioned above, the viscosity of ILs that is an important factor influencing the extraction of polysaccharides from bran depends on the anionic structure of ILs due to the bulkiness of the alkoxyl groups on the anion. Hence, alkylphosphonate with a shorter alkyl chain should be used for preparing low-viscosity ILs.

Figure 8.5 shows the degree of extraction of polysaccharides from bran by using this series of phosphonate-derived salts at 50°C. The degree of cellulose extraction after 2 hours of mixing shows the following order of magnitude: $[C_2mim][(i\text{-}PrO)(H)PO_2]$ (16%) < $[C_2mim][(n\text{-}BuO)(H)PO_2]$ (19%) < $[C_2mim][(EtO)(H)PO_2]$ (30%) < $[C_2mim][(MeO)(H)PO_2]$ (39%).

The solubility of polysaccharides depends mainly on the β value of the ILs. However, the degree of extraction of polysaccharides from bran in this

TABLE 8.10 Physicochemical Properties of [C$_2$mim]-Type ILs

Anion	T_g (°C)	T_{dec} (°C)	α	β	π^*
[(MeO)(H)PO$_2$]	−86	278	0.52	1.00	1.06
[(EtO)(H)PO$_2$]	−79	266	0.55	1.02	1.02
[(i-PrO)(H)PO$_2$]	−71	256	0.55	1.03	1.00
[(n-BuO)(H)PO$_2$]	−77	259	0.56	1.06	0.97

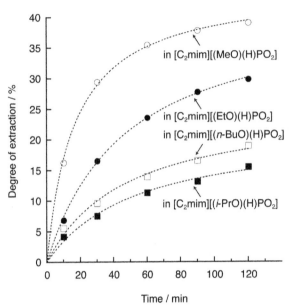

FIGURE 8.5   Degree of extraction of polysaccharides from bran using polar ILs.

experiment is not observed to depend on the β value. Among these four ILs, the β value is highest for [C_2mim][(n-BuO)(H)PO$_2$] and lowest for [C_2mim][(MeO)(H)PO$_2$]. However, [C_2mim][(MeO)(H)PO$_2$] is capable of extracting more polysaccharide than [C_2mim][(n-BuO)(H)PO$_2$]. We infer that for a series of ILs having the β value within a certain range, the amount of polysaccharide extracted from bran within a limited period is influenced more by the IL viscosity than any other factors. Among these ILs, [C_2mim][(n-BuO)(H)PO$_2$] has a larger alkyl side chain so that it has better capacity to extract polysaccharides than [C_2mim][(i-PrO)(H)PO$_2$]. The properties and interaction force estimated based on the size of the ion are not therefore primary factors that affect the degree of extraction; the IL viscosity is closely related to the degree of extraction. The ILs with lower viscosity have the best capacity to extract polysaccharides within a short period under mild conditions. Low-viscosity ILs are capable of extracting a large amount of polysaccharides at 50°C within the given time. For polar ILs with similar β values, the degree of extraction in a limited time depends strongly on the viscosity. Although [C_2mim][(MeO)HPO$_2$] and [C_2mim][(i-PrO)HPO$_2$] have similar β values (1.00 and 1.03), [C_2mim][(i-PrO)HPO$_2$] ($\eta = 65$ cP at 50°C) needs 120 min stirring at 50°C to extract ca. 15% of polysaccharides from bran, whereas [C_2mim][(MeO)HPO$_2$] ($\eta = 31$ cP at 50°C) needs only 10 min under similar conditions to extract the same quantity of polysaccharides. The solubility of cellulose in these ILs is independent of their viscosity over a very long time, but faster solubilization can be carried out using less viscous ILs. The bran is dispersed more easily in

ILs with lower viscosity so that the ILs may penetrate into the complex of cellulose and lignin more easily, which leads to the observed greater solubility under mild conditions and over short times.

8.4.3. Extraction of Cellulose from Bran with Phosphinate-Type ILs

The aforementioned results strongly suggest that ILs with lower viscosity are efficient to extract polysaccharides within a short period of time when the ILs have adequate polarity. Hence, a novel IL with high hydrogen-bonding basicity and low viscosity can be designed. Imidazolium cations with short alkyl side chains generally leading to ILs with low viscosity have been confirmed by the results obtained with a short alkyl side chain of phosphonate-derived anions. The IL, i.e. 1-ethyl-3-methylimidazolium phosphinate ([C_2mim][H_2PO_2]; Chart 8.4), is a newly designed extraction reagent. [C_2mim][H_2PO_2] is a liquid at a room temperature (T_m) of 17°C. Results of TGA measurements indicate that this IL is stable at temperatures of up to 250°C. The Kamlet–Taft parameter measurement reveals that [C_2mim][H_2PO_2] exhibits relatively strong hydrogen-bonding basicity. Compared with the alkylphosphonate series salts ($1.00 < \beta < 1.06$; see Table 8.11), the β value of [C_2mim][H_2PO_2] is relatively low at 0.97. Additionally, it has the lowest viscosity in these phosphate derivative ILs. As a result of the strong hydrogen-bonding characteristics of [C_2mim][H_2PO_2] and its lower viscosity, this IL is expected to be a superior solvent for extracting polysaccharides from bran.

As expected, [C_2mim][H_2PO_2] is demonstrated to extract polysaccharides better than any other phosphonate salts evaluated in this study. It is capable of

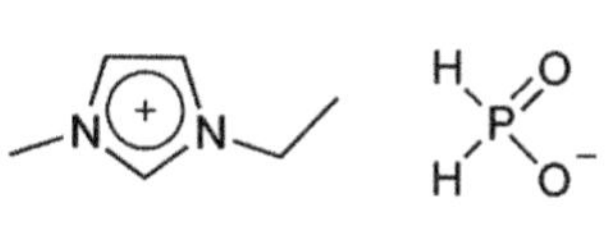

[C_2mim][H_2PO_2]

CHART 8.4 Structure of 1-ethyl-3-methylimidazolium phosphinate.

TABLE 8.11 Physicochemical Properties of Phosphinate-Derived ILs

Ionic liquid	T_g (°C)	T_m (°C)[†]	T_{dec} (°C)[‡]	Kamlet–Taft parameters at 25°C		
				α	β	π^*
[C_2mim][H_2PO_2]	–	17	260	0.52	0.97	1.09

[†]Temperature at the signal peak.
[‡]Temperature at 10% weight loss.

extracting 42% of polysaccharides from bran, whereas [C$_2$mim][(MeO)HPO$_2$] can extract only 39% of polysaccharides from bran at 50°C with gentle stirring for 2 hours. As described in the previous section, the authors successfully dissolved cellulose using a series of alkylphosphonate derivative salts without heating, and expect that [C$_2$mim][H$_2$PO$_2$] can also be used to extract polysaccharide components from bran without heating. The test results show that, using [C$_2$mim][H$_2$PO$_2$] with gentle stirring at room temperature for 5 hours, 14% of polysaccharides can be extracted. Although the efficiency of extraction is much lower than at 50°C (>35% for 2 hours), the results clearly demonstrate the possibility of extracting polysaccharides without heating. To confirm the IL stability, NMR spectra of the ILs after extraction experiments have been examined to confirm the stability of the newly designed IL, and the results confirm that the methylphosphonate salt has not been hydrolyzed during the polysaccharide extraction process. This observation strongly indicates that these newly designed polar ILs can be recycled and reused efficiently in closed systems for biomass processing.

8.5. CONCLUSION AND FUTURE ASPECTS

Some polar ILs have been designed and examined for their capability to dissolve cellulose. They are also tested for extracting cellulose from biomass under mild conditions. Only polar ILs are capable of extracting cellulose from biomass without heating. Additionally, because these ILs are thermally stable, they can be used repeatedly, suggesting that the system for processing biomass can be operated cost-effectively. The only concern is that ILs are costly on the current market; it is expected that a dramatic increase of future applications and demand for this reagent will significantly lower its cost, or novel types of ILs with higher efficiency and lower cost will be developed and made available for cost-effective extraction of polysaccharides from biomass.

REFERENCES

Abe, M., Fukaya, Y., & Ohno, H. (2010). *Green Chemistry, 12*, 1274–1280.

Anderson, J. L., Ding, J., Welton, T., & Armstrong, D. W. (2002). *Journal of American Chemical Society, 124*, 14247.

Crowhurst, L., Mawdsley, P. R., Perez-Arlandis, J. M., Salter, P. A., & Welton, T. (2003). *Physical Chemistry Chemical Physics, 5*, 2790–2794.

Fukaya, Y., Sugimoto, A., & Ohno, H. (2006). *Biomacromolecules, 7*, 3295–3297.

Fukaya, Y., Hayashi, K., Wada, M., & Ohno, H. (2008). *Green Chemistry, 10*, 44–46.

Kamlet, M. J., & Taft, R. W. (1976). *Journal of American Chemical Society, 98*, 377.

Mizumo, T., Marwanta, E., Matsumi, N., & Ohno, H. (2004). *Chemistry Letters, 33*, 1360.

Moulthrop, J. S., Swatloski, R. P., Moyna, G., & Rogers, R. D. (2005). *Chemical Communications*, 1557.

Ohno, H. (2011). *Electrochemical aspects of ionic liquids*. Wiley Interscience.

Ohno, H., & Fukaya, Y. (2009). *Chemistry Letters, 38*, 2–7.

Reichardt, C. (2005). *Green Chemistry, 7,* 339–351.

Remsing, R. C., Swatloski, R. P., Rogers, R. D., & Moyna, G. (2006). *Chemical Communications,* 1271–1273.

Rogers, R. D., & Seddon, K. R. (2003). *Science, 302,* 792–793.

Swatloski, R. P., Spear, S. K., Holbrey, J. D., & Rogers, R. D. (2002). *Journal of American Chemical Society, 124,* 4974–4975.

Welton, T. (1999). *Chemical Reviews, 99,* 2071–2083.

Wassersheid, P., & Welton, T. (2008). *Ionic liquids in synthesis.* Wiley-VCH.

Youngs, T. G. A., Holbrey, J. D., Deetlefs, M., Nieuwenhuyzen, M., Gomes, M. F. C., & Hardacre, C. (2006). *ChemPhysChem, 7,* 2279–2281.

Chapter 9

Enzymes for Cellulosic Biomass Conversion

Takashi Tonozuka, Makoto Yoshida and Michio Takeuchi

9.1. GENERAL INFORMATION ON CELLULASES

9.1.1. Cellulose

Cellulose is the major carbohydrate component of plant cell walls, and thus is one of the most abundant biomass resources on earth. The cellulose molecule is an insoluble linear homopolymer of β-D-glucopyranose units linked together by the 1 and 4 carbon atoms. Because the adjacent glucose moieties are rotated 180° about the axis of the cellulose backbone chain as shown in Figure 9.1, the basic structural repeating unit of cellulose is cellobiose rather than glucose (Blackwell, 1982; Atalla, 1983). The ends of the cellulose chain contain

Research Approaches to Sustainable Biomass Systems. http://dx.doi.org/10.1016/B978-0-12-404609-2.00009-X

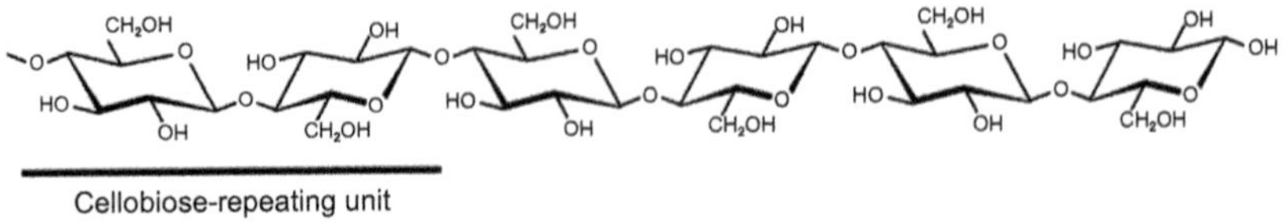

Cellobiose-repeating unit

FIGURE 9.1 Structure of cellulose.

different types of β-anhydroglucose units, i.e. the reducing end with a free hemiacetal (or aldehyde) group at C-1, and the non-reducing end with a free hydroxyl at C-4. The hydroxyl groups in cellulose are positioned in the ring plane, whereas the hydrogen atoms are in the vertical position. Therefore, a hydrophobic surface is formed in the cellulose chain due to the nature of successive β-anhydroglucose units.

The cellulose molecule does not exist as an isolated individual molecule. It forms a crystalline structure, known as cellulose microfibril (Nishiyama, 2009), through the tight packing of chains by intra- and intermolecular bonds such as hydrogen bonds and van der Waals forces. The crystalline structure gives cellulose high tensile strength and resistance to chemical or enzymatic attack. In nature, therefore, cellulose can be digested only by a limited number of organisms, most of which are bacteria and fungi.

9.1.2. General Properties of Cellulases

The word "cellulase" is a generic term used to refer to the enzymes that hydrolyze glucan chains in cellulose polymer into cellooligosaccharides or ultimately glucose. Traditionally, cellulases have been divided into two types, i.e. cellobiohydrolases (CBHs; EC 3.2.1.91) and endoglucanases (EGs; EC 3.2.1.4), based on their activity profiles. The former hydrolyzes cellulose molecules from the end of the chain with the release of cellobiose, whereas the latter randomly cleaves the internal β-1,4-glucosidic bond of cellulose. EGs preferentially act on amorphous cellulose and soluble cellulose derivatives such as carboxymethyl cellulose (CMC), resulting in a rapid decrease of the degree of polymerization in cellulose; they lack hydrolytic activity against crystalline cellulose. On the other hand, CBHs possess significant activity against the crystalline region of cellulose, although the majority of CBH-type enzymes show poor activity against CMC. These enzymes act concertedly to degrade cellulose biologically (Streamer et al., 1975; Wood and McCrae, 1978; Henrissat et al., 1985) in three sequential reactions: (1) EGs initially hydrolyze the cellulose chain in the amorphous region; (2) CBHs recognize the newly generated ends of cellulose chains and start the catalytic reaction with the release of cellobiose; and (3) successive hydrolysis of cellobiose by CBHs finally results in degradation of the crystalline region of cellulose. The sequence of these reactions is called the endo–exo mechanism (Figure 9.2).

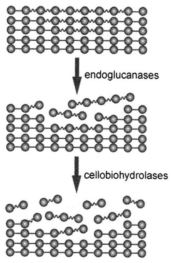

FIGURE 9.2   Endo–exo mechanism.

Recently, the endo–exo concept has been considered as insufficient to understand cellulose biodegradation, so that the concept of a processivity mechanism has been suggested to comprehend the behavior of cellulase's action (Davies and Henrissat, 1995). This concept explains the action of cellulases after the catalytic reaction (Figure 9.3), i.e. a high processive enzyme slides to the next cellobiose unit following hydrolysis of the β-1,4-glucosidic bond of cellulose, whereas a low processive enzyme easily detaches itself from the cellulose chain after the catalytic reaction. A correlation between structural features and processivity has been shown based on three-dimensional structures of cellulases. In particular, a tunnel structure around the active site is critical for expressing processive functions. This is covered in detail in Section 9.2.

Cellulase generally consists of two domains: (1) a catalytic domain, which catalyzes hydrolysis of the β-1,4-glucosidic bond of cellulose, and (2) a cellulose-binding domain (CBD), the binding domain from fungal enzymes, which is currently called family 1 carbohydrate-binding module (CBM1) and is one of the best characterized CBMs. This domain lacks any catalytic activity; it binds the surface of cellulose via three aromatic amino acids, as shown in Figure 9.4. Removal of this domain causes a significant decrease in both

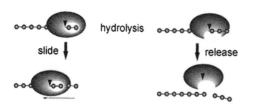

FIGURE 9.3   Concept of processivity. Arrowheads indicate the active sites of enzymes.

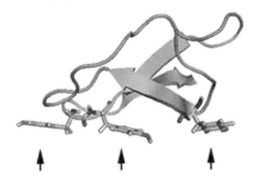

FIGURE 9.4 Structure of cellulose-binding domain of cellobiohydrolase I from *Hypocrea jecorina* (PDB ID, 2CBH). Arrows indicate three aromatic amino acid residues, which are associated with the binding on to cellulose.

binding and catalytic activities against crystalline cellulose but shows no effects on the hydrolysis of amorphous cellulose (Tomme et al., 1988; Ståhlberg et al., 1993; Linder and Teeri, 1997). These observations suggest that CBM1 is related to the degradation of crystalline cellulose.

9.2. STRUCTURE AND FUNCTION OF CELLULASES

9.2.1. The CAZy Database: A Classification System Based on Structures

A database designated as CAZy (Carbohydrate-Active enZymes; http://www. cazy.org/; Cantarel et al., 2009) is indispensable for studying the structure and function of cellulases. The CAZy database was developed by Bernard Henrissat and colleagues, and is now used worldwide as the standard classification system of carbohydrate-acting enzymes. In the CAZy database, carbohydrate-acting enzymes are classified based on their structural similarity into four major groups, i.e. glycoside hydrolase (GH), glycoside transferase (GT), polysaccharide lyase (PL), and carbohydrate esterase (CE). Each group is divided into families, which are serially numbered. For example, one of the cellobiohydrolases is classified and designated as glycoside hydrolase (GH) family 6, and the abbreviated name of the family, GH6, is normally used.

The nomenclature of cellulases has recently been developed based on the CAZy database. As described in Section 9.1.2, cellulases are divided into EG (endoglucanase) and CBH (cellobiohydrolase) in terms of their enzymatic properties. The name of cellulases is conventionally given as a combination of EG/CBH and numbers; for example, cellulases from *Trichoderma reesei* are designated as "EG I" or "CBH II". According to the recent nomenclature, the name of cellulases is a combination of "Cel" (which means cellulase), the number of the glycoside hydrolase family, and a capital letter. For example, the cellulases "EG I" and "CBH II" from *Trichoderma reesei* as described above are designated as "Cel7B" and "Cel6A" respectively.

Many carbohydrate-acting enzymes including cellulases consist of a catalytic domain and a carbohydrate-binding domain; the latter plays a crucial role in the enzymatic activity. In the CAZy database, these non-catalytic domains are classified into carbohydrate-binding modules (abbreviated as CBM).

9.2.2. Cellulases of Two Well-Studied Organisms, *Trichoderma reesei* and *Clostridium thermocellum*

Cellulases are classified into glycoside hydrolase families, GH5, GH6, GH7, GH8, GH9, GH12, GH44, GH45, GH48, GH51, GH74, and GH124. Enzymes belonging to GH61 also show activities similar to cellulases, but the GH61 enzymes have been reported not to be identified as cellulases; they are identified as copper-dependent monooxygenases (Quinlan et al., 2011). Although cellulases are found in many glycoside hydrolase families, none of the cellulase-producing organisms possesses genes encoding all of these GH members.

A filamentous fungus, *Trichoderma reesei*, is the most used microorganism in industry as a source of cellulolytic enzymes (Gusakov, 2011), and therefore numerous studies of enzymes derived from this organism have been carried out. The organism has been shown to be the anamorph of the ascomycete *Hypocrea jecorina*, but is still better known by its former name, *T. reesei*. The complete genome of *T. reesei*, which was sequenced in 2008, indicates that the organism possesses genes for GH5 (two genes), GH6 (one gene), GH7 (two genes), GH12 (one gene), GH45 (one gene), and GH61 (three genes) (Martinez et al., 2008). Among these enzymes, the most predominant cellulases secreted by *T. reesei* are Cel7A (CBH I) and Cel6A (CBH II), which account for 50–60% and ~20% respectively of the total cellulases. Likewise, a basidiomycete, *Phanerochaete chrysosporium*, which is also well known as a producer of cellulolytic enzymes, secretes cellobiohydrolase (Cel7D) of up to ~10% of the total secreted protein in liquid culture (Ubhayasekera et al., 2005).

Bacteria and fungi are important players in degrading the cellulolytic materials in nature. Although bacterial enzymes are not utilized by the industry, numerous studies on bacterial enzymes have been conducted. In particular, several anaerobic bacteria produce a multienzyme complex termed cellulosome, which is an attractive target for the structural biology. The cellulosome was first observed to exist in *Clostridium thermocellum*, and now several *Clostridium* species and *Ruminococcus* species as well as some other organisms have been reported to produce this multienzyme complex (Demain et al., 2005; Doi, 2008). The cellulosome is composed of a non-catalytic subunit called scafoldin that consists of multiple repeats of cohesin domains and cellulosomal enzymes, which typically comprise a catalytic domain and a dockerin domain. Each cohesin domain interacts with the dockerin domain of the cellulosomal enzymes, resulting in the formation of a multienzyme complex, or cellulosome

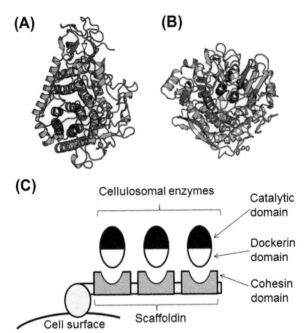

FIGURE 9.5 Cellulases from *Clostridium thermocellum*. (A, B) Ribbon representations of: (A) GH9 cellobiohydrolase (PDB ID, 1RQ5) and (B) GH48 cellobiohydrolase (PDB ID, 1L2A). (C) Schematic illustration of a cellulosome. Note that the actual structure is much more complicated.

(Figure 9.5C). Cellulases consisting of the cellulosome belong to GH5, GH8, GH9, and GH48, and the majority of the enzymes are classified as either GH9 or GH48. The catalytic domains are composed of an $(\alpha/\alpha)_6$ barrel (Figure 9.5A, B) for GH9 and GH48, and a $(\beta/\alpha)_7$ barrel and β-sandwich (Figure 9.6A, B) for GH6 and GH7. Thus, the structural folds of cellulosomal enzymes are completely different from those of cellulases from *T. reesei*.

9.2.3. The GH6 and GH7 Enzymes from *Trichoderma reesei*

Trichoderma reesei mainly produces the GH6 and GH7 CBHs, Cel6A and Cel7A, as described above. The GH6 CBH releases cellobiose units from the non-reducing end of cellulose (EC 3.2.1.91), while the GH7 CBH removes cellobiose units from the reducing end of cellulose (EC 3.2.1.176) (Igarashi et al., 2011). GH7 CBH and GH48 CBH have a unique enzymatic character because there are only a few glycosidases that hydrolyze polysaccharides from the reducing end in an exo manner. In contrast, all the exo-acting amylases degrade starch from the non-reducing end.

A cellulose fiber contains crystalline and amorphous regions. For the *T. reesei* cellulase system, attacking the crystalline region by GH6 and GH7

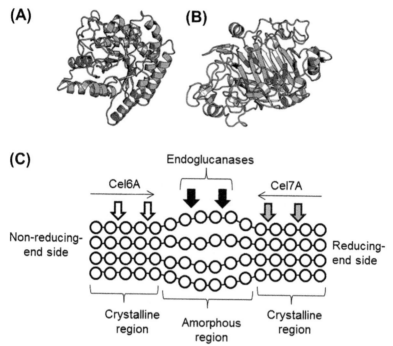

FIGURE 9.6   **Cellulases from *Trichoderma reesei*.** (A–C) Ribbon representations of: (A) Cel6A catalytic domain (PDB ID, 1QJW); (B) Cel7A catalytic domain (PDB ID, 5CEL); and (C) cellulose-binding domain of Cel7A (PDB ID, 1CBH). In (C), three tyrosine residues that are critical for the cellulose-binding are indicated. (D) Schematic illustration of the enzymatic hydrolysis of cellulose. Glucose residues are indicated by circles.

CBHs, Cel6A, from the non-reducing end and by Cel7A from the reducing end has been proposed, whereas the amorphous region is accessible to endoglucanase action (Figure 9.6C) (Teeri, 1997). Although no structural homology is found between the catalytic domains of GH6 and GH7, an enclosed tunnel is present at the active site of both GH6 and GH7 CBHs (Figure 9.7A, B). It is noteworthy that the length of the tunnel of *T. reesei* Cel7A is ~50 Å (Divne et al., 1998), which is an unusual structure among glycosidases. Some EGs are also classified as GH6 or GH7, but no enclosed tunnel is found in the structures of GH6 and GH7 EGs. *Trichoderma reesei* Cel6A and Cel7A possess a cellulose-binding domain that is grouped in the CAZy family, CBM1. The cellulose-binding domain of Cel6A is located on the N-terminal side, whereas the cellulose-binding domain of Cel7A is found on the C-terminal side (Linder et al., 1995; Rabinovich et al., 2002). The structural features of CBM1 are described in Section 9.1.2; the CBM1 cellulose-binding domain, which is smaller than the catalytic domain, consists of ~40 amino acid residues.

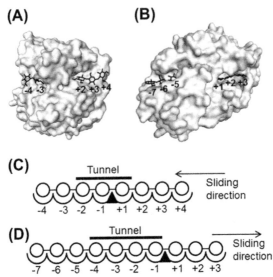

FIGURE 9.7   Subsite structures of Cel6A and Cel7A from *Trichoderma reesei*. (A) A surface model of the catalytic domain of Cel6A. The cellulose chain was constructed by combining the ligands of the structures, PDB IDs 1OCB and 1QJW, and overlaid on the surface model of 1QJW. (B) A surface model of the catalytic domain of Cel7A. The cellulose chain was constructed by combining the ligands of the structures, PDB IDs 5CEL and 6CEL, and overlaid on the surface model of 5CEL. (C, D) Schematic models of the subsite structures of Cel6A (C) and Cel7A (D). Symbols: circle, glucose residue; triangle, enzymatic cleavage point. The non-reducing end and the reducing end of the substrates are drawn on the left and the right respectively.

9.2.4. Subsite and Catalytic Residues

The term subsite refers to the structural relationship between an enzyme and a polymer substrate. In the case of glycosidase and polysaccharide substrates, the glycosidase has individual regions called subsites, each of which binds to a specific monosaccharide unit of the polysaccharide substrate. The subsites that bind to the non-reducing end are labeled "minus" and those that bind to the reducing end of the substrate are labeled "plus"; the glycosidase cleaves between the −1 and +1 subsites (Davies et al., 1997). An enclosed tunnel is present at the active site of the GH6 and GH7 CBHs (Figure 9.7A, B). Therefore, the substrate cellulose is initially considered to bind the plus subsites of the GH6 CBH, and then slides through the active center tunnel until it occupies subsites −2 and −1. In the case of GH7 CBH, the plus and minus subsites have opposite signs. For GH6 and GH7 CBHs, subsite −3 and subsite +3 are presumed unnecessary to produce the respective cellobioses, but *T. reesei* Cel7A has been reported to have 10 subsites from −7 to +3 (Figure 9.7C) (Divne et al., 1998). In *T. reesei* Cel6A, the existence of eight subsites from −4 to +4 has also been proposed (Figure 9.7D) (Varrot et al., 2003).

Knowledge of the catalytic residues and the concept of subsites is of great importance in understanding the structure–function relationship of enzymes. Reaction mechanisms of glycosidases are divided into two groups: inverting mechanism and retaining mechanism. In the inverting mechanism, the enzymatic reaction is accompanied by an inversion of the anomeric configuration, whereas in the retaining mechanism the anomeric configuration of the substrate is retained. For instance, the "inverting" CBH and the "retaining" CBH produce α-cellobiose and β-cellobiose respectively from cellulose. GH6 and GH7 enzymes act through the inverting mechanism and retaining mechanism respectively. In the inverting mechanism, two amino acid residues that act as a general acid and a general base are found to be the catalytic residues, whereas in the retaining mechanism a nucleophile residue and a general acid/base residue are identified as the catalytic residues. In both mechanisms, aspartic acid and/or glutamic acid residues normally function as the catalytic residues of glycosidases (Rye and Withers, 2000). In *T. reesei* Cel7A, the nucleophile residue is Glu212 and the general acid/base residue is Glu217 (Divne et al., 1994). In GH6 enzymes, the catalytic general acid residue is identified as a conserved aspartic acid, Asp221 in *T. reesei* Cel6A, but it is still debatable concerning which residue acts as the catalytic base of the GH6 enzymes. Recent studies show that Asp175 in *T. reesei* Cel6A is located relatively far from the cleavage site of the substrate, but has been proposed to be the catalytic base. It acts indirectly through the hydrogen bond network of water molecules in what has been called a Grotthus mechanism (Koivula et al., 2002).

In conclusion, the structures and functions of the GH6 and GH7 cellulases are unusual compared with other glycosidases such as amylases. GH6 and GH7 CBHs are characterized by an enclosed tunnel at the active site. GH7 CBHs release cellobiose units from the reducing end of cellulose, and the catalytic base of the GH6 enzymes has been proposed to act indirectly in the so-called Grotthus mechanism. It is likely that these unusual features of cellulases make them suitable for the hydrolysis of cellulose. For more detailed information see Box 9.1.

9.3. OTHER BIOMASS-DEGRADING ENZYMES

9.3.1. β-Glucosidases

β-Glucosidases (BGLs; EC 3.2.1.21) catalyze the hydrolysis of β-glucosidic bonds in various β-glucosidic compounds, and thus are involved in various metabolic pathways such as degradation of polysaccharides, cellular signaling, oncogenesis, and host–pathogen interactions. Among these pathways, the enzymes important for cellulose degradation, i.e. cellobiases, are focused in this section. These enzymes are indispensable for the production of glucose, which is then fermented into ethanol by yeast in the subsequent process of bioethanol production.

In nature, various cellulolytic microorganisms produce BGLs during the course of cellulose degradation. Fungal enzymes are especially regarded as the

BOX 9.1 Structures of the GH6 Cellobiohydrolases

As described above, cellulases belonging to GH6 are known as major cellulolytic enzymes produced by fungi. In 1990, the first crystal structure of a cellulose that is a catalytic domain of *T. reesei* Cel6A was reported (Rouvinen et al., 1990). The crystal structure of the catalytic domain of a CBH, i.e. Cel6A, obtained from the ascomycete *Humicola insolens* was also determined in the same decade (Varrot et al., 1999a). The catalytic domains of both CBHs consist of a distorted seven-stranded β/α barrel. The active sites are enclosed by the N-terminal and C-terminal loops to form a tunnel, and the two loops have been demonstrated to open and close in response to ligand binding (Varrot et al., 1999b; Zou et al., 1999).

Coprinopsis cinerea (formerly known as *Coprinus cinereus*) is known as a model basidiomycete with its genome completely sequenced (Stajich et al., 2010). While *T. reesei* produces only one GH6 enzyme, *C. cinerea* produces five GH6 enzymes, i.e. CcCel6A, CcCel6B, CcCel6C, CcCel6D, and CcCel6E (Yoshida et al., 2009). There is a high sequence identity between CcCel6A and *T. reesei* Cel6A (48%), and these two enzymes contain an N-terminal cellulose-binding domain. These facts indicate that CcCel6A is a major enzyme involved in the degradation of crystalline cellulose in *C. cinerea*, and that the expression of CcCel6A is induced by cellobiose. On the other hand, the four enzymes, CcCel6B, CcCel6C, CcCel6D, and CcCel6E, have been mapped to a region closer to the endoglucanase in the evolutionary tree, and they have no cellulose-binding domain.

The structures of CcCel6A and CcCel6C have been determined. CcCel6C is produced by *C. cinerea* at low and constitutive levels. Although CcCel6C is classified as a CBH, its activity and structural characteristics are unique in that the enzyme hydrolyzes carboxymethyl cellulose (CMC), which is a poor substrate for typical CBH (i.e. CcCel6A) (Liu et al., 2009). The closed form of CcCel6C shows that the enclosed tunnel is wider than that of CcCel6A, indicating that the wide tunnel is suitable for the binding of substrates such as CMC (Liu et al., 2010). It is likely that the enzymatic properties and the structural architecture of CcCel6C can be optimized to hydrolyze amorphous cellulose effectively to produce a small amount of cellobiose, a key molecule for CcCel6A expression.

Detailed structural analyses of the open-to-closed conformational changes of GH6 CBHs from *Coprinopsis cinerea* have been carried out (Tamura et al., 2012). The conformational change in CcCel6A is only observed in the vicinity of the enclosed tunnel (Figure 9.8A). In contrast, CcCe16C has a more drastic conformational change than CcCel6A, and the structure of CcCel6C behaves like a pair of tweezers (Figure 9.8B). Phylogenetic analyses of GH6 enzymes demonstrate that the length of loops that form the active-site tunnel in CBHs vary among subfamilies

BOX 9.1 Structures of the GH6 Cellobiohydrolases—cont'd

(Mertz et al., 2005). Further studies may provide more information on the relationship between conformational change, physiological role, and phylogenetic position of GH6 CBHs.

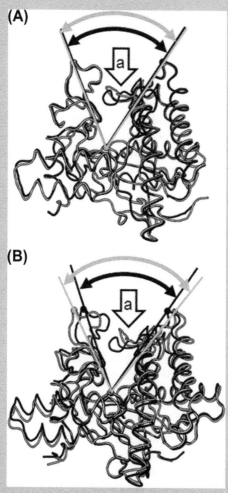

FIGURE 9.8 Conformational changes in two cellobiohydrolases, CcCel6A and CcCel6C, from *Coprinopsis cinerea* upon substrate binding. Comparison of the open form (gray) and the closed form (black) of: (A) CcCel6A (PDB IDs, 3VOG and 3VOJ) and (B) CcCel6C (PDB IDs, 3A64 and 3VOF). Arrows "a" indicate a large loop movement observed in the active site. The angles produced by the two tunnel-forming loops are shown.

most promising BGL in bioethanol production. Filamentous fungi mainly produce three types of BGLs, i.e. extracellular, intracellular, and cell wall-associated enzymes (Lynd et al., 2002); extracellular and intracellular enzymes have been well characterized so far. These enzymes, which are retaining enzymes, catalyze the hydrolysis of cellobiose and cellooligosaccharides with the release of glucose. In addition to their hydrolytic activities, the enzymes often show transglycosylation activities. Because the activity of BGL decreases at higher concentrations of glucose, improving the glucose tolerance is one of the many important targets for conducting BGL research.

The extracellular BGLs belong to the GH3 family in the CAZy database, and some of the enzymes possess CBM1 beside the catalytic domain. Because the enzymes catalyze the hydrolysis of cellobiose, which is the main product of cellulose hydrolysis by cellulases, the cellobiases are considered to have an important role in the final step of extracellular cellulose degradation (Sternberg, 1977; Eriksson, 1978). Previous studies on extracellular BGL obtained from *Phanerochaete chrysosporium* reveal interesting results. The enzyme prefers laminaribiose (disaccharide with β-1,3-glucosidic bond) instead of cellobiose, and the reaction efficiency against β-1,3-oligosaccharides increases with higher degree of polymerization of the substrates. This observation suggests that the enzyme is β-1,3-glucanase (Igarashi et al., 2003). Hence, not all extracellular BGLs from cellulolytic fungi are cellobiases. On the other hand, intracellular BGLs belong to the GH1 family, and the intracellular localization of the BGLs requires the uptake of cellobiose into the cell. In this regard, *T. reesei* is expected to be a putative diglucoside permease involved in the transport of cellobiose into the cell (Kubicek et al., 1993). In addition, intracellular BGL from *T. reesei* is reported to produce sophorose, which is a very strong inducer of cellulases (Mandels et al., 1962), via transglycosylation activities (Saloheimo et al., 2002), although the induction of cellulases by sophorose is observed only in a few limited fungal species, such as *T. reesei*.

9.3.2. Hemicellulases

Hemicelluloses, which represent about 20–35% of the plant cell wall, are heterogeneous polymers of various sugars such as xylose, arabinose, glucose, mannose, and galactose. The composition of hemicellulose depends on the source and type of plant biomass. Hardwood and graminaceous plants contain 4-O-methylglucuronoxylan and arabinoxylan as the major hemicellulose respectively, whereas hemicellulose of softwood mainly consists of glucomannan and glucuronoxylan. Xylans, such as 4-O-methylglucuronoxylan, arabinoxylan, and glucuronoxylan, have a backbone of 1,4-linked β-D-xylopyranose units. In addition to xylopyranoside residues, xylans contain various substituents, i.e. arabinoside residues, O-acetyl groups, ferulic acid, *p*-coumaric acid, and 4-O-methylglucuronic acid (Wilkie and Woo, 1977). Glucomannan is a mainly linear heteropolymer consisting of β-1,4-linked D-mannose and

D-glucose (Timell, 1967). Xyloglucan is a highly branched heteropolymer with D-glucose and D-xylose. Xylan and glucomannan are located in the secondary wall, whereas xyloglucan is contained in the primary wall. Therefore, xyloglucan accounts for only a small fraction of the plant cell wall.

The word "hemicellulase" is a generic term used to refer to enzymes that have a role in hemicellulose degradation, and thus includes not only glycosidases but also esterases (de Vries and Visser, 2001).

Endo-1,4-β-D-xylanases (EC 3.2.1.8) are enzymes that randomly hydrolyze β-1,4-xylosidic bonds of xylan backbone (Polizeli et al., 2005). They belong to GH5, GF7, GH8, GH 10, GH11, and GH43 in the CAZy database; GH10 and GH11 have been well characterized. The GH10 enzymes act preferentially on soluble substrates and can readily hydrolyze small xylooligosaccharides such as xylotriose. Unlike GH10 enzymes, GH11 enzymes are most active in the hydrolysis of insoluble xylans, although the activity is more likely to be inhibited by the presence of arabinose decorations. Because xylanase enhances the hydrolysis of cellulose by cellulases, it has been considered as a major accessory enzyme in the saccharification of plant biomass.

α-L-Arabinofuranosidase (EC 3.2.1.55) catalyzes the hydrolysis of α-1,2-, α-1,3- or α-1,5-L-arabinofuranosidic bonds in arabinose-containing polysaccharides (Numan and Bhosle, 2006). It has been classified as GH family 3, 43, 51, 54, and 62 in the CAZy database. Some α-L-arabinofuranosidases demonstrate broad substrate specificity, i.e. acting on arabinofuranosyl residues at O-2, O-5 and/or O-3 positions as a single substituent. On the other hand, α-L-arabinofuranosidases that specifically catalyze the release of αα-1,3-L-arabinofuranosyl substituents from doubly substituted xylose residues have also been found. These enzymes work synergistically on arabinoxylan degradation with xylanases.

Acetylxylan esterases (EC 3.1.1.72) release acetate from xylan and xylooligosaccharides. The enzymes have been classified as carbohydrate esterase (CE) families 1, 2, 3, 4, 5, 6, 7, and 12 in the CAZy database. CE16 is also reported to contain acetyl esterase that releases acetate from 4-nitrophenyl-β-D-xylopyranoside monoacetate (Li et al., 2008). The removal of acetyl groups by acetylxylan esterases improves the access of xylanases to the xylan backbone and facilitates the degradation of xylans.

In addition to the enzymes described above, various other enzymes such as α-glucuronidases, xyloglucanases, galactosidases, glucuronoyl esterases, feruroyl esterases, mannanases, endoglucanases, and xylosidases are required to degrade hemicellulose (de Vries and Visser, 2001). The synergistic action of these enzymes leads to the hydrolysis of the complicated structure of hemicelluloses.

9.3.3. Starch-Hydrolyzing Enzymes: Amylases

Although producing ethanol from starch adversely affects food supplies, the process is still important to convert non-edible biomass such as crop waste into ethanol. The term "amylase" is defined as an enzyme that hydrolyzes starch, and this section includes a synopsis of amylases.

Starch is a polymer of glucose units linked predominantly via α-1,4-glucosidic linkages and some α-1,6-glucosidic linkages to form branch points. Amylases are classified according to their enzymatic reactions into the following four groups (Figure 9.9A):

1. α-Amylase (EC 3.2.1.1) is an endo-acting enzyme that hydrolyzes internal α-1,4-glucosidic linkages of starch via a retaining mechanism. The enzyme belongs to GH13, GH57, and GH119 in the CAZy database, and is produced by a wide range of organisms, including archaea, bacteria, plants, and animals.

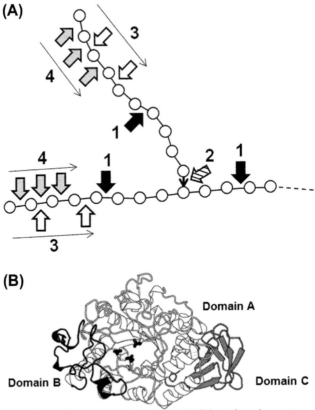

FIGURE 9.9 **Action and structure of amylases.** (A) Schematic action pattern of amylases. Symbols: circle, glucose; bold line, α-1,4-glucosidic linkage; ↓, α-1,6-glucosidic linkage; arrow 1, cleavage points of α-amylase; arrow 2, cleavage points of debranching enzymes; arrow 3, cleavage points of β-amylase; arrow 4, cleavage points of glucoamylase. (B) Structure of a typical α-amylase, *Aspergillus oryzae* α-amylase. Domains are shown in different gray scales. Three catalytic residues conserved in amylase family enzymes are indicated in a stick representation.

2. Starch-debranching enzymes that hydrolyze α-1,6-glucosidic linkages of starch are further classified as isoamylase (EC 3.2.1.68) and pullulanase (EC 3.2.1.41) based on their substrate specificities. Both isoamylase and pullulanase belong to GH13, and are produced by plants and some archaea and bacteria.

3. β-Amylase (EC 3.2.1.2) is an exo-acting enzyme that hydrolyzes the non-reducing ends of starch to produce β-maltose via an inverting mechanism. The enzyme belongs to GH14 in the CAZy database, and is produced by plants and some bacteria.

4. Glucoamylase (EC 3.2.1.3) is an exo-acting enzyme that hydrolyzes the non-reducing ends of starch to produce β-glucose via an inverting mechanism. The enzyme belongs to GH15 in the CAZy database, and is produced by some archaea, bacteria, and fungi.

In the CAZy database, numerous amylases are members of GH13, which was originally known as the α-amylase family. There are four conserved amino acid sequences in the original definition of the α-amylase family (Kuriki and Imanaka, 1999), and three catalytic residues (i.e. aspartic acid, glutamic acid, and aspartic acid residues) are located in the second, third, and fourth conserved regions. Now the α-amylase family is defined as a clan GH-H composed of three glycoside hydrolase families, including GH13, GH70, and GH77 (MacGregor et al., 2001). The essential structure of GH13 is composed of domains A, B, and C; domain A consists of a $(\beta/\alpha)_8$ barrel with a small domain, i.e. domain B, protruding from it. Domain C is located at the C-terminal side of domains A and B, and it forms a β-sandwich (Figure 9.9B) (Kuriki and Imanaka, 1999). Isoamlase, pullulanase, and many other GH13 enzymes have some extra domains other than A, B, and C.

REFERENCES

Atalla, R. H. (1983). *The structure of cellulose: Recent developments*. New York: Academic Press.

Blackwell, J. (1982). *The macromolecular organization of cellulose and chitin*. New York: Plenum.

Cantarel, B. L., Coutinho, P. M., Rancurel, C., Bernard, T., Lombard, V., & Henrissat, B. (2009). The Carbohydrate-Active EnZymes database (CAZy): An expert resource for Glycogenomics. *Nucleic acids research, 37*, D233–D238.

Davies, G., & Henrissat, B. (1995). Structures and mechanisms of glycosyl hydrolases. *Structure, 3*, 853–859.

Davies, G. J., Wilson, K. S., & Henrissat, B. (1997). Nomenclature for sugar-binding subsites in glycosyl hydrolases. *The Biochemical journal, 321*, 557–559.

Demain, A. L., Newcomb, M., & Wu, J. H. (2005). Cellulase, clostridia, and ethanol. *Microbiology and molecular biology reviews: MMBR, 69*, 124–154.

de Vries, R. P., & Visser, J. (2001). *Aspergillus* enzymes involved in degradation of plant cell wall polysaccharides. *Microbiology and molecular biology reviews: MMBR, 65*, 497–522.

Divne, C., Ståhlberg, J., Reinikainen, T., Ruohonen, L., Pettersson, G., Knowles, J. K., Teeri, T. T., & Jones, T. A. (1994). The three-dimensional crystal structure of the catalytic core of cellobiohydrolase I from. *Trichoderma reesei. Science, 265*, 524–528.

Divne, C., Ståhlberg, J., Teeri, T. T., & Jones, T. A. (1998). High-resolution crystal structures reveal how a cellulose chain is bound in the 50 Å long tunnel of cellobiohydrolase I from Trichoderma reesei. *Journal of molecular biology, 275,* 309–325.

Doi, R. H. (2008). Cellulases of mesophilic microorganisms: Cellulosome and noncellulosome producers. *Annals of the New York Academy of Sciences, 1125,* 267–279.

Eriksson, K. E. (1978). Enzyme mechanisms involved in cellulose hydrolysis by the white-rot fungus *Sporotrichum pulverulentum. Biotechnology and bioengineering, 70,* 317–332.

Gusakov, A. V. (2011). Alternatives to *Trichoderma reesei* in biofuel production. *Trends in biotechnology, 29,* 419–425.

Henrissat, B., Driguez, H., Viet, C., & Schulein, M. (1985). Synergism of cellulases from *Trichoderma reesei* in the degradation of cellulose. *Biotechnology, 3,* 722–726.

Igarashi, K., Tani, T., Kawai, R., & Samejima, M. (2003). Family 3 β-glucosidase from cellulose-degrading culture of the white-rot fungus *Phanerochaete chrysosporium* is a glucan 1,3-β-glucosidase. *Journal of bioscience and bioengineering, 95,* 572–576.

Igarashi, K., Uchihashi, T., Koivula, A., Wada, M., Kimura, S., Okamoto, T., Penttilä, M., Ando, T., & Samejima, M. (2011). Traffic jams reduce hydrolytic efficiency of cellulase on cellulose surface. *Science, 333,* 1279–1282.

Koivula, A., Ruohonen, L., Wohlfahrt, G., Reinikainen, T., Teeri, T. T., Piens, K., Claeyssens, M., Weber, M., Vasella, A., Becker, D., Sinnott, M. L., Zou, J. Y., Kleywegt, G. J., Szardenings, M., Ståhlberg, J., & Jones, T. A. (2002). The active site of cellobiohydrolase Cel6A from *Trichoderma reesei*: The roles of aspartic acids D221 and D175. *Journal of the American Chemical Society, 124,* 10015–10024.

Kubicek, C. P., Ressner, R., Gruber, F., Mandels, M., & Kubicek-Pranz, E. M. (1993). Triggering of cellulase biosynthesis by cellulase in *Trichoderma reesei*: Involvement of a constitutive, sophorose-inducible, glucose-inhibited β-diglucoside permease. *The Journal of biological chemistry, 268,* 19364–19368.

Kuriki, T., & Imanaka, T. (1999). The concept of the alpha-amylase family: Structural similarity and common catalytic mechanism. *Journal of bioscience and bioengineering, 87,* 557–565.

Li, X. L., Skory, C. D., Cotta, M. A., Puchart, V., & Biely, P. (2008). Novel family of carbohydrate esterases, based on identification of the *Hypocrea jecorina* acetyl esterase gene. *Applied and environmental microbiology, 74,* 7482–7489.

Linder, M., & Teeri, T. T. (1997). The role and function of cellulose-binding domains. *Biotechnology journal, 57,* 15–28.

Linder, M., Mattinen, M. L., Kontteli, M., Lindeberg, G., Ståhlberg, J., Drakenberg, T., Reinikainen, T., Pettersson, G., & Annila, A. (1995). Identification of functionally important amino acids in the cellulose-binding domain of *Trichoderma reesei* cellobiohydrolase I. *Protein science: a publication of the Protein Society, 4,* 1056–1064.

Liu, Y., Igarashi, K., Kaneko, S., Tonozuka, T., Samejima, M., Fukuda, K., & Yoshida, M. (2009). Characterization of glycoside hydrolase family 6 enzymes from Coprinopsis cinerea. *Bioscience, biotechnology, and biochemistry, 73,* 1432–1434.

Liu, Y., Yoshida, M., Kurakata, Y., Miyazaki, T., Igarashi, K., Samejima, M., Fukuda, K., Nishikawa, A., & Tonozuka, T. (2010). Crystal structure of a glycoside hydrolase family 6 enzyme, CcCel6C, a cellulase constitutively produced by Coprinopsis cinerea. *The FEBS journal, 277,* 1532–1542.

Lynd, L. R., Weimer, P. J., van Zyl, W. H., & Pretorius, I. S. (2002). Microbial cellulose utilization: Fundamentals and biotechnology. *Microbiology and molecular biology reviews: MMBR, 66,* 506–577.

MacGregor, E. A., Janecek, S., & Svensson, B. (2001). Relationship of sequence and structure to specificity in the α-amylase family of enzymes. *Biochimica et Biophysica Acta, 1546,* 1–20.

Mandels, M., Parrish, F. W., & Reese, E. T. (1962). Sophorose as an inducer of cellulase in Trichoderma viride. *Journal of bacteriology, 83,* 400–408.

Martinez, D., Berka, R. M., Henrissat, B., Saloheimo, M., Arvas, M., et al. (2008). Genome sequencing and analysis of the biomass-degrading fungus *Trichoderma reesei* (syn. *Hypocrea jecorina*). *Nature biotechnology, 26,* 553–560.

Mertz, B., Kuczenski, R. S., Larsen, R. T., Hill, A. D., & Reilly, P. J. (2005). Phylogenetic analysis of family 6 glycoside hydrolases. *Biopolymers, 79,* 197–206.

Nishiyama, Y. (2009). Structure and properties of the cellulose microfibril. *J. Wood Sci., 55,* 241–249.

Numan, M. T., & Bhosle, N. B. (2006). α-L-Arabinofuranosidases: The potential applications in biotechnology. *Journal of industrial microbiology & biotechnology, 33,* 247–260.

Polizeli, M. L., Rizzatti, A. C., Monti, R., Terenzi, H. F., Jorge, J. A., & Amorim, D. S. (2005). Xylanases from fungi: Properties and industrial applications. *Applied microbiology and biotechnology, 67,* 577–591.

Quinlan, R. J., Sweeney, M. D., Lo Leggio, L., Otten, H., Poulsen, J. C., Johansen, K. S., Krogh, K. B., Jørgensen, C. I., Tovborg, M., Anthonsen, A., Tryfona, T., Walter, C. P., Dupree, P., Xu, F., Davies, G. J., & Walton, P. H. (2011). Insights into the oxidative degradation of cellulose by a copper metalloenzyme that exploits biomass components. *Proceeding of the National Academy of Sciences of the United States of America., 108,* 15079–15084.

Rabinovich, M. L., Melnick, M. S., & Bolobova, A. V. (2002). The structure and mechanism of action of cellulolytic enzymes. *Biochemistry (Moscow), 67,* 850–871.

Rouvinen, J., Bergfors, T., Teeri, T., Knowles, J. K., & Jones, T. A. (1990). Three-dimensional structure of cellobiohydrolase II from. *Trichoderma reesei. Science, 249,* 380–386.

Rye, C. S., & Withers, S. G. (2000). Glycosidase mechanisms. *Current opinion in chemical biology, 4,* 573–580.

Saloheimo, M., Kuja-Panula, J., Ylosmaki, E., Ward, M., & Penttila, M. (2002). Enzymatic properties and intracellular localization of the novel *Trichoderma reesei* β-glucosidase BGLII (cell A). *Applied and environmental microbiology, 68,* 4546–4553.

Ståhlberg, J., Johansson, G., & Pettersson, G. (1993). *Trichoderma reesei* has no true exo-cellulase: All intact and truncated cellulases produce new reducing end groups on cellulose. *Biochimica et biophysica acta, 1157,* 107–113.

Stajich, J. E., Wilke, S. K., Ahrén, D., Au, C. H., Birren, B. W., Borodovsky, M., et al. (2010). Insights into evolution of multicellular fungi from the assembled chromosomes of the mushroom *Coprinopsis cinerea* (*Coprinus cinereus*). *Proceeding of the National Academy of Sciences of the United States of America., 107,* 11889–11894.

Sternberg, D., Vijayakumar, P., & Reese, E. T. (1977). β-Glucosidase: Microbial production and effect on enzymatic hydrolysis of cellulose. *Canadian journal of microbiology, 23,* 139–147.

Streamer, M., Eriksson, K. E., & Pettersson, B. (1975). Extracellular enzyme system utilized by the fungus *Sporotrichum pulverulentum* (*Chrysosporium lignorum*) for the breakdown of cullulose. Functional characterization of five endo-1,4-β-glucanases and one exo-1,4-β-glu-canase. *European Journal of Biochemistry., 59,* 607–613.

Tamura, M., Miyazaki, T., Tanaka, Y., Yoshida, M., Nishikawa, A., & Tonozuka, T. (2012). Comparison of the structural changes in two cellobiohydrolases, CcCel6A and CcCel6C, from *Coprinopsis cinerea*: A tweezer-like motion in the structure of CcCel6C. *The FEBS journal, 279,* 1871–1882.

Teeri, T. T. (1997). Crystalline cellulose degradation: New insight into the function of cellobio-hydrolases. *Trends in biotechnology, 15,* 160–167.

Timell, T. E. (1967). Recent progress in the chemistry of wood hemicelluloses. *Wood Science and Technology., 1*, 45–70.

Tomme, P., Van Tilbeurgh, H., Pettersson, G., Van Damme, J., Vandekerckhove, J., Knowles, J., Teeri, T. T., & Claeyssens, M. (1988). Studies of the cellulolytic system of *Trichoderma reesei* QM 9414. Analysis of domain function in two cellobiohydrolases by limited proteolysis. *European Journal of Biochemistry., 170*, 575–581.

Ubhayasekera, W., Muñoz, I. G., Vasella, A., Ståhlberg, J., & Mowbray, S. L. (2005). Structures of *Phanerochaete chrysosporium* Cel7D in complex with product and inhibitors. *The FEBS journal, 272*, 1952–1964.

Varrot, A., Hastrup, S., Schülein, M., & Davies, G. J. (1999a). Crystal structure of the catalytic core domain of the family 6 cellobiohydrolase II, Cel6A, from *Humicola insolens*, at 1.92 Å resolution. *Journal of biochemistry, 337*, 297–304.

Varrot, A., Schülein, M., & Davies, G. J. (1999b). Structural changes of the active site tunnel of *Humicola insolens* cellobiohydrolase, Cel6A, upon oligosaccharide binding. *Biochemistry, 38*, 8884–8891.

Varrot, A., Frandsen, T. P., von Ossowski, I., Boyer, V., Cottaz, S., Driguez, H., Schülein, M., & Davies, G. J. (2003). Structural basis for ligand binding and processivity in cellobiohydrolase Cel6A from *Humicola insolens. Structure, 11*, 855–864.

Wilkie, K. C. B., & Woo, S. L. (1977). A heteroxylan and hemicellulosic materials from bamboo leaves, and a reconsideration of the general nature of commonly occurring xylans and other hemicelluloses. *Carbohydrate research, 57*, 145–162.

Wood, T. M., & McCrae, S. I. (1978). The cellulase of *Trichoderma koningii*. Purification and properties of some endoglucanase components with special reference to their action on cellulose when acting alone and in synergism with the cellobiohydrolase. *The Biochemical journal, 171*, 61–72.

Yoshida, M., Sato, K., Kaneko, S., & Fukuda, K. (2009). Cloning and transcript analysis of multiple genes encoding the glycoside hydrolase family 6 enzyme from *Coprinopsis cinerea. Bioscience, biotechnology, and biochemistry, 73*, 67–73.

Zou, J., Kleywegt, G. J., Ståhlberg, J., Driguez, H., Nerinckx, W., Claeyssens, M., Koivula, A., Teeri, T. T., & Jones, T. A. (1999). Crystallographic evidence for substrate ring distortion and protein conformational changes during catalysis in cellobiohydrolase Cel6A from *Trichoderma reesei. Structure, 7*, 1035–1045.

Chapter 10

Ethanol Production from Biomass

Haruki Ishizaki and Keiji Hasumi

10.1. ETHANOL FERMENTATION

10.1.1. Outline of Ethanol Fermentation

In the New Stone Age, there existed alcoholic drinks produced by natural fermentation of fruits. Several thousand years ago, in the region of Mesopotamia, there were descriptions of the production of wine or beer that was offered to God to pray for a good harvest and hunting for ancient people. Nomadic tribes produced kumis from horse milk, and agricultural people produced alcohol such as wine, beer, and rice wine from grapes, wheat, and rice. In the Middle Ages, ethanol was initially developed as medicine, and during the middle of the sixteenth century, whisky and brandy made from beer and wine became alcoholic beverages. After this, ethanol production was developed worldwide to produce beverages, raw materials for medicines, fragrances, and solvents. In 2009, 468 million barrels of bioethanol were produced worldwide for use as an alternative to gasoline.

Research Approaches to Sustainable Biomass Systems. http://dx.doi.org/10.1016/B978-0-12-404609-2.00010-6

Antoine Lavoisier (France) made the first scientific studies of alcoholic fermentation. He described how sugar is converted into alcohol and carbon dioxide in 1789, and he determined the composition of both fermentable substances and the products of fermentation. In 1857, Louis Pasteur (France) showed that lactate fermentation is caused by living organisms without air. Emil C. Hansen (Denmark) first isolated *Saccharomyces cerevisiae* from wort for beer fermentation in 1883. In 1897, Eduard Buchner (German) found that cell-free extract from yeast can also transform glucose to ethanol. This was the first demonstration that enzymes bring about fermentation reactions.

10.1.2. Principles of Ethanol Fermentation

a. Embden–Meyerhof Pathway (EM Pathway)

The EM pathway is the main pathway for the anaerobic degradation of carbohydrates. D-Glucose first enters into the EM pathway via phosphorylation to be converted to D-glucose-6-phosphate by hexokinase, with one mole of ATP required as phosphate donor. D-Glucose-6-phosphate is next converted to D-fructose-6-phosphate and then catalyzed by phosphofructokinase to produce D-fructose-1,6-bisphosphate in the follow-up phosphorylation with ATP. D-Fructose-1,6-bisphosphate is subsequently split by fructose-1,6-bisphosphate aldolase into two triosephosphates, i.e. D-glyceraldehyde-3-phosphate and dihydroxyacetone phosphate. D-Glyceraldehyde-3-phosphate is converted to pyruvate by five-step enzyme reactions, and four moles of ATP are generated from two moles of the triosephosphates in this step. Pyruvate is converted to ethanol by pyruvate decarboxylase and alcohol dehydrogenase. Thus, one mole of glucose is converted to two moles of ethanol and two moles of carbon dioxide while generating two moles of ATP.

Fructokinase converts D-fructose to D-fructose-6-phosphate, and hexokinase converts D-mannose to D-mannose-6-phoshate; the latter is subsequently converted to D-fructose-6-phosphate by mannose phosphate isomerase. D-Galactose is converted to D-galactose-1-phosphate by galactokinase, and the resulting product isomerizes to D-glucose-6-phosphate via D-glucose-1-phosphate. D-Fructose-6-phosphate and D-glucose-6-phosphate are metabolized to ethanol through the EM pathway (Figure 10.1).

b. Entner–Doudoroff Pathway (ED Pathway)

Although the ED pathway is generally considered to be restricted to a limited number of Gram-negative bacteria, the pathway is now known to be present in a diverse group of organisms ranging from archaea to bacteria, to eukarya. Most bacteria utilize the ED pathway to convert glucose to pyruvate anaerobically. D-Glucose enters into the ED pathway after phosphorylation by hexokinase. The resulting D-glucose-6-phosphate is converted to D-6-phosphogluconate using one mole of NADP. D-6-Phosphogluconate is

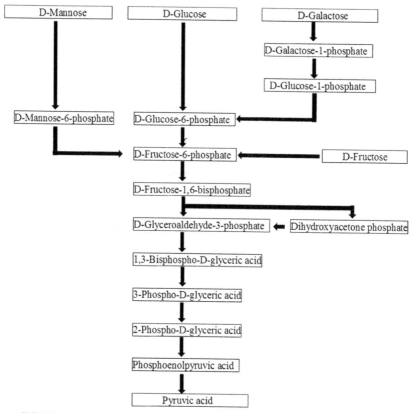

FIGURE 10.1   Pathway for hexose metabolism (the Embden–Meyerhof pathway).

oxidized to 2-keto-3-deoxy-6-phosphogluconate, which is then cleaved by aldolase to pyruvate and D-glyceraldehyde-3-phosphate. D-Glyceraldehyde-3-phosphate is oxidized to pyruvate through the EM pathway, yielding two moles of ATP (Figure 10.2). Pyruvate is converted to ethanol as in the EM pathway. Overall, one mole of glucose is converted to two moles of ethanol and two moles of carbon dioxide by consuming one mole of ATP.

c. Pentose Phosphate Pathway

In most bacteria, the D-xylose conversion proceeds via direct isomerization to D-xylulose catalyzed by xylose isomerase without any cofactor. In yeast and most filamentous fungi, D-xylose is first reduced to xylitol by xylose reductase, and the resulting xyitol is subsequently oxidized to D-xylulose by xylitol dehydrogenase. The difference in cofactor preference of xylose reductase (NADPH specific) and xylitol dehydrogenase (NAD specific) leads to redox imbalance and formation of xylitol.

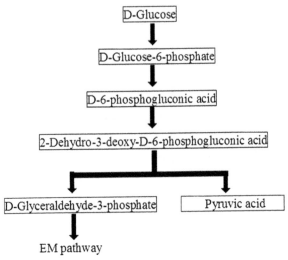

FIGURE 10.2   The Entner–Doudoroff pathway.

D-Xylulose is converted to D-xylulose-5-phosphate by xylulokinase with the consumption of one mole of ATP. D-Xylulose-5-phosphate is dimerized to D-ribulose-5-phosphate, which then isomerizes to D-ribose-5-phosphate. D-Xylulose-5-phosphate and D-ribose-5-phosphate are converted to a 7-carbon sugar, sedoheptulose-7-phosphate, and a 3-carbon sugar, D-glyceraldehyde-3-phosphate, by transketolase. The resulting 7-carbon sugar and 3-carbon sugar are converted to a 4-carbon sugar, erythrose-4-phosphate, and D-fructose-6-phosphate through the catalysis by transaldolase. Erythrose-4-phosphate and D-xylulose-5-phosphate form D-glyceraldehyde-3-phosphate and D-fructose-6-phosphate via a transketolase-catalyzed reaction. Thus, three moles of xylose is converted to two moles of D-fructose-6-phosphate and one mole of D-glyceraldehyde-3-phosphate; the latter are glycolytic intermediates and enter into the EM pathway (Figure 10.3). As a result, one mole of xylose is converted to 5/3 mole of ethanol.

10.2. ETHANOL-PRODUCING MICROORGANISMS

10.2.1. Yeast

a. Saccharomyces cerevisiae

Saccharomyces cerevisiae was first isolated in 1883 by Emil C. Hansen; it belongs to ascomycota and multiplies by budding. *Saccharomyces cerevisiae* converts glucose, fructose, galactose, maltose, and sucrose into ethanol but is not capable of converting pentoses, lactose, and cellobiose. It produces two moles of ethanol and two moles of carbon dioxide from one mole of glucose

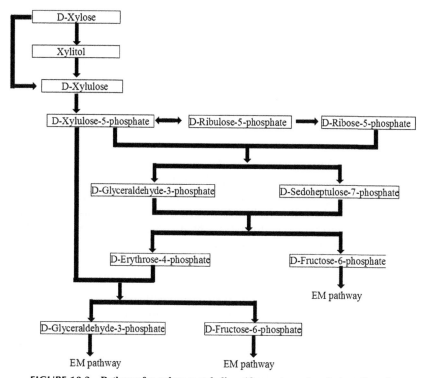

FIGURE 10.3   Pathway for xylose metabolism (the pentose phosphate pathway).

through the EM pathway. *Saccharomyces cerevisiae* is among the best known safe microorganisms, and is therefore ideal for producing alcoholic beverages such as wine, beer and Japanese sake, and for leavening.

Saccharomyces cerevisiae has the following beneficial characteristics for ethanol fermentation:

1. It is highly resistant to ethanol and toxic substances compared with bacteria, and is capable of fermenting sugars at low pH values so that the contamination risk is greatly minimized.
2. It can efficiently produce ethanol at high concentrations of ethanol from glucose and sucrose. In Japanese sake manufacturing from boiled rice, over 20% (v/v) ethanol is obtained by fermentation.
3. It has good performance for the production of beverage and bioethanol as well as for use as leavening, because each of these manufacturing processes depends on various characteristic strains.
4. It is GRAS (Generally Recognized As Safe) as a food additive.
5. It is suitable for conducting genetic engineering research and implementation because its whole genome has been elucidated.

Native strains of *S. cerevisiae* are incapable of utilizing pentoses such as xylose and arabinose, which are abundant in lignocellulosic biomass. For bioethanol production from lignocellulosic biomass, various genetic engineering approaches have been evaluated to efficiently convert pentoses to ethanol (Laluce et al., 2012). These include the expression of heterologous genes encoding xylose reductase, xylitol dehydrogenase and xylose isomerase, compensation of the imbalance of redox cofactors, and improvement of specific pentose transporter systems.

b. Schizosaccharomyces pombe

Schizosaccharomyces pombe was isolated from beer in 1893. It is a fission yeast that differentiated from budding yeasts one billion years ago in evolutionary history. In 2002, the entire nucleotide sequence of *S. pombe* was determined. *Schizosaccharomyces pombe* tolerates a high concentration of salts, but has lower ethanol resistance than *S. cerevisiae*.

c. Kluveromyces lactis *and* Kluveromyces marxianus

Kluveromyces lactis and *K. marxianus* are lactose fermenting yeasts and exhibit the activities of lactose permease and β-galactosidase. Lactose is hydrolyzed to glucose and galactose and the resulting sugars are converted to ethanol via the EM pathway. In Ireland and New Zealand, ethanol for beverages, medical and industrial uses, as well as fuels, are produced from whey by fermentation using *K. lactis* (Guimarães et al., 2010).

Kluveromyces marxianus is a thermotolerant yeast that is capable of fermenting various sugars including glucose, mannose, galactose, and xylose at 40°C or 30°C. The capability of xylose fermentation is lowered with more glucose repression at higher temperatures as compared with fermentations at 30°C. Because *K. marxianus* has thermotolerant characteristics, various genetic engineering research has been conducted for fermenting various sugars efficiently at relatively high temperatures (Rodrussamee et al., 2011).

d. Pentose-Fermenting Yeast

In the case of bioethanol production from lignocellulosic biomass, fermenting pentoses efficiently is as important as fermenting hexose. Many pentose-fermenting yeasts have been isolated; these include *Pichia stipitis, P. segobiensis, Candida shehatae, Pachysolen tannophilus,* and *Hansenula polymorpha.*

Pichia stipitis, a respiratory yeast, was isolated from decaying wood and larvae of wood-inhabiting insects. It is capable of fermenting glucose, xylose, galactose, and cellobiose under anaerobic conditions. *Pichia stipitis* has the highest native capacity of xylose fermentation among known microbes, but glucose inhibits xylose transport non-competitively. Additionally, *P. stipitis* has less ethanol resistance than *S. cerevisiae*, and controlling the dissolved oxygen at low levels is of great importance for high ethanol yield using *P. stipites* (Agbogbo and Coward-Kelly, 2008).

Hansenula polymorpha is a thermotolerant xylose fermenting yeast. It grows prolifically at 37°C and survives at temperatures of up to 48°C. While xylose fermentation by *H. polymorpha* has been studied in various metabolic engineering fields, its ethanol production rate has not yet been improved.

10.2.2. Bacteria

a. Zymomonas mobilis

Zymomonas mobilis is a Gram-negative, facultative anaerobic bacterium that was isolated from the alcoholic beverage Mexican pulque in 1928. In Mexico, the distilled spirit tequila is traditionally made from fermentation of juices from the agave plant using *Z. mobilis*. The microorganism degrades sugar to pyruvic acid via the ED pathway, and pyruvic acid is then converted to ethanol and carbon dioxide. It is GRAS as a food additive. The ED pathway yields only half as much ATP from glucose, as does the EM pathway. This low yield of ATP results in a small yield of cell mass; hence, *Z. mobilis* has higher yield per unit cell mass than *S. cerevisiae* (Weber et al., 2010). *Zymomonas mobilis* is capable of fermenting glucose, fructose, and sucrose but not other forms of sugars. Because pentoses such as xylose and arabinose, which are abundantly contained in lignocellulosic biomass, cannot be effectively fermented by *Z. mobilis*, various catabolic genes have been transferred into *Z. mobilis* to broaden the types of sugars that can be effectively fermented by *Z. mobilis* to produce bioethanol.

b. Zymobacter palmae

Zymobacter palmae, a Gram-negative, facultative anaerobic bacterium, was isolated from palm sap in Okinawa prefecture, Japan. It can ferment glucose, fructose, galactose, mannose, sucrose, maltose, melibiose, raffinose, mannitol, and sorbitol. To broaden the range of its fermentable sugar substrates, including the pentose sugar xylose, *Escherichia coli* genes encoded with the xylose catabolic enzymes such as xylose isomerase, xylulokinase, transaldolase and transketolase, have been introduced into *Z. palmae*, where their expression is driven by the *Z. mobilis* glyceraldehyde-3-phosphate dehydrogenase promoter. This new strain produces 91% of the theoretical yield of ethanol from xylose (Yanase et al., 2007).

c. Thermophilic Bacteria

Thermophilic bacteria generally possess unique thermostable enzymes for efficient hydrolyses of biomass to ferment a broad range of carbohydrates including pentoses into ethanol (Chang and Yao, 2011). Because thermophilic bacteria are capable of fermenting both hexoses and pentoses, they are suitable for producing bioethanol from lignocellulosic biomass. However, they have a much lower ethanol production rate than *S. cerevisiae*, and they exhibit low tolerance to ethanol and toxic substances. Therefore, industrial application of thermophilic bacteria is still very limited.

Clostridium thermocellum has cellulosome, and it converts celluloses and hemicelluloses into ethanol directly at high temperature (60–70°C), which reduces the risk of contamination. However, its ethanol tolerance is as low as 2% (v/v). *Thermoanaerobacterium saccharolyticum* grows at 45–65°C; it is capable of fermenting hemicellulose and xylan directly. There are many other thermophilic bacteria that have been reported to utilize lignocellulosic biomass. These include *Geobacillus thermoglucosidasius*, *Thermoanaerobacterium saccharolyticum*, *Thermoanaerobacterium ethanolicum*, *Thermoanaerobacterium pseudethanolicus*, and *Thermoanaerobacter brockii*.

d. Clostridium phytofermentans

Clostridium phytofermentans is a mesophilic, anaerobic Gram-positive bacterium. It is capable of fermenting almost all types of carbohydrates contained in lignocellulosic biomass, including hexoses, pentoses, oligosaccharides, and polysaccharides, to produce ethanol and hydrogen as the major metabolic end products (Warnick et al., 2002).

e. Escherichia coli

Escherichia coli has a near complete assimilating pathway for utilization of hexoses and pentoses; they are used as a host microorganism to improve bioethanol production by genetic engineers. *Zymomonas mobilis* genes for pyruvate decarboxylase and alcohol dehydrogenase are integrated into the *E. coli* chromosome (Ohta et al., 1991). The modified *E. coli*, known as strain KO11, is used for industrial ethanol production from waste woods. Disadvantages of the modified *E. coli* strain include its low resistance to ethanol and incapability of surviving beyond a narrow neutral pH range of 6.0–8.0.

f. Corynebacterium glutamicum

Corynebacterium glutamicum is well known as an industrial organism for the production of amino acids. A genetically engineered strain is constructed by introducing pyruvate decarboxylase and alcohol dehydrogenase genes from *Z. mobilis*. Ethanol production by this modified strain occurs in the absence of cell growth and thus is not affected by the presence of toxic substances such as furfural, 5-hydroxymethyl-2-furaldehyde, and 4-hydroxybenzaldehyde, so that the ethanol production is proportional to cell density (Inui et al., 2004).

10.3. METHODS OF ETHANOL FERMENTATION

10.3.1. Raw Materials for Ethanol Fermentation

a. Saccharides (Sucrose, Glucose, Fructose, and Lactose)

The saccharides sucrose, glucose, and fructose are converted to ethanol by using a simple process that does not involve the saccharification process so that this process is especially useful for starch-based raw materials. These saccharides are contained abundantly in crystallized sugar cane juice, beet juice, and molasses. Because sugar cane juice and beet juice cannot be stored for more than a few days, their use is seasonally limited. Additionally, because these raw materials do not contain insoluble substances, they are suitable for cell-recycle fermentation as currently implemented in Brazil to produce most bioethanol from sugar cane juice and molasses.

Lactose, β-D-galactopyranosyl-(1 → 4)-D-glucose, is a disaccharide found most notably in milk. Lactose fermentation occurs in two steps: (1) lactose is hydrolyzed to glucose and galactose by β-galactosidase, and (2) the resulting sugars are converted to ethanol via the EM pathway. Whey is a by-product during the manufacture of cheese and casein; it contains about 4% lactose that can be used as a raw material for ethanol production. In New Zealand and Ireland, industrial ethanol production from whey started in the 1970s, and fuel ethanol production has been in operation since 2005 (Guimarães et al., 2010).

b. Starch

Starchy raw materials such as corn, cassava, sweet potato, rice, and wheat have been used to produce beverages since ancient times. These raw materials are liquefied by α-amylase and then hydrolyzed to glucose by glucoamylase.

In the USA, bioethanol has been produced using corn by means of wet or dry milling, and nowadays dry milling is the preferred method of processing (Figure 10.4). In this process, ground corn kernels are blended with water and α-amylase to hydrolyze starch into smaller sugar chains at 90–110°C. These fragments are saccharified to glucose by glucoamylase at 50–60°C, and the resulting glucose is converted to ethanol by S. cerevisiae at 30°C.

c. Lignocellulosic Biomass

Lignocellulosic biomass refers to agricultural residues such as rice straw, wheat straw, corn stover, bagasse, and plant residues. These biomasses that are the most abundant sustainable raw materials worldwide and are attractive feedstock for bioethanol production because their use as raw materials for bioenergy does not deplete sources of food and animal feed.

Ethanol production from these raw materials is more complicated than that from starchy raw material. Lignocellulosic biomass must be hydrolyzed to fermentable monosaccharide, hexoses and pentoses, but there are problems

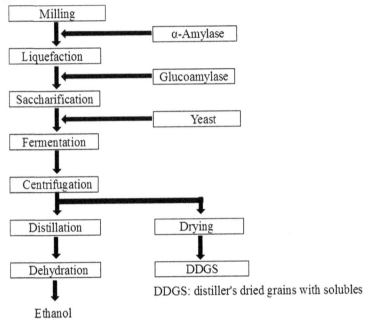

FIGURE 10.4 A flow sheet of dry-milling corn-based bioethanol production.

with this process. Cellulose and hemicellulose are densely packed by layers of lignin that protect cellulose and hemicellulose against enzymatic hydrolyses. Hence, breaking the lignin layers to expose cellulose and hemicellulose by decreasing the crystallinity of cellulose, increasing biomass surface area, removing hemicellulose, and breaking the lignin seal to facilitate subsequent enzyme action is a necessary but expensive pretreatment. Hence, an efficient, rapid, and complete enzymatic hydrolysis to pretreat the biomass is one of the major technical and economical bottlenecks in the overall bioconversion process of lignocellulose to bioethanol. There is another concern in the production of bioethanol from lignocellulosic biomass; a microorganism to produce ethanol efficiently from the second abundant sugar, xylose, has not yet been discovered. Therefore, much genetic engineering research has been devoted to finding microorganisms that are able to ferment xylose efficiently.

10.3.2. Fermentation Technology

a. Batch Fermentation

In batch fermentation, substrate and microorganism are loaded into the fermenter batchwise, and this is the most popular and simple method for ethanol production. Batch fermentation has the advantages of low investment costs, simple control and operations, and easy-to-maintain complete sterilization. However, seed culture is needed for each new batch. Bioethanol from corn in the USA is almost entirely produced using batch fermentation.

A higher initial sugar concentration is required in order to achieve more efficient ethanol production; however, a high sugar concentration will inhibit the growth and function of fermenting microorganisms due to excessive osmosis to result in a low fermentation yield with a prolonged fermentation period. The batchwise fermentation method will alleviate this drawback. Fermentation is started with a low initial sugar concentration to allow robust growth of the microorganisms at an early stage, and sugar is added periodically when it is consumed.

b. Semi-Continuous Fermentation

(i) Melle–Boinot Method

In semi-continuous processes, a portion of fermented broth is withdrawn at intervals and fresh medium is added to the system. There is no need to inoculate the reactor except during the initial startup.

The Melle–Boinot method is the most common fermentation process in Brazil to produce about 85% ethanol from cane juice or cane molasses. The initial processing stage for bioethanol production is basically the same as that for sugar production (Figure 10.5). Fermentation is carried out with sugar cane juice, water-diluted molasses, or mixtures of juice and molasses within 6–12 hours. At the end of fermentation, the ethanol concentration reaches 7–11% (v/v). The fermented broth is centrifuged to separate yeasts from liquid

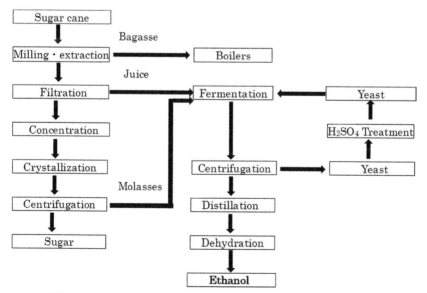

FIGURE 10.5 Semi-continuous fermentation (the Melle–Boinot method).

containing ethanol. The liquid is distilled to yield ethanol whereas the recovered yeast cells are treated with sulfuric acid (pH 2.0–2.5) to prevent bacterial contamination. After 2–3 hours of acid treatment, the yeast cells are returned to fermentation tanks for the next round of fermentation. Yeast cells are recycled 400–600 times during the sugar cane harvest season. However, there is a high risk of contamination and mutation of the yeast during the long-term cultivation and periodic handling (Amorim et al., 2011).

(ii) Ethanol Fermentation by Flocculent Yeast

The use of flocculent yeast reduces the cost of recovering yeast cells because flocculent cells are easily separated from medium without centrifugation during ethanol production in batch or fed-batch fermentation. The fermented broth is left quiescent without agitation for a short time, and yeast cells agglomerate to one another to form large flocs that settle rapidly to the bottom of the fermenter. After 70–90% volume of fermented broth is withdrawn, fresh sugar solution is added to the fermenter to start the next round of fermentation.

Flocculent yeast fermentation has a major advantage in that efficient ethanol production is achieved with a short fermentation period because a high density of yeast culture can be accumulated in the fermenter. The recovery of yeast cells is easier than the recovery of non-flocculent microbial mass from semi-batch fermentation such as the Melle–Boinot method, and the process is not as complicated as the yeast immobilization method. The mechanism of yeast flocculation, which is controlled by cell wall components, is very complicated. Flocculation is affected by many factors such as morphology of cells, ethanol concentration, nutrients, dissolved oxygen, pH, metals, and fermentation temperature, among many others.

c. Continuous-Flow Fermentation

(i) Cell Recycling

In a continuous-flow process, sugar solution is continuously fed to a fermenter while the fermenter content is simultaneously withdrawn. The withdrawn fermented broth is fed to a centrifuge to be separated from the solution; the separated yeast cells are returned to the fermenter, and the liquid is fed to a distillation process to recover ethanol. Most of the ethanol-depleted fermented broth is pumped back to the fermenter to reduce water consumption. Because fermentation and distillation are conducted simultaneously to keep sugar concentrations in the fermenter at a low level, inhibition of ethanol production due to high sugar concentrations is alleviated in this method.

(ii) Immobilized Microbial Cells

Microbial cells are immobilized by mixing them with a polysaccharide such as sodium alginate or κ-carrageenan. The mixture is dropped into salt solutions to form gel particles. Trace dissolved oxygen in the solution will enable the microorganisms to grow prolifically near the surface of the gel particles. Therefore, the overall cell density in the reactor during fermentation becomes about 10 times higher than that in batch fermentation. As a result, the fermentation period can be shortened and the productivity of ethanol becomes highly efficient. Advantages of this method are smaller reactor size, greater feasibility of continuous processing, and shorter startup time. The reactor behaves as a fluidized bed that is agitated by the carbon dioxide produced during fermentation in the reactor. However, there are risks of contamination and poor activities for the microbial cells trapped in the gel.

10.3.3. Fermentation Process

a. Separate Hydrolysis and Fermentation (SHF)

SHF is a method by which enzymatic hydrolysis and fermentation are performed sequentially. In this process, enzymatic saccharification of starchy biomass or pretreated lignocellulosic biomass is carried out first at the optimal temperature of the saccharifying enzyme. Subsequently, appropriate microorganisms are added to ferment the saccharified solution. In the SHF process, the temperatures of the enzymatic hydrolysis and fermentation can be optimized independently. Because enzymatic hydrolyses are performed at optimal temperature, this process requires a smaller quantity of saccharifying enzymes than the simultaneous saccharification and fermentation process.

Additionally, because saccharified solution containing fermentable sugar can be sterilized, the risk of contamination is reduced. However, the SHF process is carried out in two separate processes that use two independent reactors for the saccharification and fermentation processes; the capital cost is therefore higher than that for the simultaneous process.

b. Simultaneous Saccharification and Fermentation (SSF)

SSF is a method by which enzymatic hydrolysis and fermentation are performed simultaneously in the same reactor. The combination of saccharification and fermentation can decrease the number of vessels needed and thereby reduce the initial costs. Using the SSF process eliminates the inhibition of saccharifying enzyme by sugars because the resulting sugars are immediately converted to ethanol by fermentation microorganisms (see Box 10.1). However, the SSF process has disadvantages when compared with the SHF process. The optimum temperature for yeast fermentation is typically lower than that for enzymatic hydrolysis. Therefore, the saccharification process requires more enzyme than the SHF process. Thermotolerant microorganisms have been developed to ferment sugars at an elevated temperature that is optimal for saccharification. In the case of ethanol production using lignocellulosic biomass as raw material, recycling yeast is very difficult because the fermented broth contains many insoluble solids.

BOX 10.1 Bioethanol Production from Forage Rice "Leaf Star" using an SSF Process

The forage cultivar "Leaf Star", which is a rice plant bred for harvesting the whole crop silage, has high lodging resistance and biomass productivity. It is suitable as a raw material for producing bioethanol because the rice straw of Leaf Star contains large amounts of starch and saccharides such as sucrose and glucose (Table 10.1). These carbohydrates are easily converted to ethanol without complicated pretreatment prior to the fermentation of lignocellulosic raw material. The following paragraphs present the results of ethanol production from rice straw or the whole crop of Leaf Star using an SSF process.

Rice straw or the whole crop that had been milled to small size was mixed with water and α-amylase, and the mixture was held for 1 hour at 90°C to liquefy the starch. The temperature was then reduced to 30°C to start the fermentation process with simultaneous addition of glucoamylase and yeast. The results show that 135 L of ethanol and 580 kg of distillation residues can be obtained from one ton of rice straw, whereas 242 L of ethanol and 550 kg of distillation residues can be obtained from one ton of the whole crop (Table 10.2). These distillation residues are more valuable as a nutrient than the original raw materials.

TABLE 10.1 Composition of the Forage Cultivar "Leaf Star" Biomass

Composition	Biomass			
	Rice straw	Rice hull	Brown rice	Whole rice
Starch (%)	12.2	5.6	80.9	27.1
Sucrose (%)	8.5	0.2	0	6.7
Glucose (%)	2.1	0.1	0	1.6
Fructose (%)	2.1	0.1	0	1.6
Cellulose (%)	21.8	26.8	0	7.6
Xylan (%)	9.5	13.3	0	7.6
Lignin (%)	10.8	–	0	8.5
Ash (%)	16.4	25.7	1.6	13.6

To increase ethanol yield from rice straw, both the lignocellulosic fraction of rice straw and easily fermentable saccharides (EFSs) such as starch and sucrose must be used. The EFSs were recovered from the liquefaction process by centrifugation, and the solid portion treated with 1.0% NaOH for 1 hour at 100°C. The NaOH-pretreated solids were recovered by centrifugation and then washed with water. The liquid containing EFSs was added to the NaOH-pretreated solids, and fermentation was again started with the simultaneous addition of glucoamylase, cellulase, and yeast at 30°C.

BOX 10.1 Bioethanol Production from Forage Rice "Leaf Star" using an SSF Process—cont'd

TABLE 10.2 Summary of the Ethanol Production from the Forage Cultivar "Leaf Star" by SSF

	Biomass			
			Rice straw	Whole crop
			NaOH	NaOH
	Rice straw	Whole crop	treatment	treatment
Ethanol (L per t-raw material)	135	242	244	309
Residue (kg per t-raw material)	580	550	200	180

As a result, 244 L of ethanol and 200 kg of distillation residues can be obtained from one ton of rice straw, and 309 L of ethanol and 180 kg of distillation residues can be obtained from one ton of whole crop (Table 10.2). As these distillation residues have high ash content (about 45%), their nutrient values are considered to be low.

Simultaneous saccharification and co-fermentation (SSCF) of hexoses and pentoses is a process similar to the SSF process except that hexose and pentose fermentation occurs simultaneously. SSCF offers the potential of streamlined processing while reducing capital costs.

c. Consolidated Bioprocessing (CBP)

In the CBP process, enzymetic hydrolysis and fermentation take place in one vessel using a single species of microorganism or a co-culture of microorganisms. The CBP process is simple, and therefore both operation and capital costs associated with enzyme production are reduced.

Naturally occurring microorganisms are incapable of producing saccharolytic enzymes to convert the released sugars into ethanol simultaneously. Hence, engineered microorganisms need to be developed in order to make this process suitable for industrial applications.

REFERENCES

Agbogbo, F. K., & Coward-Kelly, G. (2008). Cellulosic ethanol production using the naturally occurring xylose-fermenting yeast *Pichia stipitis. Biotechnology letters, 30*, 1515–1524.

Amorim, H. V., Lopes, M. L., Castro Oliveira, J. V., Buckeridge, M. S., & Goldman, G. H. (2011). Scientific challenges of bioethanol production in Brazil. *Applied microbiology and biotechnology, 91*, 1267–1275.

Chang, T., & Yao, S. (2011). Thermophilic, lignocellulolytic bacteria for ethanol production: Current state and perspectives. *Applied microbiology and biotechnology, 92*, 13–17.

Guimarães, P. M. R., Teixeira, J. A., & Domingues, L. (2010). Fermentaion of lactose to bio-ethanol by yeasts as part of integrated solution for the valorization of cheese whey. *Biotechnology advances, 28*, 375–384.

Inui, M., Kawaguchi, H., Murakami, S., Vertès, A. A., & Yukawa, H. (2004). Metabolic engineering of *Corynebacterium glutamicum* for fuel ethanol production under oxygen-deprivation conditions. *Journal of molecular microbiology and biotechnology, 8*, 243–254.

Laluce, C., Schenberg, A. C. G., Gallard, J. C. M., Coradello, L. F. C., & Pombeiro-Sponchiado, S. R. (2012). Advances and developments in strategies to improve strains of *Saccharomyces cerevisiae* and processes to obtain the lignocellulosic ethanol – A review. *Applied biochemistry and biotechnology, 166*, 1908–1926.

Ohta, K., Beall, D. S., Mejia, J. P., Shanmugam, K. T., & Ingram, L. O. (1991). Genetic improvement of *Escherichia coli* for ethanol production: Chromosomal integration of *Zymomonas mobilis* genes encoding pyruvate decarboxylase and alcohol dehydrogenase II. *Applied and environmental microbiology, 57*, 893–900.

Rodrussamee, N., Lertwattanasakul, N., Hirata, K., Suprayogi, Limtong, S., Kosaka, T., & Yamada, M. (2011). Growth and ethanol fermentation ability on hexose and pentose sugars and glucose effect under various conditions in thermotolerant yeast *Kluveromyces marxianus*. *Applied microbiology and biotechnology, 90*, 1573–1586.

Warnick, T. A., Methe, B. A., & Leschine, S. B. (2002). 11.2.2.4 *Clostridium phytofermentans* sp.nov., a cellulolytic mesophile from forest soil. *International journal of systematic and evolutionary microbiology, 52*, 1155–1160.

Weber, C., Farwick, A., Benisch, F., Brat, D., Dietz, H., Subtil, T., & Boles, E. (2010). Trends and challenges in the microbial production of lignocellulosic bioethanol fuels. *Applied microbiology and biotechnology, 87*, 1303–1315.

Yanase, H., Sato, D., Yamamoto, K., Matsuda, S., Yamamoto, S., & Okamoto, K. (2007). Genetic engineering of *Zymobacter palmae* for production of ethanol from xylose. *Applied and environmental microbiology, 73*, 2592–2599.

Co-Generation by Ethanol Fuel

Hideo Kameyama

11.1. FUEL CELL: INNOVATIVE TECHNOLOGY FOR GENERATING POWER AND HEAT

11.1.1. What is a Fuel Cell?

The fuel cell converts fuel and air directly into energy through quiet, efficient, solid-state electrochemical reactions. It generates power more efficiently than internal combustion engines (ICEs) because it converts chemical energy directly into electrical current rather than via an inefficient mechanical intermediate phase. When loaded partially, fuel cells can be operated at maximum efficiency whereas most ICE generators are very ineffective, and the efficiency of fuel cells is largely unaffected by cell size. In addition, the modular design allows fuel cells to be stacked with capacities to match the specific output power needed.

The simple nature of their design and operation makes fuel cells highly reliable. If pure hydrogen is used as the input fuel, the outputs are electricity, heat, and harmless water vapor.

Research Approaches to Sustainable Biomass Systems. http://dx.doi.org/10.1016/B978-0-12-404609-2.00011-8

11.1.2. Domestic Fuel Cells: Polymer Electrolyte Membrane Fuel Cells (PEMFCs)

There are more than a dozen distinct fuel cell technologies currently developed by academia and industries, as shown in Table 11.1; however, only a select few are suitable for domestic micro-CHP (combined heat and power) applications. The fuel cell stack must have at least the potential of being low cost in production and having a long operating lifetime under suboptimal conditions, particularly with regard to impurities in the hydrogen fuel. There are also concerns about the safety and practicality of using pressurized hydrogen vessels in fuel cells in addition to cost-effectiveness because of the high initial and operating costs.

The PEMFC type of cell is suitable for stationary and mobile applications. The high power density of the PEMFC enables a compact design, thus allowing

TABLE 11.1 General Operating Characteristics of Each Fuel Cell Technology

	PEMFC (proton exchange membrane fuel cell)	SOFC (solid oxide fuel cell)	PAFC (phosphoric acid fuel cell)	AFC (alkaline fuel cell)
Electrodes	Pt, Ru, C, PTFE	Ni, LSM	Pt, C, PTFE	Pt or Ni, C, PTFE
Electrolyte	Solid polymer (PFSA)	Ceramics: YSZ, LSM	Liquid H_2SO_4	Liquid KOH
Interconnect	Graphite, steels	Chromium alloys, steels	Graphite	Graphite, metal or plastic
Operating temperature (°C)	30–100	500–1000	200–250	50–200
Fuels	H_2	H_2, CO	H_2	H_2
Fuel tolerance Sulfur	<0.1 ppm	<1 ppm	<50 ppm	?
CO	<10–100 ppm	Fuel	<0.5–1%	<0.2%
CO_2	Diluent	Diluent	Diluent	<100–400 ppm or <0.5–5%
CH_4	Diluent	Fuel/diluent	Diluent	Diluent
NH_3	Poison	<0.5%	<4%	?

reduced volume and weight as compared with other alternatives. The PEMFC is operated with hydrogen and air as the inputs; the hydrogen is derived from natural gas in an upstream process known as reforming. As in a gas-fired boiler, the fuel cell heating equipment requires a gas supply pipe, combustion ventilation, and a flue pipe. PEM cells reach a constant operating temperature of approx. 90°C in only a few minutes and the heat generated can be utilized for central heating and domestic hot water (DHW) via a heat exchanger.

Domestic CHP systems based on PEMFCs have undergone intense research and commercial development over the last decade. There are at least a dozen major companies actively pursuing this market, and products have been deployed in large-scale field trials throughout Japan, South Korea, and Germany.

11.1.3. Commercialization Requirements for the Reformer

Although domestic fuel cells have been commercialized to some extent in Japan since 2005, more than 40,000 pieces of PEMFC domestic systems were sold in 2006. Reformer durability has already been confirmed to be 40,000 h (startup and shutdown, 4000 times). Nevertheless, some effort should be made to improve the performance and reduce the cost of PEMFC systems in order to meet the expected popularity of domestic fuel cells in Japan and worldwide, as shown in Table 11.2.

The minimum requirements of future PEMFC systems should include low cost, long durability and reliability, and quick startup.

a. Low Cost

There is considerable competition between domestic fuel cells and traditional cost-effective energy facilities. Present fuel cells cost as much as 3 million yen

TABLE 11.2 Roadmap for the Domestic PEMFC System in Japan

	Now (2010)	Initial popularity (2015)	Expanding popularity (2020)	Real popularity (2030)
Effect (%)	33–37	33–37	33–37	36–40
Durability (h) (startup and shutdown)	40,000	60,000 (4000 times)	90,000 (4000 times)	90,000 (4000 times)
Highest work temperature (°C)	70	90	90	90
Cost of the system (million yen)	200–250	50–70	40–50	<40

because they are made manually. The fuel cell commercialization conference of Japan (FCCJ) predicted that mass production of fuel cells with 100,000 or more sets would greatly lower the production cost. It has been reported that fuel cells would be applied extensively in the urban regions of the USA if the cost of fuel cells could be reduced to the same level as that of traditional energy.

b. Long Durability and Reliability

To reach the 2020 final goal a fuel processor should suffer no degradation in its normal performance for 10 years, or it should be operated to exceed 90,000 hours with more than 4000 startup and shutdown cycles without changing any catalysts, including that for desulferizer.

c. Quick Startup

For residential fuel cell utilization, the startup time presents one of the many problems that affect the performance of fuel cells. Until now, the startup time of a domestic fuel cell system has been reported to be more than 50 min, and the reformer plays an important role in reducing the startup time.

11.2. HYDROGEN PRODUCTION THROUGH ETHANOL STEAM REFORMING

11.2.1. Introduction

Within the past few decades, hydrogen has been considered as the least polluting fuel that has the potential to be used in internal combustion engines or fuel cells for electricity generation. Natural gas is still the main source of hydrogen due to its abundance. However, developing alternative methods or using renewable sources to produce hydrogen is of great importance based on economic and environmental considerations. In this context, a low concentration of about 30% bioethanol, which has been derived from biomass, is used to produce hydrogen through the ethanol steam reforming reaction (Figure 11.1).

Nickel-impregnated alumina is suggested as the most suitable metal catalyst for ethanol steam reforming because of its high activity and low cost. However, the main problem associated with using Ni catalysts is that the catalyst is easy to deactivate by sintering and carbon deposition or coking. With these

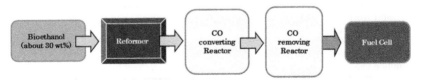

FIGURE 11.1　Hydrogen production and fuel cell system.

concerns, the objective of the present research is to study the effect of adding Ce on Ni catalysts to inhibit sintering and coking.

The steam reforming of ethanol can be represented as:

$$CH_2H_5OH + 3H_2O \rightarrow 2CO_2 + 6H_2 \qquad (11.1)$$

This research is unique in that it uses an alumite catalyst, which was patented by Kameyama (1996) with superior heat transfer as well as flexibility in shape and structure. This catalyst is supported by an aluminum-clad-based material made of nickel–chromium alloy with super pure aluminum compressed on both surfaces of the alloy. Laboratory studies verified that the catalyst surface temperature reaches 800°C in a few seconds by passing electricity through the base metal (Tran et al., 2004), and this property helps to solve the problem of long startup times. Results of many other studies have suggested Ni as the most suitable metal for ethanol steam reforming using metals with the following order of performance: Rh > Ni, Co > Pt >> Cu.

Although γ-Al$_2$O$_3$-supported Rh shows the best performance (Cavallaro, 2000; Cavallaro et al., 2003), Ni has been studied further because of its low cost, availability, and industrial applications (Haryanto et al., 2005); it has been proven to be the most suitable metal for ethanol steam reforming. The main by-products over the Ni/γ-Al$_2$O$_3$ catalyst are CH_4, CO and C_2H_4, and C_2H_4 is formed through the dehydration of ethanol over the acid sites of alumite supports (Breen et al., 2002; Fierro et al., 2005):

$$C_2H_5OH \rightarrow C_2H_4 + H_2O \qquad (11.2)$$

In order to suppress the acidity of alumite supports, silica is coated over anodized aluminum plates, and this method has also been applied to particle-type catalysts (Zhang et al., 2009). As a result, the formation of C_2H_4 is reduced remarkably with increasing selectivity of the ethanol steam reforming reaction.

11.2.2. Experimental

a. Catalyst Preparation

Al-clad plates were anodized in 4 wt% oxalic acid solution with an electric current density of 65 A m^{-2} at 20°C for 6 hours to form porous alumina films (ca. 100 μm thick) on the outside surface. The anodized plates were rinsed, air dried, and then subject to pore widening treatment (PWT) by immersing the plate in an aqueous solution of oxalic acid for 2 hours. The plate was later rinsed with deionized water, dried, and calcinated at 350°C for 1 hour in open air. Subsequently, hot water treatment (HWT) of the calcinated plate in deionized water at 80°C was performed for 2 hours. Finally, γ-alumina films were formed by calcinating the plates in air at 500°C for 3 hours.

A 4 wt% colloidal silica solution (pH 1) was prepared by adding oxalic acid to commercial colloidal silica sol (SI50, Catalyst & Chemicals Ind. Co., LTD).

An anodized aluminum plate was immersed in the sol solution at 15°C for 20 hours. Then, the plate was rinsed with deionized water, dried, and calcinated at 400°C in air for 3 hours. The resulting silica-coated support was labeled SiAl.

The catalysts were prepared by using the impregnation method. The adherence of Ni on alumina-clad plates was carried out by dipping the plate in hot Ni solution, $Ni(NO_3)_2 \cdot 6H_2O$ (pH 10), with the solution temperature varied from 25 to 40°C for 1 hour, 3 hours, and 5 hours. After the catalysts were dried, the plate was calcinated at 400°C for 3 hours in air. On completion of impregnation, the resulting catalyst was labeled NiSiAl.

The amount of metal impregnated on the alumina-clad plates was measured using atomic adsorption spectrometry (AA-680, Shimadzu Corp.). The surface area of the catalysts was analyzed using a nitrogen adsorption method (BET) with an SA3100 Surface Area and Pore Size Analyzer. The surface of the catalysts after activity tests and durability tests was examined using scanning electron microscopy (SEM).

b. Activity Tests

The system used to conduct this research is shown schematically in Figure 11.2. The ethanol steam reforming reaction was carried out in a fixed-bed flow reactor at atmospheric pressure. The plate-type catalysts were cut into smaller pieces (3×3 mm), mixed with 10 g of quartz sand, and packed in a quartz tube (i.d. $= 10$ mm, L $= 333$ mm). Before the activity test, the catalyst was reduced

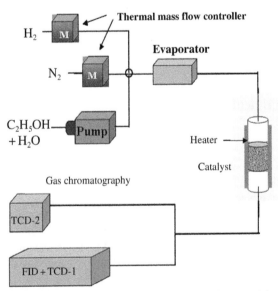

FIGURE 11.2   **Experimental apparatus for evaluating the surface reactivity of the prepared catalyst.**

in a 10% H_2/N_2 mixture (10 mL min^{-1} H_2 and 90 mL min^{-1} N_2) at 500°C for 3 hours. After the reduction treatment, the catalyst was exposed to a stream of N_2 flowing at 45 mL min^{-1} for 1 hour in order to remove the remaining H_2 from the reactor. The reaction temperature was set between 450 and 600°C at atmospheric pressure. A solution of ethanol and water in a 1:6 ratio was pre-mixed and then fed into an evaporator (170°C) using an injection pump at a flow rate of 25 μL min^{-1}. Then, the vaporized reactants were mixed with inert N_2 at a flow rate of 25 mL min^{-1} regulated by using a mass flow controller. The steam-reformed gases collected at the reactor outlet were analyzed online using a gas chromatograph equipped with two thermal conductivity detectors (TCDs) and a flame ionized detector (FID).

The ethanol conversion rate C_{EtOH} (%) and product selectivity S_i (dimensionless) were calculated using:

$$C_{EtOH} = \frac{F_{EtOH,in} - F_{EtOH,out}}{F_{EtOH,in}} \times 100 \tag{11.3}$$

$$S_i = \frac{X_i}{F_{EtOH,in} - F_{EtOH,out}}, \tag{11.4}$$

where $F_{EtOH,in}$ and $F_{EtOH,out}$ are the inlet and outlet ethanol flow rates (mol s^{-1}) respectively; X_i is the flow rate, which is expressed as the number of moles formed per mole of ethanol consumed, calculated by multiplying the product i (mol s^{-1}) and selectivity S_i for product compound i.

c. Reaction Tests

In order to evaluate the effective quantity of Ni impregnated on the supporting plate, the quantity of catalyst impregnated was measured using an inductively coupled plasma (ICP) spectrometer. The results can be used to calculate the amount of active catalyst due to Ni impregnation on the plate. The catalyst prepared by impregnating the plate for 3 hours in the 30°C solution was found to produce the largest amount of H_2 (approximately 2.7 mol per mol of ethanol) with less CO and CH_4 as compared with the results obtained using the catalysts prepared under other conditions. Accordingly, at these conditions, i.e. 30°C solution temperature and 3 h calcinating time, the resulting catalyst has the largest specific surface area that allows the greatest contact between ethanol and catalyst, so that the ethanol steam reforming reaction is significantly enhanced to speed up the production of hydrogen gas.

The NiSiAl catalyst obtained by immersing the plate in 30°C solution for 3 h was further subject to durability tests. The results show that at low space velocity (SV), ethanol conversion is 100% with no apparent degradation of the catalyst observed after 50 hours of operation (Figure 11.3).

Figure 11.4 shows that a longer cerium impregnation time leads to an increasing amount of cerium on the catalyst with a reduced amount of carbon

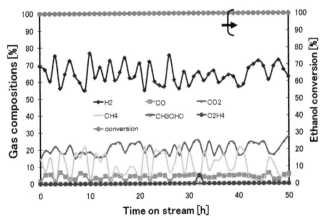

FIGURE 11.3　Relationship between gas-phase composition and ethanol conversion in durability tests at 550°C (SV = 32,000 h^{-1}, S/C = 3).

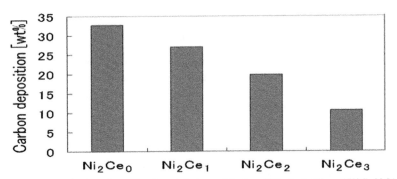

FIGURE 11.4　Carbon deposition in catalysts used in durability tests (T = 550°C, 20 h).

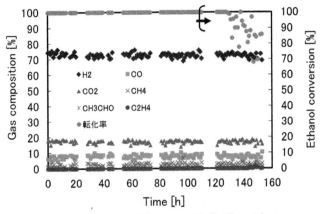

FIGURE 11.5　Durability tests with Ce$_2$Ni$_2$ catalyst.

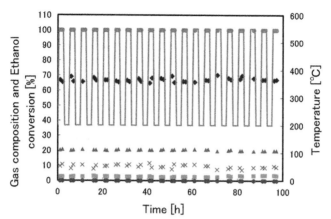

FIGURE 11.6   DSS test with PtNi₂Ce₂ catalyst.

deposition. This observation indicates that the presence of cerium inhibits carbon formation. Additionally, results of durability tests show that the Ni_2Ce_2 catalyst usage time of 85 hours is the longest. The major depletion of Ni_2Ce_3 was found to be associated with the weak $NiAl_2O_4$ inner layer that causes quicker growth of nickel particles on the surface catalyst in a process known as "sintering".

Durability tests using different levels of Ni and Ce show that Ce_2Ni_2 yields the longest useful time of 146 hours and the highest H_2 selectivity of 72.5% due to less carbon formation (about 13.9%) (Figure 11.5).

The addition of Pt to Ni_2Ce_2 catalyst leads to good results in DSS tests (Figure 11.6) with 20 startup and shutdown cycles, and the conversion of ethanol is always 100% with almost no depletion of catalyst observed.

REFERENCES

Breen, J. P., Burch, R., & Coleman, H. M. (2002). Metal-catalysed steam reforming of ethanol in the production of hydrogen for fuel cell applications. *Applied Catalysis B: Environmental, 39*(1), 65–74.

Cavallaro, S. (2000). Ethanol steam reforming on Rh/Al₂O₃ catalysts. *Energy Fuels, 14,* 6.

Cavallaro, S., Chiodo, V., Vita, A., & Freni, S. (2003). Hydrogen production by auto-thermal reforming of ethanol on Rh/Al₂O₃ catalyst. *Journal of Power Sources, 123*(1), 10–16.

Fierro, V., Akdim, O., Provendier, H., & Mirodatos, C. (2005). Ethanol oxidative steam reforming over Ni-based catalysts. *Journal of Power Sources, 145*(2), 659–666.

Haryanto, A., Fernando, S., Murali, N., & Adhikari, S. (2005). Current status of hydrogen production techniques by previous steam reforming of ethanol: A review. *Energy Fuels, 19,* 2098–2106.

Kameyama, H. (1996). *Production method of thermal conductive catalyst.* Japanese Patent No. 2,528,701.

Tran, T. P., Koyama, S., Zhang, Q., Sakurai, M., & Kameyama, H. (2004). Developing a new electrically heated catalyst supported by anodized alumina layers with high heat resistance. *Proceedings of the 10th APPChE Congress*, 1G-05, Kitakyushu, Japan.

Zhang, L., Li, W., Liu, J., Guo, C., Wang, Y., & Zhang, J. (2009). Steam reforming of ethanol over $Al_2O_3 \cdot SiO_2$-supported Ni–La catalysts. *Fuel, 88*(3), 511–518.

Chapter 12

Thermochemical Transformation of Biomass

Kenichi Yakushido, Yuichi Kobayashi and Hitoshi Kato

12.1. NEED FOR BIOMASS UTILIZATION TECHNOLOGY IN JAPAN

In the EU, the main demand for biomass energy is for heat, and biomass resources are plentiful; hence, traditional technologies can be used to obtain heat energy from biomass. In Japan, in contrast, biomass energy is used mainly for the production of electricity as well as both liquid and gaseous fuels. However, geographic conditions in Japan make it difficult to transport biomass; technologies that can accomplish high-efficiency energy transformation on a small scale are needed. In addition, the expenditures in terms of funds and energy needed for collecting and transporting the widespread and sparsely distributed biomass can be saved if a resource-circulating society is developed in Japan (Figure 12.1).

Research Approaches to Sustainable Biomass Systems. http://dx.doi.org/10.1016/B978-0-12-404609-2.00012-X

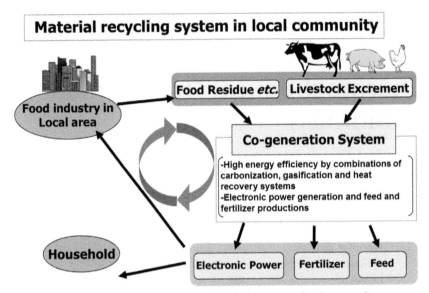

FIGURE 12.1 Promotion of low utilized biomass in a local community.

Figure 12.2 shows material flows in a full-scale model of a co-generation system that uses livestock urine and feces as well as food wastes as inputs. A herd of 1300 head of beef cattle generates 34 t of urine and feces per day. These materials can be used to generate power and heat in a co-generation system that simultaneously performs carbonization and gasification to supply 4000 kWh per day of electricity. The heat captured by the co-generation system can dry 7 t of food residues to produce 1.6 t of feed for feeding 500 pigs. Additionally, 1.7 t of carbide can be generated from the gasification operation of the co-generation system to be used as fertilizer.

12.2. SOLIDIFICATION OF BIOMASS FUEL

Through photosynthesis, plants store light energy in plant tissues that can be combusted to release the energy as heat. The combustion also generates ash, which contains various essential and trace elements. The calorific value of plant material depends on both the species and individual plants. In general, slow-growing trees contain less ash and more energy than fast-growing grasses.

Compression of biomass into pellets or briquettes concentrates the energy density by increasing the bulk mass density. The compression process facilitates storage and transportation; the uniformity of shape and quality of the compressed biomass facilitates the control of combustion, improves heat utilization efficiency, and manages contaminant emissions. Major methods

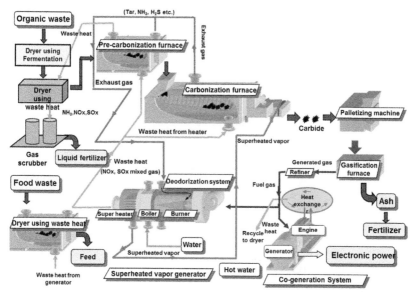

FIGURE 12.2   Multi-stage carbonization and gasification of livestock excrement in a co-generation system.

for molding pellet-form fuels include a ring-die system (Figure 12.3) and a flat-die system (Figure 12.4). In the ring-die system, a column rotating along the inner circumference extrudes pellets from the cavity. In the flat-die system, a column rolling on a disk with a cavity extrudes pellets from the top downward.

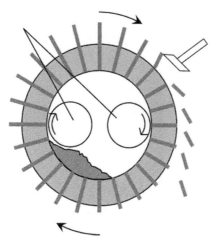

FIGURE 12.3   Ring-die system.

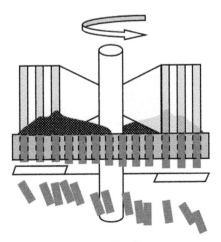

FIGURE 12.4   Flat-die system.

12.3. COMBUSTION

12.3.1. Combustion System

Combustion systems include the fluidized-bed furnace, fire-grate furnace, rotary kiln furnace, and vortex combustion furnace (JSME, 1996). Fire-grate furnaces can be divided into upflow and downflow systems. In an upflow system, the gas generated by thermal decomposition of biomass in a fire grate goes upwards to be burned. The combustion rate increases with biomass input because the heat produced by combustion decomposes the biomass efficiently. In a downflow system, gas generated by biomass combustion in a fire grate is forced downwards to burn beneath the grate (Figure 12.5). The gas can be

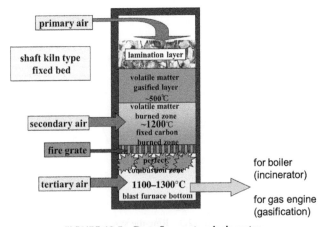

FIGURE 12.5   Downflow system incinerator.

burned at a constant rate regardless of the biomass input rate because only a small portion of the combustion heat is used for thermal decomposition of the biomass. The downflow fire-grate furnace generates extra heat by providing tertiary air. If the supply of this tertiary air is blocked, the furnace becomes an imperfect gasification furnace.

12.3.2. Problems in Biomass Combustion

Although biomass is typically burned at 600–700°C, agricultural biomass must be burned at temperatures higher than 800°C to reduce the emission of dioxins from fertilizers and chlorinated compounds derived from the soil. Unlike wood, which consists of 99.7% organic matter, agricultural biomass contains appreciable quantities of ionic compounds and silicic acid that will melt to a lava-like slurry to choke the furnace. For example, rice straw from Hokkaido can melt at temperatures exceeding 1000°C, whereas rice straws from Kyushu to Kanto cause no problem at the same temperature. In addition, wild bamboo does not melt but fertilized bamboo grown for edible shoots melts easily. The melting point of agricultural biomass depends on fertilizer application, harvest time, and cultivar. Therefore, the melting temperature of a material should be determined before it is used to assess the material's suitability for combustion.

12.4. GASIFICATION WITH METHANOL SYNTHESIS

12.4.1. Gasification Method

Organic matter decomposes at high temperature to generate combustible gases such as carbon monoxide (CO), hydrogen (H_2), and methane (CH_4) (Nishizaki, 2008) in addition to generating tar that is then decomposed to form gas at temperatures above 900°C. In traditional internal combustion, incomplete combustion occurs in a conventional furnace. In a new external combustion system, the furnace is externally heated to $\sim 600°C$ (Figure 12.6) to gasify the biomass (new technology). The temperature is further raised to $\sim 1000°C$ by adding air and oxygen in order to remove the residual tar.

In a new suspension/externally heated system developed for instant gasification, water vapor and biomass particles of ~ 3 mm size are converted to gas in a reaction tube with temperature maintained at 800–1000°C (Figure 12.7). The reaction tube that is heated by hot gas drawn from a furnace is used to convert the organic matter into high-calorific gas fuel (~ 16 MJ Nm^{-3}); the resulting gas is typically composed of 35–50% H_2, 20–30% CO, 7–15% CH_4, 1–4% ethylene (C_2H_4), and 10–20% CO_2 (Table 12.1) (Sakai, 1998, 2011). The efficiency of gasification, including external combustion heat, is 85%. In other words, 85% of the biomass energy can be converted into a gas fuel similar to household natural gas. Because there is no internal

[Conventional Technology]
Partial Combustion Gasification Using O_2 and Steam

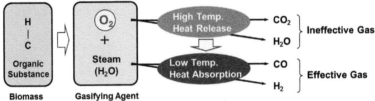

Biomass   Gasifying Agent

[New Technology]
Suspension/External Heat Type Gasification Using Steam

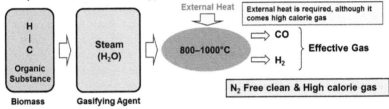

Biomass   Gasifying Agent

FIGURE 12.6   Principle of the gasification method.

combustion involved, the final product is a high-calorific gas with a high hydrogen content. The mixed gas can be used for electricity generation with a gas engine to achieve an efficiency of 15–30% for producing electricity from a few kilowatts to several hundred kilowatts. The method can also co-generate heat with the electricity.

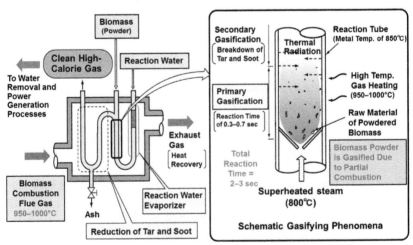

Powdered biomass is heated by an external source, and reacts with steam. Dioxin is not composed due to the absent of air.

FIGURE 12.7   Diagram of suspension/external heat-type high-calorie gasification system.

TABLE 12.1 Components of High-Calorific Gas Converted from Biomass

Gas component	Japanese cedar	Sorghum	Bagasse	Rice straw
H_2 (%)	44–52	47–51	34–44	42–47
CO (%)	14–25	24–28	24–28	24–28
CO_2 (%)	5–21	4	4–6	3–6
CH_4 (%)	7–11	9–10	9–11	9–11
C_2H_4 (%)	0–1	0	0–1	0–1
N_2 (%)	14–19	16–17	19–27	15–20
O_2 (%)	0–1	0	0–1	0

12.4.2. Methanol Synthesis

The biomass gas contains H_2 and CO as the main ingredients in a ratio of 2:1; it has attracted attention as a source for synthesizing methanol at pressure of 1–2 MPa using copper and zinc as catalysts:

$$2H_2 + CO \rightarrow CH_3OH$$

This exothermic reaction occurs at 200–250°C. When cooled to lower than 60°C, ~96% of methanol and ~4% of hydrocarbon-type light fuel and water are produced.

Figure 12.8 outlines a demonstration plant designed to use materials such as tree thinnings, sawdust, bark, rice straw, Napier grass, and sweet sorghum as

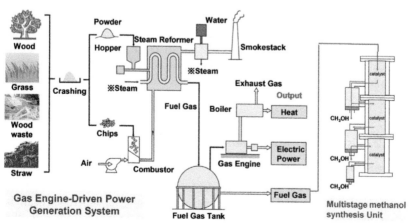

FIGURE 12.8 Schematic diagram of gas engine-driven power generation and multi-stage methanol synthesis system.

inputs to produce liquid methanol fuel. For 1 t of dried rice straw, 340 L of methanol can be produced in a multi-stage synthesizing tower operating at a pressure of less than 1 MPa. Unreacted gas is passed to the generator.

12.5. ENERGY PRODUCTION FROM LIVESTOCK WASTES

12.5.1. Introduction

Since the commencement of the Livestock Excrement Law of Japan in 1999, the production of livestock manure compost has increased rapidly. Because of oversupply, some compost is exported or is diverted to carbonization or combustion for energy.

Composting is cheaper than carbonization and combustion. When the unit cost of composting is given a value of 1, the cost of processing per unit of input is 1.7–2.0 for carbonization, 3.7 for the production of activated carbon, and 2.0 for combustion. The cost of processing per unit of output is 1.5–2.5 for carbonization, 4.3 for activated carbon, and 8.0 for combustion. Therefore, composting is the most cost-effective use of livestock manure. Other uses of livestock manure are considered only when the cost of its distribution is too high for the composting practice to be cost-effective.

Since the enforcement of the Act on Special Measures Against Dioxins in 1999, an alternative to the traditional chicken manure boiler has emerged in the form of a large-scale power generation plant fired by chicken manure. The plant, which can burn more than 100 t day^{-1}, is designed to reduce dioxin emissions to alleviate adverse environmental effects. The plant produces high-temperature, high-pressure vapor to generate electricity using turning steam turbines. Currently, such plants with a total capacity of 300 t day^{-1} are operated in the Miyazaki and Kagoshima prefectures; the plant in Miyazaki prefecture is a co-combustion system that burns cattle, swine, and chicken manure. Both prefectures have a surplus of manure, and thus a lower demand for chicken manure compost than cattle and swine manure composts. The difficulties in the distribution of broiler manure compost have favored the introduction of the plants. In addition, small- to medium-scale livestock waste boilers that can be used to heat livestock barns and to produce hot water for washing eggs have been commercialized.

12.5.2. Use of Livestock Manure by Combustion

a. Combustion Method

Various combustion methods are available. In the stoker furnace, manure is burned in a fixed fire grate. In the fluidized-bed furnace, superheated air is blown through a sand layer to fluidize the sand to be mixed with the combustible input. Although the fluidized-bed furnace consumes a great deal of power to fluidize the sand, it has been used in three plants because the temperature is easy to control. Large-scale power plants use these two systems. In the rotary kiln furnace, combustibles are burned in a cylindrical rotating furnace; the rotary kiln furnace is appropriate for small-scale plants to process animal manure for heating.

The stoker furnace that is often used to burn wood chips has been built in Miyazaki prefecture to process broiler chicken manure.

b. Considerations in Combustion of Livestock Manure

Because of its high content of potassium and sodium, manure melts easily during combustion. For example, beef cattle manure starts to melt at 900–1000°C, broiler manure melts at 1200–1250°C, and layer manure melts at more than 1350°C. Therefore, to prevent melting, the temperature of the fluidized bed must be maintained at 700–750°C, and inflammable gases are used in order to keep the furnace temperature higher than 900°C. In addition, because floating ash can melt on to the wall of the furnace and the inside of the heat-exchange pipes as clinker, the heat-exchange unit must be designed for ease of cleaning.

c. Treatment of the Exhaust

Because livestock manure is an industrial waste, it is regulated under the Act on Special Measures Against Dioxins. To reduce dioxin emissions, the furnace exhaust is burned at 800°C for more than 2 s in a secondary combustion unit and then rapidly cooled. Particles of ash contained in the exhaust are removed by using a cyclonic filter or bag filter.

The exhaust can also contain NO_x, SO_x, and HCl. Although their concentrations usually meet environmental standards, equipment for water washing and calcium hydroxide processing should be added when the capacity of the system is larger or the limits become stricter.

d. Use of Ash After Combustion

The ash of combusted manure contains phosphate, potassium, and calcium, but little or no nitrogen. It can be incorporated into fertilizer or added to compost.

12.5.3. Future Issues

A manure-fired power plant processing 300 t of manure per day can generate ~10 MW of electricity at an efficiency of 20–25%. Even when vapor is used, the total thermal efficiency is only about 50%, and the rest of the energy is lost as waste heat. The effective use of this waste heat remains to be addressed, perhaps through a heat utilization annex to power plants.

REFERENCES

Japanese Society of Mechanical Engineers (JSME). (1996). *Combustion handbook*. Tokyo: Maruzen (p. 5).

Nishizaki, J. (2008). *Creative energy engineering science*. Tokyo: Kodansha (pp. 84–102).

Sakai, M. (1998). *The 21st century energy that biomass breaks up*. Tokyo: Morikita (pp. 48–57).

Sakai, M. (2011). *Present conditions and development of suspension/external heat type high-calorie gasification system*. Tokyo: CMC (pp. 70–77).

Chapter 13

Biomass Production and Nutrient Cycling

Shoji Matsumura, Takuya Ban, Shuhei Kanda,
Aye Thida Win and Koki Toyota

Chapter Outline

Research Approaches to Sustainable Biomass Systems. http://dx.doi.org/10.1016/B978-0-12-404609-2.00013-1

13.1. CROP PRODUCTION AND CYCLING OF NUTRIENTS

Shoji Matsumura

The primary purpose of agriculture is to produce edible plant and animal products. However, some non-dietary plant fibers and wastes from dietary products such as corn stover (or cob) can be utilized as bioenergy. Figure 13.1 shows the cycling of some nutrients in the production system of food products and bioenergy. At present, bioethanol is mainly produced from corn grain and sugar cane molasses, whereas biodiesel fuel (BDF) is obtained from rape and sunflower seeds. The leftover pomace and oil cake after extraction of the liquid portion still contain fibers and energy source materials that can be consumed by ruminants like cows and sheep. Excreta from these ruminants also contain energy source materials that can be fermented by methanogenic bacteria to produce methane gas under anaerobic conditions. The liquid waste after methane fermentation contains many nutrients; it is a valuable nutrient source for growing crops. Such a system, in which both food materials and energy can be sustainably produced, is a more environmentally friendly system than the agricultural systems currently used.

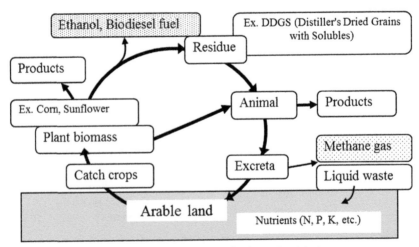

FIGURE 13.1 **Bioenergy production and nutrients.**

13.1.1. Reasonable Pathway of Nutrient Cycling

In general, crop production is mostly in open fields, with some crops grown in greenhouses. In an open field (Figure 13.2), mineral (chemical) and organic fertilizer (As), compost (manure), dissolved nutrients in rain (Rs), nitrogen fixed by microorganisms (Fxs), and the residue of precedent crops (Rc) are supplied to soil as input, whereas the outputs of nutrients include crop uptake (Us) and the loss through leaching and volatilization (Ls). Similarly, as far as the crop plant is concerned, the nutrient uptake from soil (Us) is the only input, and the outputs are harvested parts of crops for consumption by human (Pc) and domestic animals (Fa), as well as the crop waste residues (Rc). For livestock, the only input is the feed (Fa), and outputs are livestock product (Pa), manure and loss through leaching and volatilization (La).

The mass balance of the whole production system can be expressed by the following equation:

$$Pc + Pa = As + (Fxs + Rs) - (Ls + La).$$

This equation shows that raising As and Fxs while decreasing ($Ls + La$) is necessary to increase products ($Pc + Pa$), because Rs is constant.

However, real-world situations are not as simple as depicted in this equation. The fertilizer use efficiency (FUE) expressed in yield per unit of nutrient gradually decreases based on Mitscherlich's Law of Diminishing Returns. When fertilization rate As is raised, the crop product Pc will respond to the increase but not proportionally to the increasing rate of As, especially when the rate is outside the optimum range. Increasing ($Ls + La$) and adverse effects such as crop lodging and degradation of product quality are also expected to such an extent that there must be a reasonable input level, and As cannot be increased in an unlimited manner to give higher yields.

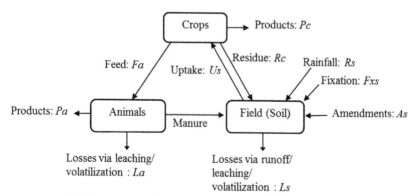

FIGURE 13.2 **Pattern diagram of cycling pathways and the balance of nutrients in a typical agricultural system.** Nutrient balance for the whole system: $(Pc + Pa) = As + (Fxs + Rs) - (La + Ls)$.

In today's eco-friendly agriculture, a combination of high yields and adequate environmental preservation is essential. To accomplish these ends, new technologies that will provide sufficient nutrient supply for crop growth while minimizing nutriment loss must be developed. One of the solutions is the utilization of catch crops.

13.1.2. Fates of Nutrients Applied to Crop Fields – Nutrient Balance and Water Pollution

Sequi et al. (2008) showed the circumstances of nitrogen balance in several EU countries for large-scale agricultural activities (Table 13.1). In four countries, i.e. the Netherlands, Belgium, Denmark, and the UK, nitrogen input is superfluous in comparison with the plant uptake based on output of crops from the field, and the nitrogen recovery is less than half of the total nitrogen input that includes animal manure. Especially in the former three countries, nitrogen balance per area exceeds 100 kg N ha^{-1}. What is the fate of the excess nitrogen? A portion is converted by ammonia volatilization and denitrification and then discharged to

TABLE 13.1 Soil Surface Nitrogen Balance Estimated in Several EU Countries (1995−1997)

Country	N input (A) (1000 Mg)	N output (B) (1000 Mg)	B/A (%)	N balance A − B (1000 Mg)	per area (kg N ha^{-1})
Netherlands	960	447	47	513	262
Belgium	443	196	44	247	181
Denmark	611	287	47	323	118
UK	2865	1387	48	1478	86
Germany	3442	2390	69	1052	61
France	4550	2965	65	1585	53
Spain	2086	885	42	1202	41
Italy	1909	1424	75	485	31
Poland	1881	1348	72	533	29
Turkey	2716	2216	74	500	12
ED-15 countries	19,789	11,709	59	8080	58

Sources: Sequi et al. (2008); OECD (2001). Environmental Indicatiors for Agriculture: Methods and Results, vol. 3. Several countries have been extracted by the author.

the atmosphere as nitrogen gas, whereas a considerable amount permeates underground to contaminate groundwater, or is discharged into surface water bodies to cause serious eutrophication. Additionally, the agriculture of these countries is characterized by abundant domestic animals with numbers exceeding the human population, so that more animal manure is generated than the total capacity of all farmland for treating and final disposal of these wastes.

By contrast, the recovery rates of nitrogen input in Germany and France of 69% and 65% respectively are high in comparison with the aforementioned four countries. Because the crop yield level in both Germany and France is high, the production systems are the most progressive when both high crop yield and environmental preservation are considered. Therefore, high yield at a recovery rate of 60–70% is a realistic level that can be achieved by applying present-day technology, and the development of future innovative technologies should aim at recovery rates higher than 60–70%.

The water pollution of major rivers and their apportionment in several EU countries is cited by Nishio (2006) from the European Environment Agency Report (Table 13.2). In this table, the amount of nitrogen discharged from agriculture, domestic wastewater, industrial effluent, among many others, to the environment and the source apportionments are evaluated. Particular

TABLE 13.2 Nitrogen Loading to Watershed and the Source Apportionment in Several EU Countries

Countries and watersheds	Total N loading (kg N ha^{-1})	Non-particular source (%) Background	Non-particular source (%) Agriculture	Particular source (%)
Netherlands	31	71		29
Belgium	34	64		36
Denmark	18	12	79	9
England/Wales	36	66		34
Germany	19	14	64	22
Poland	11	14	74	12
Danube river	9	23	45	32
Elbe river	16	12	55	33
Po river	36	10	54	36
Rhine river	29	14	54	32

Source: European Environment Agency Report No 7/2005. Cited in Nishio (2006). Several countries have been extracted by the author.

source denotes the source that can be observed, including domestic waste-water and industrial effluents. Background source refers to the non-point source of nitrogen flowing out of the natural soil, and the remaining portion is considered to be derived from agricultural sources. Background and agriculture-derived nitrogen levels have not been differentiated by country. Table 13.2 shows that the percentage of agriculture-derived nitrogen in Denmark, Germany, and Poland is high, at 79%, 64%, and 74% respectively. Although not supported by any data included in Table 13.2, the agriculture-derived nitrogen contributes more than 50% of the total nitrogen pollution in the watersheds of the four major rivers in Europe, excepting the Danube. This observation suggests that agriculture is the most significant source of nitrogen pollution in major European rivers. Hence, preventive measures to reduce nitrogen leaching from the farmland need to be implemented in order to alleviate nitrogen pollution of aquatic environments. Similarly, development of such control technologies is urgently needed in Asia because the abundant rainfall is expected to accelerate nitrogen leaching from the farm-land, which causes more serious problems than in Europe, where there is a comparatively small amount of rainfall.

13.1.3. Balancing Crop Production and Environmental Conservation – Recycling of Nutrients

As mentioned in the above sections, high crop yields require appropriate input of nutrients and adequate reduction of losses caused mainly by leaching. Nitrogen and phosphorus are often considered as an environmental issue, whereas other nutrients like potassium, calcium, and magnesium do not cause environmental concerns. This is because, once discharged into surface water bodies, nitrogen and phosphorus lead to serious eutrophication in aquatic systems to contaminate drinking water sources. In addition to water pollution problems, evidence reveals that other positive ions may leach out along with the negative nitrate ions from the soil layer. If the soil originally contains relatively small quantities of positive ions in relation to pedogenic parent materials, as in Japan, leaching of these positive ions should also be a concern from the viewpoint of soil fertility maintenance and economic efficiency. Intensive agriculture is characterized by multiple cropping and heavy fertilization. At present, the interval of two adjacent crops has become shorter than classical crop rotation such as the three-field system, but a period of bare field still exists in Japan. For example, in a typical double-cropping system of vegetables, the first crop (spring sown) is completed by the end of July, and the second crop (autumn sown) begins with fertilization at the end of August. Nutrient uptake by crops is very little for the 2 months when the second crop is still in the seedling stage during September. Therefore, heavy rain during this period will bring about serious leaching of nutrients (Matsumura, 2010). Then, the utilization of catch crops such as sunflower and crotalaria that take up water-soluble nutrients from

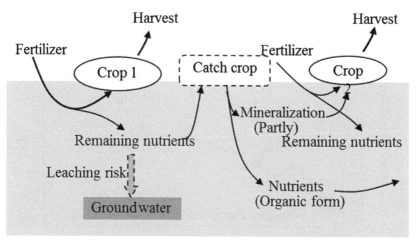

FIGURE 13.3   Image of environmentally sound crop production.

soil during this period will be effective in reducing leaching loss considerably
(Figure 13.3). Askegaarda et al. (2011) have examined the effects of cropping
pattern, catch crops, and manure application on nitrate leaching over 12 years
in Denmark, where the rainfall is light in April to June and heavy in autumn and
winter. They found that plowing in autumn increases nitrate leaching (55 kg-N
ha^{-1}) whereas catch crop in autumn minimizes nitrate leaching (20 kg-N ha^{-1}).
Thorup-Kristensen et al. (2012) also recognized that growing catch crops during
autumn after the main crop as their main source of soil fertility significantly
reduces nitrogen leaching loss. Results of field tests conducted in Japan by the
authors using sunflower as a catch crop reveal that accumulations of nitrogen,
phosphorus, potassium, calcium, and magnesium in sunflower plants 63 days
after seeding are 210 kg-N ha^{-1}, 20 kg-P ha^{-1}, 570 kg-K ha^{-1}, 100 kg-Ca ha^{-1},
and 30 kg-Mg ha^{-1} respectively (Matsumura and Kanazawa, 2011).

However, other observations indicate that catch crops have no such
outstanding benefits (Doltra et al., 2011; Nett et al., 2011). Doltra et al. (2011)
tested the effects of nitrogen availability on cereal yield, and nitrogen content in
the crop for fertilizer application, catch crops, and crop rotation. Their results
suggest that only fertilizer application shows positive results, whereas the influence of catch crops is insignificant on cereal yield and nitrogen content.

13.1.4. A Good Example of Improved Nutrient Balance

As mentioned above, Mitscherlich's Law of Diminishing Returns that relates
fertilization level to crop yield is true for cultivation of all scales from a pot
experiment to a huge serial field. Maene et al. (2008) have studied the relationship between fertilization level and cereal yield worldwide for 40 years
(1961–2001), with the studied agricultural areas divided into three groups, i.e.

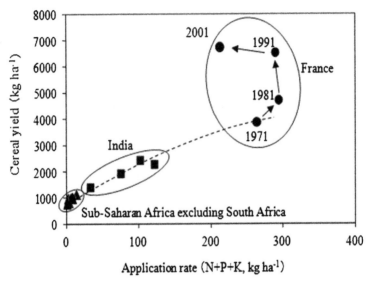

FIGURE 13.4   Improvement of recent fertilizer use efficiency in France.

advanced agricultural technology nations (e.g. France), industrializing (e.g. India), and stagnating countries (e.g. SSA or Sub-Saharan Africa excluding South Africa). Their data illustrate the fertilizer use efficiency (FUE) of these three groups over 40 years (Figure 13.4).

Although FUEs in SSA and India follow Mitscherlich's Law of Diminishing Returns for 40 years, as seen by the dotted line shown in Figure 13.4, the FUE in France before 1971 fits the dotted line. Since 1971, the FUE in France has exceeded the limitation depicted by this law. For 10 years from 1981 to 1991, cereal yield in France increased about 50% (4500–6500 kg ha^{-1}) without additional application of fertilizer input and, moreover, for the next 10 years from 1991 to 2001, the fertilization level has been drastically reduced without decreasing the yield. This seems to be the result of practicing "integrated nutrient management" (Maene et al., 2008), which should be disseminated worldwide.

Beaudoin et al. (2005) showed the positive result of using catch crops to reduce nitrate leaching in intensive agriculture in France. They examined the effect of good agricultural practice (GAP), which consists of applying accurate amounts of nitrogen, planting catch crops before summer crops, and recycling all crop residues, on nitrate leaching in northern France for 8 years. The results indicate that leaching of nitrate is affected mainly by soil properties, and the nitrate concentration in the percolated water is reduced by half for catch crops. They concluded that GAP is an essential agricultural practice, although not a perfect technique for intensive agriculture. The authors have developed an alternative farming practice, in which a corn production is accompanied by the incorporation of hairy vetch (Vicia villosa Roth.) and manure (see BOX 13.1).

BOX 13.1   High-Yield Corn Production by Applying a Combination of Hairy Vetch and Manure without Synthesized Fertilizer

Plant growth requires reasonable amounts of numerous nutriments such as nitrogen, phosphorus, potassium, etc. Synthesized chemical fertilizers are usually used in conventional crop production, but there have been serious concerns about their usage for biomass energy production because chemical fertilizers are energy-intensive products. Hence, the possibility of biomass production without using fertilizers has been examined for 5 years.

Figure 13.5 shows the nitrogen mineralization rate of the incorporated green manures (CR: crotalaria; F: ammonium sulfate as a control; HV: hairy vetch; SF: sunflower; UR: upland rice straw). The diagram suggests that HV and SF are easily decomposed in soil, and about 40% of the nitrogen contained in HV and SF is mineralized in 90 days. From this result, HV, a winter cover crop, lends itself well to summer corn production.

In 2007−2009, a nitrogen supply test was conducted on corn using HV on a TUAT farm in Tokyo by applying: no amendment (A), HV incorporation (B), HV surface mulching (C), and conventional compound fertilizer (D). The results show that the nitrogen accumulated by HV reaches about 280 kg ha^{-1}, as shown in Table 13.3 (Sharifi et al., 2011). About 70% of the accumulated nitrogen is estimated to be from nitrogen fixation (Provorov and Tikhonovich, 2003). Thus, HV cultivation is capable of supplying a sufficient amount of nitrogen for corn crops comparable to

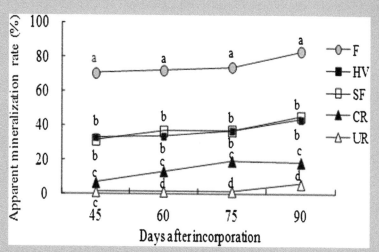

FIGURE 13.5   Patterns of apparent mineralization of incorporated green manure during 90 days. CR, crotalaria; F, ammonium sulfate as a control; HV, hairy vetch; SF, sunflower; UR, upland rice straw. Different letters mean significant differences at $P<0.05$ (Fischer's PLSD). (Sharifi et al. 2009)

BOX 13.1 High-Yield Corn Production by Applying a Combination of Hairy Vetch and Manure without Synthesized Fertilizer—cont'd

applying conventional fertilizer, with almost the same corn yields in the plots of B—D (Table 13.4).

As mentioned above, a considerable amount of nitrogen is expected to be supplied internally to crops through nitrogen fixation by symbiotic rhizobia with HV, but the other non-nitrogen nutrients must be supplied externally to replenish those depleted by the crop. Thus, the combined application of HV and manure to grow corn was examined during 2010—2011 by changing the nutrient addition to plot C from HV-mulching to HV plus manure for evaluating the balance of the five nutrients (Adachi et al., 2011).

The results indicate that both the above-ground biomass (AGB) and the grain yield of corn from plot C of 24 and 13.9 Mg ha^{-1} respectively are the highest. Plot C has significantly greater average crop yields for 2 years than plots A, B, and D based on increasing total nutrient input with manure. As the nutrient balance is concerned, plots A, B, and D have much greater outputs of nitrogen, phosphorus, potassium, calcium, and magnesium content in the harvested corn than inputs, whereas the budget of nutrients for plot C is positive for all nutrient elements except nitrogen. This suggests that when unavoidable nutrient leaching via precipitation is considered, the nutrient balance achieved with plot C is of great significance, especially for rain-abundant Japan.

In conclusion, combining HV and manure applications is confirmed to be a sustainable agricultural system based on the observation that many essential nutrients needed for crop growth are supplied in sufficient quantities by recycling the remaining nutrients in soil so that the problem of nutrient leaching to cause

TABLE 13.3 Dry Matter Production, Nitrogen Concentration, and Nitrogen Accumulation of Hairy Vetch at the Flowering Stage in 2007—2009

Year	Dry matter (A) (Mg ha^{-1})	N concentration (B) (g kg^{-1})	N accumulation (A × B) (kg ha^{-1})
2007	8.13[a]	33.0[ab]	264[a]
2008	7.95[a]	37.9[a]	301[a]
2009	9.92[a]	27.0[b]	269[a]
Mean	8.67	32.6	278

Values within columns followed by the same letter are not significantly different by Fisher's PLSD ($P = 0.05$).

TABLE 13.4 Above-Ground Biomass (AGB) Grain Yield and Nitrogen Uptake of Corn in a 3-Year Experiment

Treatments	2007			2008			2009		
	AGB (Mg ha^{-1})	Grain yield (Mg ha^{-1})	N uptake (kg ha^{-1})	AGB (Mg ha^{-1})	Grain yield (Mg ha^{-1})	N uptake (kg ha^{-1})	AGB (Mg ha^{-1})	Grain yield (Mg ha^{-1})	N uptake (kg ha^{-1})
A	22.2[a]	14.4[a]	183[a]	20.2[c]	13.6[b]	236[c]	23.8[a]	11.8[a]	341[a]
B	22.0[a]	11.6[a]	208[a]	25.3[ab]	16.2[a]	334[ab]	25.2[a]	12.9[a]	390[a]
C	20.5[a]	11.2[a]	205[a]	28.2[a]	15.2[ab]	374[a]	25.3[a]	12.7[a]	380[a]
D	20.8[a]	10.4[a]	203[a]	21.9[b]	15.4[ab]	280[b]	25.7[a]	12.7[a]	409[a]

A: Control; B: HV incorporation; C: HV mulching; D: Fertilizer.
Values within columns followed by the same letter are not significantly different by Fisher's PLSD ($P = 0.05$).

BOX 13.1 High-Yield Corn Production by Applying a Combination of Hairy Vetch and Manure without Synthesized Fertilizer—cont'd

environmental pollution problems can be significantly alleviated. Of course, a monitoring system for monitoring nutrient balance is necessary in the implementation of this practice.

In this system, the input and output of nutrients required for corn are well balanced, ensuring compatibility between corn production and soil fertility maintenance without the need for synthetic amendment.

13.2. BIOFERTILIZER

Takuya Ban

13.2.1. What is Biofertilizer?

It is not clear when humans began to use fertilizers for crop production. However, people in the distant past were aware of the advantages of enhancing crop production by applying human and animal wastes, plant ash, carcasses, and many other forms of organic matter. Some of these natural organic fertilizers are still used in today's agriculture. John Lawes conducted experiments in 1840 with the objective of using artificial chemicals, i.e. bones treated with sulfuric acid, as a fertilizer (Perry, 1992). He discovered that substances made from non-fertilizer materials have similar fertilizer values as natural bones. These artificial substances are capable of providing a readily available source of phosphorus for crop growth. The Haber–Bosch process, which is a chemical process to convert atmospheric N_2 to NH_3, was discovered in 1913 and provides an unlimited supply of nitrogen fertilizer for crop production (Galloway and Cowling, 2002). After this process was discovered, crop productivity and the global human population began to increase rapidly. However, excessive use of chemical fertilizer on crops is realized nowadays to cause serious environmental problems, including water pollution, salt accumulation in crop fields, and emission of excessive greenhouse gases.

Research on building effective cyclic food production systems with low nutrient and energy inputs has been undertaken globally in recent years. Biofertilizer plays a very important role in making such food production systems practical; it is defined as preparations containing microbial living or latent cells that enhance plant growth via the nature of their interactions in the rhizosphere (Mishra and Dadhich, 2010). In Japan, fertilizers that originate from biomass are defined as biofertilizers in a broad sense. Nitrogen-fixing bacteria, some fungi that form mycorrhizae on host plants, and phosphate solubilizers contribute to the natural production of biofertilizer (Ohyama, 2006).

In this chapter, an overview of biofertilizer usage in crop production is provided. The findings concerning (1) the development of crop production systems with applications of green manure and compost, and (2) application of solid-phase methane fermentation for crop production system are discussed in Sections 13.1 and 13.3.

The area of agricultural field where biofertilizers are actually applied for crop production has increased at the forefront of East Asian economies. In Japan, biofertilizers have also been commercialized. The application of biofertilizers has been extended to growing various agricultural products such as flowers, vegetable, fruit, field crops, and feed crops. The beneficial effects of bio-fertilizers include: (1) accelerated organic compound degradation in the soil; (2) improved aeration properties of the soil; (3) promoted root system development of the crop; (4) propagation of useful microbes for plant development in the soil; and (5) inhibited propagation of harmful microbes (Table 13.5). If more farmers use biofertilizers for crop production, the use of chemical fertilizers will decline and, as a result, the emission of greenhouse gases during the production and application of chemical fertilizers will be significantly alleviated.

13.2.2. Use of Biofertilizers in Crop Production

The availability of biofertilizers in crop production is obvious. However, there has been little information published in the literature concerning the effects of biofertilizer use during crop production on environmental quality and economic development. In this chapter, the beneficial effects of biofertilizers during crop production on crop yields, greenhouse gas emissions, and fertilizer costs are discussed.

TABLE 13.5 Substances of Commercially Supplied Biofertilizers in Japan

Distribution source	Fertilizer component (%)			Rhizobacteria contained	Remarks
	N	P	K		
Company A	2.4	11	4.8	Undisclosed	Bacterial number: 10^{11} per gram
Company B	1.2	3.9	1.3	Bacillus	C/N ratio: 21.2
Company C	—	—	—	Bacteria and fungi	
Company D	—	—	—	Azospirillum	
Company E	—	—	—	Arbuscular mycorrhiza	

Case 1: Soybean Production with Manure Compost

One of the aims of the Field Science Center, Tokyo University of Agriculture and Technology, is to develop new cultural techniques based on recent scientific advances. To examine the effects of organic manure usage on the productivities of main crops and cultivation environments, a selected field was fertilized with organic manure for more than 20 years. The total harvested area of soybean in 2011 was 1.45 ha, with a total yield of 1.85 Mg (Table 13.6). Chemical fertilizers and organic manure were also used side by side for soybean production. The total monetary values of fertilizers applied in 2011 were 15,139 yen per 10a for chemical fertilizers and 16,828 yen per 10a for organic manure. Based on the price of these fertilizers, the level of carbon dioxide emissions in manufacturing these fertilizers was estimated (NIAES, 2003) to be 71.9 kg-CO_2 for chemical fertilizer, which is higher than the 40.4 kg-CO_2 for organic manure. Little difference in crop yield was observed between the chemical fertilizer and organic manure applications. These results indicate that, for cultivation of soybean, organic manure emits less greenhouse gas than chemical fertilizers.

Case 2: Forage Rice 'Leafstar' Production with Nitrogen-Fixing Bacteria

Torii (2012) investigated the environmental factors that affect the inoculation effects of *Bacillus pumilus* TUAT-1 on the growth of the forage rice 'Leafstar'. He concluded that inoculating *Bacillus pumilus* TUAT-1 into 'Leafstar' improves the water and nutrition absorption capacity of the plant root system and promotes the absorption of nitrogen from soils during the early growth period of the plant (Table 13.7). The increased nitrogen absorption promotes tillering, increases the number of ears, and holds the photosynthesis activity at the late growth period of the plant so that the crop yield is boosted. There are no differences among the yields for the control, low-input, and low-input with inoculation samples (Figure 13.6). The total monetary values of chemical fertilizers applied in control, low-input, and low-input with inoculation samples were 3,179, 2,225, and 2,225 yen per 10a respectively. Based on the cost of fertilizer, the amount of carbon dioxide emission in the manufacture of the fertilizer was calculated (NIAES, 2003) to be 21.3 kg-CO_2 for the standard practice, higher than the 15.4 kg-CO_2 for low-input treatment. These results indicate that cultivation of 'Leafstar' inoculated with nitrogen-fixing bacteria emits less greenhouse gas than standard practice.

Case 3: Corn Production with Green Manure

Sharifi et al. (2011) examined the effect of crop rotation with hairy vetch as a green manure on corn production. They showed that the incorporation of hairy vetch residue or placement of hairy vetch residue on the soil surface increases above-ground biomass, grain yield, and nitrogen uptake of the corn plants. From these results, they concluded that incorporation or surface placement of

TABLE 13.6 Inhibitory Effect of Organic Manure Usage During Soybean Production on the Emission of Greenhouse Effect Gases

Type of fertilizer	Chemical fertilizer usage			Organic manure usage		
	Application level (kg per 10a)	Price (¥)	Amount of CO_2 (kg)*	Application level (kg per 10a)	Price (¥)	Amount of CO_2 (kg)
Chemical fertilizer[†]	30	3,465	20.4	—	—	—
Organic manure	—	—	—	2,150	10,750	25.8
Phosphorus iodide	80	8,104	19.4	60	6,078	14.6
Potassium sulfate	20	3,570	32.1	—	—	—
Total		15,139	71.9		16,828	40.4

*Emitted during the production of fertilizers.
[†]16–0–16% as $N–P_2O–K_2O_5$.

TABLE 13.7 Effect of *Bacillus* Inoculation on Carbon Dioxide Emission During Forage Rice 'Leafstar'

Type of fertilizer	Standard practice			Low-input treatment					
				Inoculation of Bacillus			Non-inoculation of Bacillus		
	Application level (kg per 10a)	Price (¥)	Amount of carbon dioxide (kg)*	Application level (kg per 10a)	Price (¥)	Amount of carbon dioxide (kg)*	Application level (kg per 10a)	Price (¥)	Amount of carbon dioxide (kg)*
Ammonium sulfate	10	851	7.7	7	595	5.4	7	595	5.4
Phosphorus iodide	10	1,013	2.4	7	709	1.7	7	709	1.7
Potassium chloride	10	1,315	11.2	7	921	8.3	7	921	8.3
Total		3,179	21.3		2,225	15.4		2,225	15.4

*Emitted during the production of the fertilizers.

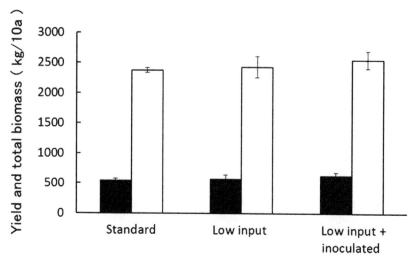

FIGURE 13.6   Effect of *Bacillus* inoculation on the yield and total biomass of forage rice 'Leafstar'.

hairy vetch residue is effective in supplying nitrogen to corn plants. As shown by Cases 1 and 2, the use of green manure reduces the total amount of chemical fertilizers and carbon dioxide emissions without affecting crop production. They also noted that the main crop is capable of utilizing the nitrogen released from the green hairy vetch manure, but an appropriate supply of minerals is needed to ensure successful growth of hairy vetch to sustain corn production over a long period of time.

13.2.3. Problems and Future Perspectives of Biofertilizer Use

Numerous publications confirm the effectiveness of biofertilizer use on plant growth under laboratory conditions; however, there are many problems in applying biofertilizers to practical fields to grow crops. Plant growth promotes rhizobacteria that accelerate the development of plant root systems, but the inoculation effect may be masked depending on the cultural environment. Many farm fields use chemical fertilizers that are easily absorbed and used by crop plants. Hence, the plants do not need to depend on nutrients derived from the coexisting rhizobacteria. For example, when grown under horticultural conditions with rich nutrients from chemical fertilizer applied by the grower, blueberry plants have lower colonization rates of ericoid mycorrhiza than those growing in the wild (Yang et al., 2002). The maximum effect of biofertilizers needs to be investigated by analyzing the functions of rhizobacteria in order to study how environmental factors affect their inoculation.

When using compost manures, one has to address the optimization of manure distribution channels. In Japan in particular, livestock are not always grazed in the neighborhood of the crop field where the manure is to be applied. Hence, the transportation of manure from the source to the application point will also lead to additional emissions of carbon dioxide.

13.3. METHANE FERMENTATION AND USE OF DIGESTED SLURRY

Shuhei Kanda

13.3.1. Methane Fermentation

a. General Outline

Increasing consumption of fossil fuels, e.g. natural gas, petroleum and coal, leads to rapidly increasing emissions of greenhouse gases such as carbon dioxide (CO_2) that cause global warming problems. Reducing the use of fossil fuels by promoting the use of renewable energy will alleviate the problems brought about by consuming fossil fuels.

Nowadays, European countries like Germany and Denmark attempt to utilize biomass resources in addition to promoting the production of clean energy from biomass. Small-scale methane fermentation that was popularly used in the past has been revived to produce biofuel, especially in some Asian countries like China and Vietnam.

In Japan the utilization of renewable energy from disposal of biomass, including animal wastes, unused biomass, and inedible crops, is stipulated by the Ministry of Agriculture Forestry and Fisheries (2003) in "Biomass Japan Strategy". Recently, the implementation of methane production techniques using animal wastes has brought remarkable achievements because of substantially increasing production of both heat and electricity from biogas, as well as use of by-product of biogas for fertilizer.

b. Current Situation of Methane Fermentation in Europe

Concerning the welfare of future generations, Germany has decided to refrain from the use of nuclear power to avoid the risks of nuclear pollution due to accidents and inappropriate treatment and disposal of radioactive waste.

In Sweden, renewable energy like biogas and bioethanol is produced by using various types of biomass resources. Large-scale methane fermentation plants now operate in 10 areas. Biogas plants using animal wastes as raw material are managed by local governments producing methane gas and digested slurry to benefit rural farms.

In Denmark, large-scale biogas plants are managed by cooperating communities. Profits from the production of methane gas were insignificant when the techniques were still at the developmental stage. Because most costs were defrayed by farmers, agricultural groups formed the "Auction Program" in 1987 for rebuilding biogas plant enterprises (Research Association of Hokkaido

Biogas, 2002), with 30–40% of the initial capital cost paid by central government. Wastes from food industries and household kitchens are now also added as raw materials to produce biogas. Renewable energy such as biogas and wind power, among others, supplies almost 100% of energy requirements in Denmark.

Therefore, these nations support the development of natural energies like solar, wind and geothermal, and many others by advancing current technologies. However, many problems associated with the distribution of renewable energy exist. For example, rural regions may be more suitable than urban regions to produce renewable energy but the demand there is much less. How to transport the excess energy from rural producers to urban consumers cost-effectively needs to be studied.

c. Current Situation of Methane Fermentation in Japan

(i) General Outline

In Japan the first group of methane fermentation plants was put into operation in 2000. Some of these plants have been closed for economic reasons.

The infamous earthquake in the Tohoku district that caused the accident at the Fukusima First Nuclear Power Station has motivated the nation to emphasize the development of renewable energy sources, including solar, wind, geothermal, and biogas, as a priority to meet future energy demands. Since 2011, the Japanese government has subsidized the construction of commercialized biogas plants and products as well.

(ii) Present Standing of Methane Fermentation Plants

The present situation regarding methane fermentation plants in Japan is shown in Table 13.8. More than 60 plants, including experimental pilot plants, have been installed; 40% of these plants were financially supported by local governments and 65% were constructed on animal husbandry farms.

In Hokkaido in particular, 33 plants have been built in order to effectively use the anaerobically digested slurry (ADS) in addition to preventing bad odors. Many mega dairy farms in Hokkaido experienced the problem of bad odor when making slurry manure before they adopted anaerobic equipment. Using methane fermentation equipment has helped dairy farms to solve these problems, as well as producing clean methane gas energy and ADS as liquid fertilizer.

13.3.2. Methane Fermentation Equipment

a. Mechanisms of Methane Fermentation

In a general model of the two-stage methane fermentation process, raw materials are decomposed biologically into organic acids in the first-stage

TABLE 13.8 Number of Biogas Plants (Methane Fermentation) in Japan (2011)

	Number	%
Financier		
Local governments	26	42
Universities, etc.	5	8
Experimental stations	3	5
Dairy farms	22	37
Pig farms	1	2
Factories	4	7
Total	**61**	**100**
Kinds of materials		
Animal waste	39	65
Kitchen waste	19	31
Factory residue	3	5
Total	**61**	**100**
Details animal waste		
Cattle waste	35	90
Others	4	10
Total	**39**	**100**
District		
Hokkaido	33	55
Others	28	45
Total	**61**	**100**

Sources: *Dairy Journal* special edition (2002). Animal wastes: Effective usage of biogas systems. *Dairy Journal* special edition (2006). Social and economic evaluations of dairy biogas systems.

acid-producing tank as the first step. The acids are then sent to the second-stage fermentation tank where they are fermented anaerobically into methane gas (Eckenfelder, 1966). Two separate tank reactors are needed in this two-stage methane fermentation process for smooth and continuous operation (Ghosh, 1987).

b. Single-Tank System for Methane Fermentation

Using a one-tank system will save the initial cost and make the operation more cost-effective. Two types of single-tank system have been developed by Mori group & Taguchi (M&T) (Kanda, 2005) and P-Works (2003) in Japan.

(i) M&T Methane Fermentation Plant

The M&T methane fermentation plant is shown schematically in Figure 13.7, with the dimensions and capacities of the plant listed in Table 13.9.

The major features of the fermentation plant are as follows:

- Fixation of methanogenic bacteria. The bacteria stall, which has a brush made of polypropylene material, is capable of reproducing anaerobic bacteria and enhances their retention abilities.
- Horizontal flow control boards. These can improve the symbiotic environment of anaerobic bacteria, and maintain the slow flow that makes effective fermentation possible.
- Automatic gas lifted stirring system (Figure 13.8). Biogas in the methane fermentation tank is stored temporarily in a holding tower. When the gas pressure rises to a certain high level, the control valve opens automatically, allowing the biogas to flow toward the methane fermentation tank, and the rising biogas stirs the fermentation slurry in the tank. This method allows the fermentation slurry to be stirred efficiently and cost-effectively.

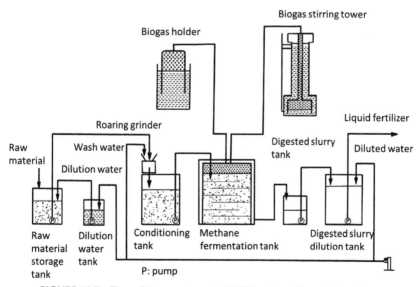

FIGURE 13.7   Flow of the experimental M&T methane fermentation plant.

TABLE 13.9 General Specification of M&T Methane Fermentation Experimental Plant*

Item	Material	Size (mm)	Volume (L)
Raw materials input tank	Steel	Diameter 600 × 700H	198
Mascroider (roaring grinder)		1.4 kW	
Raw materials stock tank	Steel	Diameter 1040 × 1475H	1252
M&T fermentation tank	Steel	1400L × 700W × 1200H	1176
Stealing tower	Steel	Diameter 150 × 4500H	79
Gas holder	Steel	Diameter 1400 × 2200H	3385
Digested slurry dilution tank	Resin	Diameter 1040 × 1475H	1252
Sprinkling liquid tank	Resin	Diameter 1040 × 1475H	1252

Designed by Mr Akira Taguchi (Science Information System and Co., Mori group).

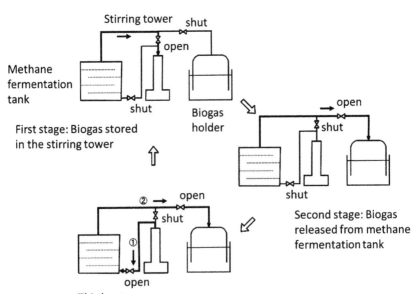

Stirring tower   shut

open

Methane fermentation tank

shut

Biogas holder

First stage: Biogas stored in the stirring tower

open

shut

shut

Second stage: Biogas released from methane fermentation tank

② open

shut

①

open

Third stage:
1. Biogas sent to methane fermentation tank for stirring
2. Biogas sent from methane fermentation tank to biogas holder

FIGURE 13.8   M&T automatic biogas stirring system.

TABLE 13.10 Features of Methane Fermentation Experimental Plant

Item	Treatment
Input	Cattle wastes 20 kg + water 20 L
Biogas production	$0.5\ m^3\ day^{-1}$
Fermentation condition	Middle temperature 35–40°C

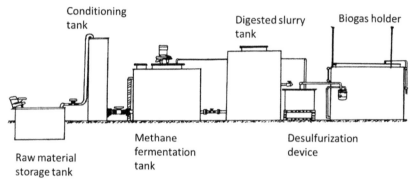

FIGURE 13.9 Outlook of the experimental methane fermented plant.

(ii) P-Works Methane Fermentation Plant

An outline of the methane fermentation experimental plant developed by P-Works is shown in Tables 13.9 and 13.10 and Figure 13.9. The plant consists of five directly connected units: (1) raw material storage tank, (2) conditioning tank, (3) methane fermentation tank, (4) digested slurry tank, and (5) desulfurization device and biogas holder (Table 13.11). The second, third, and fourth tanks are interconnected with 100-mm pipes and control valves; a PVC pipe is used to connect the head of the methane fermentation tank to the biogas holder through a desulfurization device.

The methane fermentation plant is easy to operate because of the simple design. This plant needs an additional heating unit to maintain the tank temperature in the optimum mesophilic range to keep the methane fermentation process operating normally during winter months, when the ambient temperature decreases below the optimum temperature range.

13.3.3. Use of Anaerobically Digested Slurry in Japan

In several cases, after the anaerobically digested slurry (ADS) is treated, the supernatant liquid is decanted to be discharged into a surface water body, and

TABLE 13.11 Retention Time of Each Tank Process

Item	Volume (m³)	Retention (days)
Raw material storage tank	0.50	12
Conditioning tank	0.13	3
Methane fermentation tank	2.00	50
Digested slurry tank	1.40	35
Biogas holder	2.83	3.6
Desulfurization device	0.18	—

FIGURE 13.10   Slurry injectors. *(Source: Watanabe, 2002.)*

the odorless treated ADS may be used as a fast-acting fertilizer by neighborhood farmers to grow crops.

Mega dairy farms (200–300 head) in Hokkaido used to make slurry manure, but the bad odor adversely affected the environmental quality of the surrounding regions. This problem can be solved by using methane fermentation plants that also have the benefit of producing methane gas as renewable energy. About 40 dairy farms in Hokkaido had built methane fermentation plants to produce odorless ADS as liquid fertilizer by 2006. The digested slurry is injected into the surface soil layer of the farmland with a slurry injector, as shown in Figure 13.10. Farmers are subsidized by government for up to 75% of the initial costs for building their methane fermentation plants. The use of ADS as fertilizer in paddy fields in Japan needs to be evaluated in future studies.

13.4. INFLUENCE OF FODDER RICE PRODUCTION USING BIOGAS SLURRY FOR BETTER NUTRIENT CYCLING ON ENVIRONMENTAL QUALITY

Aye Thida Win and Koki Toyota

13.4.1. Use of Biogas Slurry (BS) for Rice Production

Biogas plants convert many types of organic wastes, such as agricultural waste, animal waste, kitchen waste, sewage sludge, organic sludge, residues, etc., into biogas and slurry fertilizer known as biogas slurry (BS) that is rich in valuable nutrients for crop production. On the other hand, direct discharge of this slurry to surface water bodies will cause serious eutrophication of lakes and rivers. Thus, recycling these wastes is very important to avoid polluting the environment while maintaining sustainable farming systems. Many studies have revealed that BS application shows no adverse effects on rice yield but increases biomass production. For example, 450–600 kg-N ha^{-1} of biomass production was reported by Sunaga et al. (2009). Lu et al. (2012) also observed that applications of more BS lead to higher rice grain yields when fields were treated with biogas slurry at 270, 405, and 540 kg-N ha^{-1}. Because biogas slurry is available in large quantities almost anywhere for field application, its use will significantly reduce the need for commercial chemical fertilizers and the associated costs. Oki town in Fukuoka prefecture is the most advanced municipality in terms of BS recycling in Japan.

Unlike chemical fertilizers that are almost homogeneous in composition, BS contains large amounts of carbon and undesirable elements. Thus, there are concerns about several negative environmental affects caused by BS applications. One of the major characteristics of BS is its high pH (pH 8–9) that enhances NH_3 volatilization under high pH conditions. Loss of ammonia due to volatilization accounts for 13.3% of the total nitrogen contained in the soil applied with BS; this amount is much higher than the value of 1.9% for fields with CF (chemical fertilizer) application (Win et al., 2009). This high loss of NH_3 due to volatilization with BS application can be mitigated by acidifying the sludge with acid, e.g. wood vinegar or sulfuric acid, or by keeping deeper floodwater in the paddy field to decrease the NH_4^+ concentration in the floodwater. Animal feeds are often enriched with metallic elements such as Zn and Cu among many others, as well as higher levels of heavy metals. Accumulation of these heavy metals in the rice crop becomes a major concern about rice production fertilized with BS, although application of BS even at a rate of 1080 kg-N ha^{-1} has been shown not to increase the levels of heavy metals (Cu, Hg, Cd, and Pb) in the soil (Lu et al., 2012). The use of BS as fertilizer for rice may not cause heavy metal contamination in the short term; the consequences of long-term BS applications should be carefully evaluated in future studies.

13.4.2. Agriculture and Global Warming Potential

The global agricultural field constitutes about 35% of total land area (Betts et al., 2007). Because of its scale and intensity, farming land emits a significant quantity of greenhouse gases (GHGs) into the atmosphere. In agriculture, CO_2 is released largely from microbial decomposition of soil organic matter, whereas methane is produced by microbial decomposition of organic materials in soil under anoxic conditions, especially in flooded rice fields. Nitrous oxide is generated by the microbial transformation of nitrogen in soil via nitrification and denitrification processes.

There are a variety of management practices for mitigating GHG emissions in agricultural fields. The most prominent options include those that improve crop and land management (e.g. improved agronomic practices, nutrient use, tillage, and residue management), restore wetland that has been drained for crop production and degraded lands, improve water and rice management, change the land use pattern such as the conversion of cropland to grassland, and improve livestock and manure management, among many other practices (UNFCCC, 2010). According to the Intergovernmental Panel on Climate Change (IPCC) (1996), agriculture-related CH_4 emissions can be reduced by 15–56%, mainly through the improved nutrition of ruminant domestic animals and better management of paddy rice, which may also reduce N_2O emissions by 9–26%.

13.4.3. Emission of Methane and Nitrous Oxide in Rice Paddy Fields

Methane emission from rice fields is controlled by several factors such as carbon source, temperature, soil moisture regime, soil redox potential, time and duration of drainage, soil type, rice variety, fertilizer, and cultural practices as well as climate (Conrad, 2002). Methane production is negatively correlated with soil redox potential and positively correlated with soil temperature, soil carbon content, and rice growth (Neue, 1993). The water management system is one of the most important factors affecting CH_4 emissions. Easily degradable crop residues including fallow weeds and soil organic matter are the major sources for initial CH_4 production during the early stage of crop growth. During the later stages of growing rice, methane production is influenced more by root exudates, decaying roots, and aquatic biomass (Sass et al., 1991).

Methane emissions in paddy fields can be reduced by practicing various methods such as draining the field once or several times during the rice growing season to keep root exudation low, keeping the soil as dry as possible in the rice off-season, adjusting the timing of organic residue additions such as by incorporating organic materials during the dry period rather than in the flooded periods, or by composting the residues before incorporation. The application of organic manure like BS can enhance methane emission. Sasada et al. (2010) found that the global warming potential (GWP) based on the CH_4 and N_2O emissions from a forage rice soil applied with BS is 860 ± 310 g-CO_2 m^{-2} (exclusively contributed by CH_4 emission), which is more than twice the emission (400 ± 44 g-CO_2 m^{-2}) from fields with CF application. However, changing the type of BS applied, such as from cow manure to pig manure, leads to a 23% reduction of GWP (Figure 13.11). Although BS application can enhance CH_4 emission in rice soil, it may improve soil carbon contents, and the benefit will compensate for the disadvantage of the resulting enhanced CH_4 emission (Win et al., 2010).

In rice fields, N_2O flux is controlled mostly by nitrogen fertilization and also by the type of fertilizers, soil water regime, and soil texture (Hua et al., 1997).

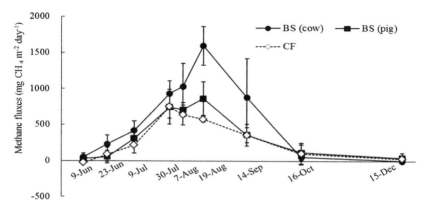

FIGURE 13.11   Effect of chemical fertilizer (CF) or different types of biogas slurry (BS) on CH₄ emissions. Mean value ± standard deviation ($n = 3$). *(Redrawn from Sasada et al., 2010.)*

In Japan, Minami (1987) found that N_2O emission from paddy fields ranges from 0.33% to 0.55% of nitrogen fertilization. However, N_2O emission from flooded rice fields is usually low and only becomes significant when the field becomes dry after the final drainage for harvest.

13.4.4. Carbon Sequestration in Rice Soil

Carbon sequestration in soil is an option for mitigating the global warming potential. The IPCC (1996) estimated that 400–800 Mt-C y^{-1}, equivalent to about 1400–2900 Mt-CO_2 y^{-1}, can be sequestered globally in agricultural soil. Various management techniques can be used to enhance carbon sequestration in soils (Janzen et al., 1999). Pan et al. (2003) reported that significant application potential for soil carbon sequestration has been found in rice paddies because the long period of flooding causes the soil to accumulate organic carbon even after the above-ground biomass has been completely removed.

REFERENCES

Adachi, M., Tsuji, A., & Matsumura, S. (2011). Nutrient balance in corn production by combined application of hairy vetch and manure. *Abstract of the Annual Meeting of the Japanese Society of Soil Science and Plant Nutrition, 57,* 142 [in Japanese].

Askegaarda, M., Olesena, J. E., Rasmussenb, I. A., & Kristensenc, K. (2011). Nitrate leaching from organic arable crop rotations is mostly determined by autumn field management. *Agriculture, Ecosystems and Environment, 142,* 149–160.

Beaudoin, N., Saad, J. K., Van Laethem, C., Machet, J. M., Maucorps, J., & Mary, B. (2005). Nitrate leaching in intensive agriculture in Northern France: Effect of farming practices, soils and crop rotations. *Agriculture, Ecosystems and Environment, 111,* 292–310.

Betts, R. A., Falloon, P., Goldewijk, K. K., & Ramankutty, N. (2007). Biogeophysical effects of land use on climate: Model simulations of radiative forcing and large-scale temperature change. *Agricultural and Forest Meteorology, 142*, 216–233.

Conrad, R. (2002). Control of microbial methane production in wetland rice fields. *Nutrient Cycling in Agroecosystems, 64*, 59–69.

Doltra, J., Lægdsmand, M., & Olesen, J. E. (2011). Cereal yield and quality as affected by nitrogen availability in organic and conventional arable crop rotations: A combined modeling and experimental approach. *European Journal of Agronomy, 34*, 83–95.

Eckenfelder, W. W. Jr (1966). *Industrial water pollution control*. McGraw-Hill. Interpreted by Ichikawa, K. & Maeda, Y. (1970). *Treatments of industrial water pollution*. Tokyo: Kouseikaku.

Galloway, J. N., & Cowling, E. B. (2002). Reactive nitrogen and the world: 200 years of change. *Journal of the Human Environment, 31*, 64–71.

Ghosh, S. (1987). Improved sludge gasification by two-phase anaerobic digestion. *Journal of Environmental Engineering, 113*, 1265–1284.

Hua, X., Xing, G. X., Cai, Z. C., & Tsuruta, H. (1997). Nitrous oxide emissions from three rice paddy fields in China. *Nutrient Cycling in Agroecosystems, 49*, 23–28.

Ichikawa, O., Nakahara, J., & Hoshiba, S. (2006). *Sociological and biological assessment of dairy biogas system*. Rakuno Gakuen University extension center, pp. 181.

IPCC. (1996). *SAR, climate change 1995: The science of climate change. Contribution of Working Group I to the Second Assessment Report of the Intergovernmental Panel on Climate Change (IPCC)*. Cambridge: Cambridge University Press.

Janzen, H. H., Desjardins, R. L., Asselin, J. M. R., & Grace, B. (1999). *The health of our air: Towards sustainable agriculture in Canada*. Publication 1981/E. Ottawa, ON: Agriculture and Agri-Food Canada.

Kanda, S. (1988). Animal waste treatment and utilization by using methane fermented device. *Symposium of Kanto district of Japanese farm work research*, 13–22.

Kanda, S. (2005). Utilization of cattle waste by using methane fermented device. *Farm work research, special edition*, pp. 87–88.

Lu, J., Jiang, L., Chen, D., Toyota, K., Strong, P. J., Wang, H., & Hirasawa, T. (2012). Decontamination of anaerobically digested slurry in a paddy field ecosystem in Jiaxing region of China. *Agriculture, Ecosystems and Environment, 146*, 13–22.

Maene, L. M., Sukalac, K. E., & Heffer, P. (2008). Global food production and plant nutrient demand: Present status and future prospects. In M. S. Aulakh, & C. A. Grant (Eds.), *Integrated nutrient management for sustainable crop production* (pp. 1–28). New York: CRC Press.

Matsumura, S. (2010). Field estimation of nitrogen balance. In T. Ohyama, & K. Sueyoshi (Eds.), *Nitrogen assimilation in plants* (pp. 33–49). Kerala, India: Research Signpost.

Matsumura, S., & Kanazawa, M. (2011). Recycling of soil-remaining nutrients by using cover crops. *Abstract of the Annual Meeting of the Japanese Society of Soil Science and Plant Nutrition, 57*, 142.

Mishra, B. K., & Dadhich, S. K. (2010). Methodology of nitrogen biofertilizer production. *Journal of Advances in Developmental Research, 1*, 3–6.

Minami, K. (1987). Emission of nitrous oxide from agro-ecosystem. *JARQ, Japan Agricultural Research Quarterly, 21*, 21–27.

National Institute for Agro-Environmental Sciences (NIAES). (2003). *Manual for life cycle assessment of agricultural practices in Japan* (pp. 1–51). NIAES.

Nett, L., Feller, C., George, E., & Fink, M. (2011). Effect of winter catch crops on nitrogen surplus in intensive vegetable crop rotations. *Nutrient Cycling in Agroecosystems., 91*, 327–337.

Neue, H. U. (1993). Methane emission from rice fields: Wetland rice fields may make a major contribution to global warming. *BioScience, 43*, 466–473.

Nishio, M. (2006). *Reports of environmental conservation typed agriculture, No. 48: Agriculture is the primary cause for water pollution in the EU.* <http://lib.ruralnet.or.jp/libnews/nishio/nishio048.htm>.

Ohyama, T. (2006). Introduction. In *Biofertilizer manual* (pp. 1–2). Japan Atomic Industrial Forum.

Pan, G. X., Li, L. Q., Wu, L. S., & Zh, X. H. (2003). Storage and sequestration potential of topsoil organic carbon in China's paddy soils. *Global Change Biology, 10*, 79–92.

Perry, W. E. (1992). The utilization of by-products and waste products in the production of commercial fertilizers. *Fertilizer Research, 32*, 111–114.

Provorov, N. A., & Tikhonovich, I. A. (2003). Genetic resources for improving nitrogen fixation in legume–rhizobia symbiosis. *Genetic Resources and Crop Evolution, 50*, 89–99.

Research Association of Hokkaido Biogas. (2002). *Animal wastes effective usage of biogas system.* Dairy Journal special edition. Rakuno Gakuen University extension center.

Sasada, Y., Win, K. T., Nonaka, R., Win, A. T., Toyota, K., Motobayashi, T., Dingiiang, C., & Lu, J. (2010). Environmental impact in a whole crop rice cultivation fertilized with anaerobically digested cattle or pig slurry. *Biology and Fertility of Soils, 47*, 948–956.

Sass, R. L., Fisher, F. M., Lewis, S. T., Jund, M. F., & Turner, F. T. (1991). Methane emission from rice fields as influenced by solar radiation, temperature, and straw incorporation. *Global Biogeochemical Cycles, 5*, 335–350.

Sequi, P., Johonston, A. E., Francaviglia, J. R., & Farina, R. (2008). Integrated nutrient management: The European experience. In M. S. Aulakh, & C. A. Grant (Eds.), *Integrated nutrient management for sustainable crop production* (pp. 221). New York: CRC Press.

Sharifi, M. Z., Matsumura, S., Hirasawa, T., & Komatsuzaki, M. (2009). Apparent nitrogen mineralization rates of several green manures incorporated in soil and the application effects on growth of komatsuna plants. *Japanese Journal of Farm Work Research, 44*(3), 163–172.

Sharifi, M. Z., Matsumura, S., Ito, T., Hirasawa, T., & Komatsuzaki, M. (2011). Improvement of nitrogen balance by rotating corn and hairy vetch. *Japanese Journal of Farm Work Research, 46*, 167–177.

Sunaga, K., Yoshimura, N., Hong, H., Win, K. T., Tanaka, H., Yoshikawa, M., Watanabe, H., Motobayashi, T., Kato, M., Nishimura, T., Toyota, K., & Hosomi, M. (2009). Impacts of heavy application of anaerobically digested slurry to whole crop rice cultivation in paddy environment on water, air and soil qualities. *Japanese Journal of Soil Science and Plant Nutrition, 80*, 596–605 [in Japanese with English summary].

Thorup-Kristensen, K., Dresbøll, D. B., & Kristensen, H. L. (2012). Crop yield, root growth, and nutrient dynamics in a conventional and three organic cropping systems with different levels of external inputs and N re-cycling through fertility building crops. *European Journal of Agronomy, 37*, 66–82.

Torii, A. (2012). Analysis of field factors resulting fructuations of yield and nutritional uptakes of forage rice Leaf Star with inoculation of an endophytic nitrogen fixing bacteria TUAT1. *Master thesis of Graduate school of agriculture. Tokyo University of Agriculture and Technology.*

UNFCCC. (2010). *United Nations Framework Convention on Climate Change, United Nations Climate Change Conference Cancun – COP 16/CMP 6, Land use, land-use change and forestry.*

Watanabe. (2002). *Effective use of animal wastes by biogas system – usage techniques of methane digested liquids.* Dairy Journal special edition 2002.

Win, K. T., Toyota, K., Motobayashi, T., & Hosomi, M. (2009). Suppression of ammonia volatilization from a paddy soil fertilized with anaerobically digested cattle slurry by wood vinegar application and floodwater management. *Soil Science and Plant Nutrition, 55*, 190–202.

Win, K. T., Nonaka, R., Toyota, K., Motobayashi, T., & Hosomi, M. (2010). Effects of option mitigation ammonia volatilization on CH_4 and N_2O emissions from a paddy field fertilized with anaerobically digested cattle slurry. *Biology and Fertility of Soils, 46*, 589–595.

Yang, W. Q., Goulart, B. L., Demchak, K., & Yadong, L. (2002). Interactive effects of mycorrhizal inoculation and organic soil amendments on nitrogen acquisition and growth of highbush blueberry. *Journal of the American Society for Horticultural Science, 127*, 742–748.

Chapter 14

Evaluation of Biomass Production and Utilization Systems

Chihiro Kayo, Seishu Tojo, Masahiro Iwaoka and Takeshi Matsumoto

Research Approaches to Sustainable Biomass Systems. http://dx.doi.org/10.1016/B978-0-12-404609-2.00014-3
309

14.1. LIFE CYCLE ASSESSMENT (LCA) ON BIOENERGY

Chihiro Kayo

14.1.1. The Concept of LCA

Human society consumes various products and services; their production, use, and final disposal gradually exhausts natural resources including minerals and fossil fuels, in addition to emitting pollutants in gaseous, liquid, and solid forms to deteriorate environmental quality. The root cause of the global environmental problems that we are facing today is the excessive burden imposed on the global environment due to mass production, mass consumption, and mass disposal to support human activities. To move toward a sustainable society requires that we minimize inputs from the natural environment to human society, and outputs from human society to the environment as well. Measuring the environmental burdens of products and services during their life cycle, and correctly ascertaining their impacts on the global environment are important initial steps toward a sustainable society.

Life cycle assessment (LCA) is a technique that considers the entire life cycle of particular products and services from their resources to their use and final disposal. This is also known as the cradle-to-grave concept that measures the quantity of resources consumed and the emission of pollutants, and assesses environmental impacts during the steps of resource extraction, transportation, material production, part manufacture, product assembly, use, and final disposal. LCA is characterized by considering not only the visible environmental impacts of products and services but also the invisible elements when the products and services are created until they are finally disposed of.

Using the LCA concept to assess the environmental impact of a product was initiated in the USA and some European countries in the 1970s (SETAC, 1993), and the term "ecobalance" became popular in the 1980s. The 1992 Earth Summit in Rio de Janeiro (Brazil) led to the development of international standards for environmental management in 1993; the term "life cycle assessment" was then adopted internationally. LCA international standards (ISO) had been discussed by Technical Committee 207 on Environmental Management (TC207), and were published as ISO 14040–14049. Under this set of standards, the four steps in the general procedures for conducting a life cycle assessment are: "Goal and scope definition", "Life cycle inventory", "Life cycle impact assessment", and "Interpretation". General discussions of these four steps are provided in the following sections.

14.1.2. Goal and Scope Definition

The first step of goal and scope definition is to define the reason for initiating the study, how the results will be utilized, and who will benefit from the reported results in the objectives for conducting an LCA. Specifically, the product or service to be studied is determined, and the impacts to be assessed are clearly defined. If one is going to assess "the impact of wood pellets with a calorific value of 500 MJ on global warming" in accordance with this goal, what pollutants should be measured must be decided beforehand, and the scope within which they are collected must then be defined.

LCAs assess the "functions" of products. If the product is wood pellets, then the function is "heat use". The unit that indicates the magnitude of this function is called the "functional unit", such as "500-MJ calorific value". When comparing multiple products in an LCA, the functional unit must be unified. If we are comparing wood pellets, coal, and oil, for example, the comparison must be based on the same functional unit of 500-MJ calorific value.

The boundary with the natural environment that subsumes the entire life cycle process of the assessed product or service is called the "system boundary". Material flows that cross the system boundary from the natural environment into each process are "input flows", whereas material flows from each process to the natural environment are "output flows". Figure 14.1 is a schematic diagram of a system boundary, and the input and output flows. International LCA standards require that the scope of an LCA be clearly described as the system boundary.

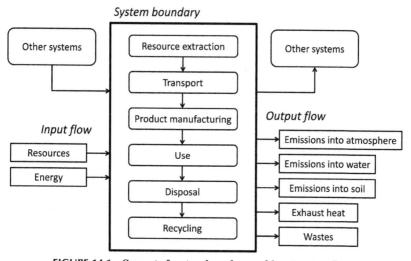

FIGURE 14.1 Concept of system boundary and input–output flows.

14.1.3. Life Cycle Inventory

A life cycle inventory (LCI) is the step to collect data on the input flows of both resources and energy entering the entire LCA system and the output flows of the products and discharged emissions (output flows) across the same LCA system boundary. When conducting an LCI, one must first gather data on the manufacture, use, and disposal of the products being assessed. The data that are directly related to the products are generally called "foreground data". Subsequently, one must investigate quantities of the input and output flows, e.g. pollutants discharged or emitted in producing power that is consumed when producing the materials to be used in the products, and when using the products. These data are indirectly related to the assessed products; hence, they are generally called "background data".

Gathering foreground data involves conducting the material and energy balance in each of the life cycle processes in the manufacturing, transport, use, disposal, and recycling of a product. Although quantifying pollutants discharged or emitted from system boundaries into the atmosphere is desirable, gaseous emissions such as those of the primary greenhouse gas CO_2 are often monitored infrequently. The equivalent greenhouse gas emissions can be obtained indirectly by determining the amount of energy consumed; the greenhouse gas emission intensity can then be estimated for each type of energy consumed, e.g. fuel oil or kerosene (Ministry of the Environment of Japan, 2012).

Collecting background data can be roughly divided into "process analyses", in which calculations are based on physical quantities such as the consumed amounts of resources, and "I-O analyses", in which the ad valorem environmental burden amounts are estimated based on the input–output tables published every 5 years by the Ministry of Internal Affairs and Communications. Background data obtained by using process analyses are produced by using physical numerical quantities such as the actual amounts of materials and fuel consumed by a factory, but the method used to collect data does not differ greatly from that for collecting foreground data. The data are easy to understand because they are presented as physical quantities so that figuring out detailed data based on the characteristics of the analyzed products is possible. However, the data are often insufficient even if one has searched for all possible background upstream or downstream data sets; the preparation of new data often becomes necessary. Therefore, process analysis is considered a labor-intensive method. For carrying out I-O analysis, a nation's economy is divided into a number of sectors (e.g. 400 sectors for Japan's economy) to represent the annual flows of goods and services among sectors in monetary units. Input–output tables can be used to determine the amounts of environmental burdens, such as the CO_2 that is directly or indirectly emitted in conjunction with production activities (Nansai and Moriguchi, 2009). Although detailed analyses of individual products are difficult because input–output tables only present macroscopic statistical data for the whole nation, this method is suitable for analyzing average or nationwide emissions of environmental burdens. In general, LCIs commonly involve using process analyses to obtain foreground and other important data sets while adopting hybrid analyses to make use of input–output tables for deriving data of less importance.

14.1.4. Life Cycle Impact Assessment

Life cycle impact assessment (LCIA) refers to the steps that assess the type and extent of environmental impacts that may arise quantitatively based on data collected in the LCI. The LCIA procedure primarily consists of: (a) categorizing, (b) classifying, (c) characterizing, (d) normalizing, (e) grouping, and (f) integrating environmental impact (Itsubo et al., 2007). Figure 14.2 shows schematically how these steps are related.

Categorizing impacts determines which technique will be used to assess what kinds of environmental problems (impact categories) such as global warming, ozone layer depletion, and acidification (SETAC, 1996). Classifying impacts is the task of sorting inventory data into their related impact categories and results in several substances being grouped into one impact category. For example, CO_2, CH_4, and N_2O are grouped into global warming, whereas NO_x and SO_x are grouped into acidification. Characterizing impacts involves assessing the environmental impacts of impact categories. Here, characterization factors that have been created for each environmental problem in the impact category are designated. For example, global warming potential (GWP; IPCC, 1995) is often used as the characterization factor for global warming so that the type of greenhouse gases contributing to global warming and to what extent can be analyzed. Normalizing impacts is a process that normalizes the assessment results obtained by characterizing each impact category in order to make relative comparisons. Grouping impacts typifies the impact categories resulting from characterization and normalization according to certain fixed conditions. Integrating impacts assigns weights to impact categories to achieve a single index weighting. Integration methods are now a subject of research under worldwide discussion. For example, one question is at which stage, e.g. from the emission of environmental burdens to the point where actual harm is done, is the integration performed? A method that directly integrates impact categories is called "midpoint assessment", whereas "endpoint assessment" is another method that calculates the actual magnitude of harm for items in each impact category, and remakes them into safeguard subjects such as "human health" or "ecosystem conservation" to be integrated later. In this way, although converting a number of approaches and methodologies into a single index weighting has been proposed, the resulting single index is often an economic or dimensionless indicator. Either way, the integration process requires that the different components of the natural environment, such as human health, the ecosystem, or social assets be weighed. The fairness and transparency of judgment criteria must be guaranteed because subjective value judgments are sometimes unavoidable. One must also fully keep in mind that results may change depending on how the environmental planners are tackling the problems, and how they implement the aforementioned integration method.

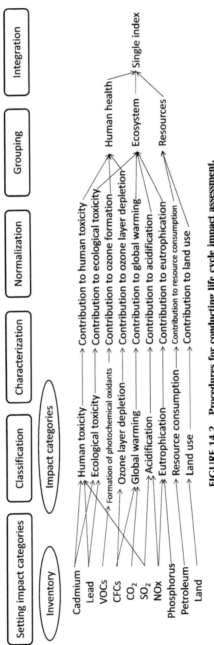

FIGURE 14.2 Procedures for conducting life cycle impact assessment.

The LCIA methodology and scientific framework are still in development, and no specific item has been widely adopted. For this reason, among the international standards being implemented currently, (a)–(c) above are mandatory whereas (d)–(f) are optional. In other words, when carrying out an LCIA, one must perform all tasks up to and including characterization. Whether to carry out normalization and integration depends on the final objective because the difficulties of comprehensively judging the importance of different impact categories and formulating a single index are recognized.

14.1.5. Interpretation

The interpretation step objectively and rationally examines the LCI and LCIA results obtained up to this point. Different results may be derived depending on, for example, differences in scope, definition of the system boundary in LCI, or the selection of characterization factors in LCIA, among many others. Hence, the influence of these factors on the results of the interpretation step must be examined. Furthermore, many of the data used in LCI, including those for measuring and estimating errors, and assessing how these errors affect the results, are important as well. One must perform "sensitivity analyses" and "uncertainty analyses" to take errors into account. However, there is a need for future research to advance knowledge related to subjects in this area because current specific methodologies lack uniformity.

14.1.6. LCA Criteria on Biomass

Some examples of using biomass for energy are presented in this section to demonstrate several important criteria for performing a biomass LCA.

a. Life Cycle

When using biomass as energy, considering the energy efficiency of the energy conversion technology is important. However, one must consider the energy efficiency and environmental burden throughout the entire life cycle. In the case of woody biomass, the environmental burdens of a series of life cycle processes, such as biomass production, preprocessing, and energy conversion, must be assessed, as shown in Figure 14.3.

b. Energy Balance

Energy balance basically involves comparing and assessing the biomass energy produced and the energy input for its production. As shown in Figure 14.3, considering the entire life cycle connected with production of biomass feedstock, its transport, and energy conversion is important in determining the invested energy. However, there are several ways of doing this, such as limiting the energy input to that from fossil fuels, or including the energy in the biomass

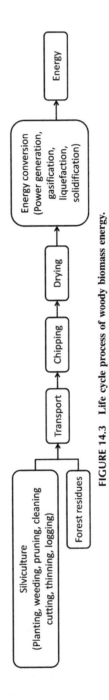

FIGURE 14.3 Life cycle process of woody biomass energy.

feedstock itself. In the former case one assesses how much more energy is produced than invested, whereas in the latter case one assesses the extent to which the feedstock energy is being effectively used (e.g. Box 14.1).

c. Carbon Neutrality

When used as energy, biomass emits CO_2, but the biomass energy is often assessed to be "carbon neutral", meaning zero emissions. This is based on the idea that because the CO_2 emitted from biomass has originally been absorbed through photosynthesis during the biomass growth process, the net carbon emission is thus considered to be zero. Another way of visualizing this concept is that the net emissions are zero because the CO_2 emitted is eventually absorbed by the next-generation biomass crop. However, there are concerns with such reasoning and assessment methods. First, fossil fuels including coal and oil are similar in that they had been formed by the absorption of CO_2 eons ago, thereby making it necessary to clearly specify that the carbon neutrality concept does not apply to fossil fuels. The second concern is the uncertainty of CO_2 reabsorption. Unless the next-generation crop is grown, zero CO_2 emissions are not warranted. Additionally, CO_2 reabsorption takes time, and future absorption is uncertain at the time when the energy is used. This leads one to perform assessments that include uncertainty. In view of these concerns, the matter of how to assess the CO_2 emission of using biomass energy itself is a major research topic that requires future discussion and study with regard to LCAs.

d. Greenhouse Gas Emissions Associated with Land Use Changes

A very large amount of carbon has been reported to be stored in soil as part of the global carbon cycle (IPCC, 2007). Changes of land use due to biomass production by forestry, agriculture, and other industries cause the carbon stored in soil emitted as CO_2 and other greenhouse gases. A representative example is the carbon emissions caused by land use changes when producing palm oil in the peat soil of tropical rainforests. However, appropriate soil management will also make it possible to keep the stored carbon intact as much as possible, thereby alleviating greenhouse gas emissions. In biomass LCAs, conducting assessments that consider the carbon emissions occurring in conjunction with land use changes is important as well.

14.2. SUSTAINABILITY INDICATORS FOR BIOENERGY

Seishu Tojo

14.2.1. Importance of Sustainability Indicators

Concerns about potential negative effects of large-scale biomass production and export on environmental quality by deforestation and reduction of

BOX 14.1 Energy Analyses on the Production of Wood Pellets from Wood Bark

1 Wood Pellets

Wood pellets are small cylindrical pieces 10–20 mm long with diameters varying from 5 to 10 mm produced from fine-ground wood bark. They are usually used as fuel by feeding into burner automatically because of their small and fixed form. On the other hand, energy is consumed in the process of drying and grinding wood barks for making wood pellets. Therefore, wood pellets have a smaller ratio of output energy to input energy than wood chips; however, wood pellets can be used in residential houses to be burnt efficiently in relatively small stoves. The energy efficiency of wood pellet production system must be improved to reduce their greenhouse gas (GHG) emission for expanding the use of wood pellets as a major energy source.

2 Materials and Methods

A mill that has been established as a guild of sawmills to process wood barks, which are a gavage from the sawmilling process, was evaluated twice on the production of bark pellets. This mill mainly produces about 500 t bark pellets per year. The first evaluation was carried out just after the establishment of the mill, and the ratio of output energy to input energy was 2.64. This ratio was considered low because the mill operation had not been stable due to a short period of operation since its establishment. The second research was carried out 5 years later; the ratio was expected to be improved because of a more efficient process and stable operations.

The output energy is defined as the high calorific value of the pellet as calculated from available shipments and the unit high calorific value. The available shipment is smaller than production because a portion of products is used as fuel for drying the ground bark in the mill. It was calculated as a reminder of subtracting stock of the previous month from the amount of sales and stocks of the current month. The output energy E_o was calculated using the following equation:

$$E_o = P_r \times \frac{100 - u}{100} \times e_p, \qquad (14.1)$$

where P_r is the available shipment (t), u is water content of pellets (%), and e_p is high calorific value of pellets (GJ t^{-1}).

The input energy can be divided into four categories: (1) electric power consumed inside of the mill; (2) energy of fuels consumed by a forklift inside of the mill; (3) transportation energy to collect barks from outside the mill; and (4) transportation energy to ship pellets outside the mill. The electric power consumption and the fuel consumption by the forklift can be obtained from the invoices. The transportation energy to collect barks and to ship pellets can be calculated using the mass-distance method based on the average loadage, number of transportations, and transportation distances. The average loadage and the number of transportations can be acquired from the invoices, whereas the transportation distance is measured using the Map Fan Web (http:// www.mapfan.com), and the input energy E_i (GJ) is calculated using the following equation:

$$E_i = F_l \times e_l + F_u \times e_u, \qquad (14.2)$$

BOX 14.1 Energy Analyses on the Production of Wood Pellets from Wood Bark—Cont'd

where F_l is fuel consumption, F_u is electric power consumption, e_l is high calorific value of fuels, and e_u is the conversion factor from electric power to energy.

The GHG emission can also be calculated using the following equation:

$$G_i = F_l \times g_l + F_u \times g_u, \tag{14.3}$$

where g_l is GHG emission factor of fuels and g_u is GHG emission factor of electric power.

This pellet mill consumes the pellet product as the fuel to dry ground wood barks as described above. The energy consumption of drying is calculated based on the quantity of consumed pellets and the quantity of pellets dried. The water contents of pellets and barks are determined using the collected pellet and bark samples.

3 Results

The average wood pellet production was 2637.6 kg per day by consuming on average 50.26 kg an hour of wood pellets during the study period. For 7.9 hours operational period per day, the total amount of consumed pellets was 397.05 kg per day, which amounted to 20% of total wood pellet production.

Results of the mass-distance method indicate that fuel consumption of the bark collection process is 2153.6 L per year to produce 478.5 t of wood pellets per year that is available for shipment. Therefore, the wood bark collection process consumes 0.17 GJ t^{-1} (Figure 14.4) and emits 11.66 kg-CO$_2$ t^{-1}. The pellet shipping process consumes 742.5 L fuel a year that is equivalent to 0.06 GJ t^{-1} with the emission of 4.42 kg-CO$_2$ t^{-1}. The invoice investigation reveals that the average electric power consumption is 4.99 GJ t^{-1} and the average CO$_2$ emission from the electric power is 240.5 kg-CO$_2$ t^{-1}. The average energy consumption to operate the forklift is 0.12 GJ t^{-1} with 8.08 kg-CO$_2$ t^{-1} emitted. Hence, electric power is the most consumed energy; it amounts to 93% of the total energy consumption. Moreover, most of CO$_2$ emission is from the electric power, and it amounts to 91% of the total CO$_2$ emission.

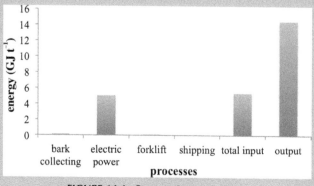

FIGURE 14.4   Input and output energy.

BOX 14.1 Energy Analyses on the Production of Wood Pellets from Wood Bark—Cont'd

Based on the data collected in this study, the ratio of output energy to input energy for processing wood pellets is 2.70 that conforms to the value observed in previous studies. This indicates that improving the production process to make the operation more stable does not contribute to improving the ratio of output energy to input energy or the energy efficiency of this process. However, the wood pellet process has a higher ratio than the biodiesel fuel from rapeseed but a smaller ratio than the ethanol fuel from sweet potato.

The process of drying the ground bark before processing into wood pellets is powered by burning the produced pellets that is equivalent to consuming 15% of total production (2.98 GJ t^{-1}). If kerosene is used for drying the ground bark, the CO_2 emission will increase from 264.6 to 394.4 kg-CO_2 t^{-1}. Hence, using the processed wood pellets to dry the bark will reduce CO_2 emissions by 129.8 kg-CO_2 t^{-1}, or a reduction of 70 t-CO_2 emissions per year.

Masahiro Iwaoka

food production by diverting food resources for energy production have led to the necessity of developing sustainability criteria and certification systems for managing and regulating biomass production and trade (Lewandowski and Faaij, 2006). There is a need to establish minimum sustainability criteria for importing bioethanol to avoid unwanted negative impacts or leakage effects (Corbière-Nicollier et al., 2011). Sheehan (2009) stresses the complex issue of sustainability for biofuels that certainly requires a holistic vision as well as the recognition of its ethical nature.

Identifying what is sustainable is difficult because sustainability as a social value is controversial by nature (Buchholz et al., 2009). For instance, some people value the social, economic, and ecological factors of sustainability equally, whereas some others support the view of nested components of sustainability to stress that sustainability can only be achieved when its social and economic factors do not violate ecological limits (Gowdy, 1999). Since the multiple perspectives encompassed in the concept of sustainability are based on normative values, the concept requires specific measurements (Buchholz et al., 2009). Thirty-five sustainability criteria that are regularly included in discussions on bioenergy have been identified. The criteria that had been identified were grouped into the broad categories of 15 social criteria, four economic criteria, and 16 environmental criteria based on four attributes including relevance, practicality, reliability, and importance using the following definitions:

- **Relevance:** How relevant is the criterion to the concept of sustainable bioenergy systems? Does its assessment contribute to a better understanding of the sustainability of the bioenergy system?

- **Practicality:** Are there existing scales and/or measurement units? Are there measurable threshold values? How easily can data be obtained? Is measuring the indicator cost, time, and/or resource effective?
- **Reliability:** How reliable is the result of assessing the criterion? Is there a high uncertainty attached to the criterion? Are results reproducible? How easily can consensus be achieved?
- **Importance:** How important is the criterion for assessing the sustainability of the bioenergy system? Is it critical, i.e. is it assessed according to opinion in a mandatory way, to include it in a sustainability assessment of bioenergy systems?

According to Mayer (2008), the determination of the sustainability of human–environment systems is based on the consideration of three main characteristics: resilience to disturbances, both natural and anthropogenic; desirability to human societies; and temporal and spatial scale boundaries. Resilience and desirability determine policy goals, and the scale determines the system to be monitored and managed to reach those goals. In relation to biofuels, such means are (Silva Lora et al., 2011):

- To remain carbon neutral, considering the necessity of fossil fuel substitution and global warming mitigation.
- Not to affect the quality, quantity, and rational use of available natural resources as water and soil.
- Not to have undesirable social consequences such as starvation because of high food prices.
- To contribute to society's economic development and equity.
- Not to affect biodiversity.

14.2.2. Sustainability Indicators Developed by the Global Bioenergy Partnership

The Partners and Observers of the Global Bioenergy Partnership (GBEP) under FAO (Food and Agriculture Organization) developed the sustainability indicators for bioenergy to provide a framework for assessing the relationship between production and use of modern bioenergy with sustainable development. These indicators offered policymakers and other stakeholders a set of analytical tools for developing national bioenergy policies and programs, and monitoring the impact of these policies and programs. A set of 24 GBEP sustainability indicators for bioenergy, as shown in Table 14.1, is developed under three pillars, i.e. environmental, social and economic, with the relevant themes listed as follows:

- **Environmental:** Greenhouse gas emissions, Productive capacity of the land and ecosystems, Air quality, Water availability, Use efficiency and quality, Biological diversity, Land-use change, including indirect effects.

TABLE 14.1 Sustainability Indicators (GBEP, 2011)

Pillars		
Environmental	**Social**	**Economic**
Indicators 1. Life-cycle GHG emissions	9. Allocation and tenure of land for new bioenergy production	17. Productivity
2. Soil quality	10. Price and supply of a national food basket	18. Net energy balance
3. Harvest levels of wood resources	11. Change in income	19. Gross value added
4. Emissions of non-GHG air pollutants, including air toxics	12. Jobs in the bioenergy sector	20. Change in consumption of fossil fuels and traditional use of biomass
5. Water use and efficiency	13. Change in unpaid time spent by women and children collecting biomass	21. Training and re-qualification of the workforce
6. Water quality	14. Bioenergy used to expand access to modern energy services	22. Energy diversity
7. Biological diversity in the landscape	15. Change in mortality and burden of disease attributable to indoor smoke	23. Infrastructure and logistics for distribution of bioenergy
8. Land use and land-use change related to bioenergy feedstock production	16. Incidence of occupational injury, illness and fatalities	24. Capacity and flexibility of use of bioenergy

- **Social:** Price and supply of a national food basket, Access to land, Water and other natural resources, Labor conditions, Rural and social development, Access to energy, Human health and safety.
- **Economic:** Resource availability and use efficiencies in bioenergy production, Conversion, Distribution and end-use, Economic development, Economic viability and competitiveness of bioenergy, Access to technology and technological capabilities, Energy security/diversification of sources and supply, Energy security/infrastructure and logistics for distribution and use (e.g. Box 14.2).

BOX 14.2 Evaluation of Bioethanol Production from Rice Straw

Due to rapid growth in population and industrialization, worldwide ethanol demand is increasing continuously. The production of conventional crops such as corn and sugar cane that have been cropped for food and feed is insufficient to meet the new global demand of bioethanol production.

Therefore, lignocellulosic substances such as agricultural wastes are attractive alternative sources for bioethanol production because they are cost-effective, renewable, and abundant. Bioethanol from agricultural waste emerges as a promising technology, although the process has several challenges and limitations such as biomass transport and handling, and efficient pretreatment to achieve a nearly complete delignification of lignocelluloses (Sarkar et al., 2012). Considering the evolution and need of second-generation biofuels, rice straw appears a promising and potent candidate for production of bioethanol due to its abundant availability and attractive composition (Binod et al., 2010). Bioconversion of lignocellosics to bioethanol is difficult due to: (1) the nature of biomass to resist breakdown; (2) identification of natural microorganisms or genetic creation of new species of organisms that are capable of fermenting the variety of sugars released from the degraded hemicellulose and cellulose polymers not being completed currently; (3) relatively high costs for collection and storage of low-density lignocellosic materials (Balat, 2011).

According to Park et al. (2011b), significant amounts of soft carbohydrates ($62-303$ g kg^{-1}) have been detected in all rice cultivars, with $58.9-86.0\%$ of the total soft carbohydrates being starch and sucrose. In addition to cellulose, soft carbohydrates are regarded as important sources of hexoses that can be fermented by using the commercially available ethanol-producing yeast, *Saccharomyces cerevisiae*. Soft carbohydrates constitute about $20.8-133\%$ of cellulose that signifies the importance of utilizing soft carbohydrates for ethanol production from rice straw.

Many researchers reported effective pretreatment, saccharification, and ethanol fermentation processes to enhance ethanol production from rice straw. These processes include acid hydrolysis, liquid hot water extraction, steam explosion, dilute acid–steam explosion, ammonia fiber explosion, lime pretreatment, pretreatment by using aqueous ammonia solution (Ko et al., 2009), the combined use of ammonia and ionic liquid treatment (Nguyen et al., 2010), and the simultaneous saccharification and ethanol fermentation process (SSF). Shinozaki and Kitamoto (2011) reported that (i) silage made from whole-plant rice can be used for bioethanol production, and (ii) proper selection and combination of commercially available enzymes can make the SSF process more cost-efficient by eliminating the pretreatment step. New strategies to ferment the mixture of hexose and pentose have been developed using co-cultures of *Saccharomyces cerevisiae* and *Pichia stipites* (Yadav et al., 2011), and with sequential application of *Saccharomyces cerevisiae* and *Pichia stipitis* (Park et al., 2011a).

Roy et al. (2012a) evaluated the life cycle of bioethanol production from rice straw with three indicators, i.e. net energy consumption, CO_2 emission, and production costs estimated for bioethanol produced from the most common variety of rice straw in Japan (*Oryza sativa* L. cv. Koshihikari). Their studies were based on the three scenarios of a basic case, an innovative case, and a futuristic case using data collected from their pilot plant and taken from the literature. The net energy

(Continued)

BOX 14.2 Evaluation of Bioethanol Production from Rice Straw—Cont'd

consumption, CO_2 emission, and production costs are $10.4-11.6$ MJ L^{-1}, $1.10-1.14$ kg L^{-1}, and $0.88-1.37$ US$ L^{-1} respectively. They also reported on bioethanol from sugar-rich straw of variety "Leaf Star" with energy scenario (Table 14.2). The net energy consumption, CO_2 emissions, and production costs vary from 10.0 to 17.6 MJ L^{-1}, -0.47 to 1.58 kg L^{-1}, and 0.86 to 1.44 US$ L^{-1} respectively. A shift in energy scenarios in the type of primary energy not only reduces emissions and production costs, but may also alleviate the fluctuation in production costs over time, thus decreasing the risk on investment in the bioethanol industry (Roy et al., 2012b). The rice straw collection/transportation cost accounts for 68% of the total cost of the ethanol production system. In terms of cost-effectiveness, bioethanol is not competitive with gasoline. The bioethanol, however, can be competitive for the end user if it is tax exempt from the gasoline tax in Japan (Yang and Sagisaka, 2009).

Rice straw management has significant impacts on CH_4 emissions from paddy fields. When rice straw was left in the field and mixed into the soil, total CO_2-equivalent greenhouse gas emission is $25.5-28.2$ t-CO_2 ha^{-1}. However, a method that is effective in mitigating CH_4 emission is to remove rice straw from the paddy fields after rice harvest (Koga and Tajima, 2011). Thus, rice straw removal from paddy fields as bioethanol feedstock has significant impacts on alleviating greenhouse gas emissions.

TABLE 14.2 Life Cycle Inventory of Indicators for Bioethanol from Rice in Japan

Cultivar	Net energy consumption (MJ L^{-1})	CO_2 emission (kg-CO_2 L^{-1})	Production costs (US$ L^{-1})	Reference
Fukuhibiki*	13.4			Saga et al. (2007)
Rice[†]	12.7	0.67	1.81	Yang and Sagisaka (2009)
Koshihikari[‡]	11.6	1.14	1.37	Roy et al. (2012a)
Leaf Star[§]	10.0	0.93	1.05	Roy et al. (2012b)

*Rice cultivar for livestock feed "forage rice".
[†]Cultivar is not identified, standard cultivar.
[‡]The heat generation from the residual lignin is used to offset some of the energy consumption, CO_2 emission, and production cost.
[§]Rice cultivar for livestock feed "forage rice". The lignin and hemicellulose recovered in the waste management process is used for heat or electricity generation.

14.3. MANAGEMENT OF FOREST LAND FOR BIOMASS PRODUCTION

Masahiro Iwaoka

14.3.1. Sustainable Forest Management

Forest land needs to be properly managed in order to maintain sustainability. The United Nations held a conference on environment and development at Rio de Janeiro in June 1992 to reach an agreement known as "Agenda 21"; moreover, the declaration of a forest principle was also agreed upon. Sustainable forest management is described in the declaration as "Forest resources and forest lands should be sustainably managed to meet the social, economic, ecological, cultural and spiritual needs of present and future generations. These needs are for forest products and services, such as wood and wood products, water, food, fodder, medicine, fuel, shelter, employment, recreation, habitats for wildlife, landscape diversity, carbon sinks and reservoirs, and for other forest products." The term "sustained yield" that was once popular in the forestry industry is defined in Webster's Dictionary as "production of a biological resource (as timber or fish) under management procedures which ensure replacement of the part harvested by regrowth or reproduction before another harvest occurs" by Merriam-Webster. "Sustained yield" mainly concerns the yield, whereas "sustainable forest management" is a broader concept including environment quality in addition to forest resources.

There are several methods to implement sustainable forest management; forest certification is one of these methods based on the consumer's intention. The forest certification schemes will be described in the following sections.

14.3.2. Forest Certification Schemes

a. Definition of Forest Certification Scheme

Forest certification is a system that certifies a well-managed forest and guarantees timber or non-timber production in the forest. A well-managed forest is "environmentally appropriate, socially beneficial, and economically viable" (Forest Stewardship Council), and uses "global and local management to ensure that all of us can enjoy the environmental, social and economic benefits that forests offer" (Program for the Endorsement of Forest Certification, PEFC). Forests will gain some benefits when consumers prefer the forest products labeled as environmentally conscious products to other products with similar quality and price. The benefits may not only come from direct income but also cover intangible benefits such as lower non-tariff barrier or less consumer resistance to non-environment-conscious products.

Forest certification can usually be divided into two parts: certification for forest management (FM) and certification for the supply chain, known as chain of custody (CoC). The FM certification guarantees forest management systems whereas the CoC certification guarantees timber or non-timber production for FM certified forests. For achieving these purposes, the FM requires a system for monitoring, checking, and improving forest management, whereas the CoC requires another separation system for certifying forest-related products. Every component of a supply chain involved in forest-related products and services must be certified as CoC in order for the final products and services to be certified as "environment-conscious products".

b. International Forest Certification Schemes

The Forest Stewardship Council A.C. (FSC) and the Program for the Endorsement of Forest Certification (PEFC) are the two most popular forest certification schemes worldwide, although there are numerous other international or local forest certification schemes. The FSC Secretariat opened in Oaxaca, Mexico and the FSC was established as a legal entity in Mexico in February 1994, 2 years after the United Nations Conference on Environment and Development – the Earth Summit was held in Rio de Janeiro in 1992. Although producing no legally binding commitments on forest management, the Earth Summit resulted in Agenda 21 and the non-legally binding Forest Principles that provide a crucial forum for congregating many non-governmental organizations and pooling support for the innovative idea of a non-governmental, independent, and international forest certification scheme. Following intensive consultations in 10 countries to build support for the idea of a worldwide certification system, the FSC Founding Assembly was held in Toronto, Canada in 1993. The mission of FSC is to promote environmentally appropriate, socially beneficial, and economically viable management of the world's forests. To perform the mission, the FSC has developed 10 principles and 56 criteria; each principle is supported by several criteria that provide a way of judging whether the principle has been met in practice. All principles and criteria must be applied by a forest management unit before the latter can receive FSC certification. The Principles and Criteria apply to all forest types and to all areas covered by the management unit included in the scope of the certificate. These principles and criteria are not specific to any particular country or region, they are applicable worldwide and are relevant to various forest areas and different ecosystems, as well as cultural, political, and legal systems. The FSC's principles and criteria are characterized by three unique principles. The first is an indigenous people's rights that require identifying and upholding indigenous people's rights of ownership and use of land and resources. The second is about maintenance of high conservation value forests that requires maintaining or enhancing the attributes that define such forests. The third is about plantations that require planning and managing plantations in accordance with FSC Principles and Criteria. FSC accredits certification bodies and the accredited certification body certifies each forest. The FSC has an integrated accreditation program that systematically checks the forests that have been certified by the FSC. The area of FSC certified forest amounts to

152,157,408 ha in 80 countries as of June 2012. In Japan, 35 certificates have been issued by the FSC covering 392,989 ha of forest.

The PEFC, which is an international non-profit, non-governmental organization dedicated to promoting sustainable forest management (SFM) through independent third-party certification, is an umbrella organization that endorses national forest certification systems. It was established in 1999 by national organizations from 11 countries representing a wide range of interests to promote sustainable forest management, especially among small forest managers. The PEFC recognized the first national system in 2000, enabling forest owners and managers in Finland, Sweden, Norway, Germany, and Austria to certify their responsible forest management practices. In 2001, in an effort to integrate social concerns more fully in its activities, the PEFC became the first global forest certification organization to require compliance with all the fundamental ILO conventions in forest management. In 2004, Australia and Chile became the first non-European nations with national standards endorsed by the PEFC. With the endorsement of the Canadian standard in 2005, the PEFC became the world's largest forest certification system with more than 100 million hectares of certified forest area. The mission of the PEFC is "to give society confidence that people manage forests sustainably" to realize their vision of "a world in which people manage forests sustainably". The PEFC bases its understanding of SFM on the definition adopted by the Food and Agriculture Organization (FAO) and originally developed by Forest Europe. SFM is defined by FAO and Forest Europe as: "The stewardship and use of forests and forestlands in a way, and at a rate, that maintains their biodiversity, productivity, regeneration capacity, vitality and their potential to fulfill, now and in the future, relevant ecological, economic and social functions, at local, national, and global levels, and that does not cause damage to other ecosystems." To achieve sustainability under this definition, the PEFC states that forest management practices must result in outcomes that are economically viable, ecologically sound, and socially just. The PEFC collaborates with national forest certification systems tailored to local conditions and involving a wide range of stakeholders to encourage the delivery of sustainably sourced products to the marketplace because the diversity of both forests and communities that depend upon the forests for their livelihoods means that a "one-size-fits-all" standard is not the solution. A national certification system that has developed standards in line with PEFC requirements can apply for endorsement to gain access to global recognition and market access through PEFC International. To achieve endorsement, the national system needs to meet the PEFC's rigorous Sustainability Benchmark. This "bottom-up" approach provides a high degree of independence of national processes, and allows for the development of standards tailored to the political, economic, social, environmental, and cultural realities of their respective countries, yet in compliance with rigorous international benchmarks. The area of PEFC-certified forest is 242,317,994 ha in 27 countries as at June 2012.

c. Japanese Local Forest Certification Scheme

The Sustainable Green Ecosystem Council (SGEC) established in 2003 is the only Japanese local forest certification scheme in the context of Japanese forestry. It was established based on the idea that "sustainable forest management (SFM)" can be achieved effectively through forest certification systems. SGEC is becoming popular because it promotes SFM, encourages cyclical use of forest resources, raises forest management level, contributes to global warming prevention efforts, and supports people's lives in both country and urban settings. One year before the SGEC was established, the investigative committee led by the Japan Forestry Association proposed the establishment of the SGEC based on the investigation of a local forest certification system suitable for Japanese forestry. The purpose of the SGEC is to spread the idea of SFM throughout Japanese societies, promote effective use of wood products from "sustainable forest management (SFM)", construct a cyclical society so as to support a pleasant lifestyle, and help maintain a natural environment. To achieve these objectives, the SGEC has initiated four important activities:

1. Operation of SGEC certification system that contains a labeling system for identifying products from SGEC-certified forests.
2. Accreditation of certification bodies and registration of the consultation bodies.
3. Corporation with overseas forest accreditation bodies.
4. Investigation on and dissemination of SFM.

The procedure of SGEC forest certification system starts with an application to a certification body by an individual or a deputized consultation body. According to forest certification standards, the certification body assesses the management of a forest being certified to report the findings to SGEC. If approved by the SGEC audit committee, SGEC issues a SGEC certificate to the applicant who can then claim himself as owner/manager of the certified forest. Subsequently, the audit committee investigates the certified forests once a year in order to confirm that the certified forest continues appropriate forest management practices. The SGEC labeling system follows the same procedure as those stated in the above sections. That is, SGEC issues a certificate to an eligible applicant who may be called a certified entity. The certification body, which is independent of the applicant, should conduct accurate examination for making fair and neutral judgments and decisions. Additionally, the certification body should operate independently from other forest certification schemes. A consultation body should be registered on condition of its ability to give advice to applicants in a precise, quick, and efficient manner. The SGEC promotes both separation control and label control so that forest products from SGEC-certified forests (hereinafter referred to as "certified wood products") are properly supplied to consumers. The SGEC has established seven standards; Table 14.3 lists a comparison of SGEC standards to principles of the FSC and criteria of the SGEC. The SGEC standards are arranged in order of

TABLE 14.3 Comparison of SGEC Standards, FSC Principles, and PEFC Criteria

Standards of SGEC	Principles of FSC	Criteria of PEFC
Standard 1: Identification of forests and their management policies	Principle 1: Compliance with laws and FSC principles	Criterion 1: Maintenance and appropriate enhancement of forest resources and their contribution to the global carbon cycle
Standard 2: Conservation of biological diversity	Principle 2: Tenure and use rights and responsibilities	Criterion 2: Maintenance of forest ecosystem health and vitality
Standard 3: Conservation and maintenance of soil and water resources	Principle 3: Indigenous people's rights	Criterion 3: Maintenance and encouragement of productive functions of forests (wood and non-wood)
Standard 4: Maintenance of productivity and health of forest ecosystem	Principle 4: Community relations and worker's rights	Criterion 4: Maintenance, conservation and appropriate enhancement of biological diversity in forest ecosystems
Standard 5: Legal and institutional framework for SFM	Principle 5: Benefits from the forest	Criterion 5: Maintenance and appropriate enhancement of protective functions in forest management (notably soil and water)
Standard 6: Maintenance and promotion of societal and economic benefits	Principle 6: Environmental impact	Criterion 6: Maintenance of other socio-economic functions and conditions
Standard 7: Monitoring and disclosure of information	Principle 7: Management plan	Criterion 7: Compliance with legal requirements
	Principle 8: Monitoring and assessment	
	Principle 9: Maintenance of high conservation value forests	
	Principle 10: Plantations	

management definition, subjects, and measures. In contrast, the FSC principles are arranged in order of relationships between the forest and people or society around the forest, management subjects and measures, whereas the PEFC criteria are arranged according to management subjects. The arrangement of SGEC standards is similar to FSC principles; however, each SGEC standard is similar to a corresponding PEFC criterion. These differences in arrangements cause discrepancies between the numbers of environmental conservation items, and larger numbers of SGEC standards than numbers of FSC criteria. The total SGEC-certified forest area is 888,779.55 ha and the number of total CoC-certified business entities is 381 in June 2012.

14.4. MANAGEMENT OF FARMING LAND FOR BIOMASS PRODUCTION

Seishu Tojo

14.4.1. Environmental Indicators for Agriculture

Many OECD countries have established goals to reduce discharges of agricultural surface runoffs, particularly from livestock farming, which are rich in nitrogen and phosphorus, into the environment. The EU (European Union) Nitrate directive (EU Council Directive 676/91) is one of the measures introduced to comply with EU drinking water standards; it is implemented by limiting the use of fertilizer to reduce nitrogen inputs to designated nitrate vulnerable zones. Many countries have also introduced similar programs to alleviate acidification, including agricultural ammonia emissions into the atmosphere resulting from livestock farming, and the use of inorganic fertilizers.

The OECD published environmental indicators for agriculture in 2001 (OECD, 2001) based on the work carried out by the OECD Joint Working Party of the Committee for Agriculture and the Environment Policy Committee. These indicators are primarily useful for policymakers and the general public, in both OECD and non-OECD countries. Specific indicators for the environmental impact of agriculture are categorized as soil quality, water quality, land conservation, greenhouse gases, biodiversity, wildlife habitats, and landscape (Table 14.4).

According to the report published by OECD, changes of cropping pattern such as exploiting the potential of agricultural land as a source of biomass instead of foodstuff production will have considerable environmental consequences. An important aspect concerning the link between biodiversity and agriculture is the relationship between biomass production from agriculture (i.e. crops and forage) and species diversity. There are also important links between biomass production in agriculture and productivity, such as the possibilities of growing extra plant biomass through technology to provide energy in addition to food stuff. Research into the relationship between biomass production and biodiversity, however, is still at an early stage of development.

TABLE 14.4 Complete List of OECD Agri-Environmental Indicators

I. Agriculture in the broader economic, social, and environmental context

1. Contextual information and indicators

- Agricultural GDP
- Agricultural output
- Farm employment
- Farmer age/gender distribution
- Farmer education
- Number of farms
- Agricultural support

- Land use
  - Stock of agricultural land
  - Change in agricultural land
  - Agricultural land use

2. Farm financial resources

- Farm income
- Agri-environmental expenditure
  - Public and private agri-environmental expenditure
  - Expenditure on agri-environmental research

II. Farm management and the environment

1. Farm management

- Whole farm management
  - Environmental whole farm management plans
  - Organic farming

- Nutrient management
  - Nutrient management plans
  - Soil tests
- Pest management
  - Use of non-chemical pest control
  - Methods
  - Use of integrated pest management

- Soil and land management
  - Soil cover
  - Land management practices

III. Use of farm inputs and natural resources

1. Nutrient use

- Nitrogen balance
- Nitrogen efficiency

2. Pesticide use and risks

- Pesticide use
- Pesticide risk

3. Water use

- Water use intensity
- Water use efficiency

(Continued)

TABLE 14.4 Complete List of OECD Agri-Environmental Indicators—Cont'd

III. Use of farm inputs and natural resources

1. Nutrient use	2. Pesticide use and risks	3. Water use
		● Water use technical efficiency
		– Water use economic efficiency
		● Water stress

IV. Environmental impacts of agriculture

		4. Greenhouse gases
		● Gross agricultural greenhouse gas emissions

1. Soil quality
● Risk of soil erosion by water
● Risk of soil erosion by wind

3. Land conservation
● Water retaining capacity
● Off-farm sediment flow (soil retaining capacity)

7. Landscape
● Structure of landscapes
  – Environmental features and land use patterns
  – Man-made objects (cultural features)
● Landscape management
● Landscape costs and benefits

2. Water quality
● Water quality risk indicator
● Water quality state indicator

6. Wildlife habitats
● Intensively-farmed agricultural habitats
● Semi-natural agricultural habitats
● Uncultivated natural habitats
● Habitat matrix

5. Biodiversity
● Genetic diversity
● Species diversity
  – Wild species
  – Non-native species
● Eco-system diversity
(see Wildlife Habitats)

EU Directive 2009/28/EC specifies three criteria for the land to provide biofuel from feedstock. First, biofuels shall not be made from raw materials obtained from land with high biodiversity value, which includes primary forest and other wooded land, areas designated for nature protection or the protection of rare, threatened or endangered ecosystems or species, and highly biodiverse grassland. Second, biofuels shall not be made from raw materials obtained from land with high carbon stock, namely wetlands, continuously forested areas, or land spanning more than one hectare with a certain minimum canopy cover. Third, biofuels shall not be made from raw materials obtained from peatland, unless evidence is provided to show that the cultivation and harvesting of that raw material does not involve drainage of previously undrained soil.

In Brazil, the expansion of soybean production could adversely affect rain-forests. In Indonesia 27% of palm oil concessions are on peat forest, whereas in Malaysia 10% of plantations are on former peat forest. Therefore, the biodiesel fuel from some Brazilian soybean as well as Malaysian and Indonesian palm oil may not conform to the EU Directive (Lendle and Schaus, 2010).

14.4.2. Nutrient Runoff from Biomass Production Fields

The sustainable development of alternative renewable fuel can offer many benefits but will demand comprehensive understanding of how our choice of land use affects the ecological systems around us. The Mississippi-Atchafalaya River Basin covers almost half of the USA in an enormous swath of land before the river finally empties into the Gulf of Mexico. The area includes predominant wheat- and corn-growing fields of the mid-west region that supplies a large share of feedstock for current production of starch-based biofuels (Figure 14.5).

History reveals that land-use choices in these upstream agricultural belts have far-reaching effects on the environment. The use of nitrogen- and phosphorus-rich fertilizer over several decades has led to abundant nutrient-rich surface runoffs that eventually make their way into the Mississippi to cause algal blooms and murky water. The algae die and decompose rapidly, causing serious depletion of dissolved oxygen, creating areas known as the "dead zone" where only a few species of organisms can survive.

The Advisory Panel of the Environmental Protection Agency's Science Advisory Board recommended the promotion of environmentally sustainable approaches to biofuel crop production, such as no-till farming, the reduced use of fertilizer, and the use of riparian buffers in targeted areas of the basin (Dale et al., 2010).

The use of dedicated energy crops generally requires less nutrient applications than corn or wheat. Switchgrass, which is currently recommended as a potential biomass crop, is harvested in quantities of about one-third of the quantity of corn cropped in the Southeast USA (Garland, 2008). The grass also contains less phosphorus than does corn. Reviewing the literature on potential water quality resulting from the use of switchgrass as a potential source of

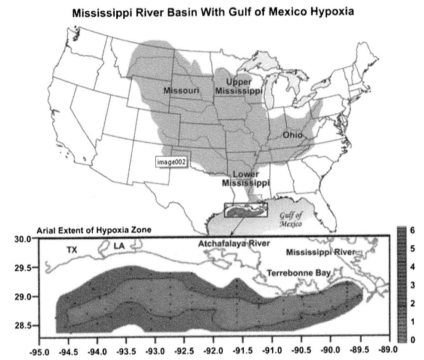

FIGURE 14.5 Mississippi river basin with Gulf of Mexico hypoxia. *(Source: Programmatic environmental impact statement, Biomass crop assistance program; USDA, 2010.)*

biofuel, Simpson et al. (2008) summarized the benefits as fewer nutrients existing in runoff and less drainage by approximately 50–90% than corn–soybean rotations that will have a significant positive impact on water quality.

14.4.3. Abandoned Agricultural Land in Japan

Urban sprawl and other changes in land use including abandonment of cultivation had caused a loss of agricultural land area by 922,700 ha from 1973 to 2001 (Takata et al., 2011). The Japanese government announced that the abandoned agricultural land increased rapidly in the 1990s and reached a high level of 396,000 ha or 10.6% of total arable land in 2010, as shown in Figure 14.6. Yoshikawa et al. (2004) warns that a large-scale soil loss of $90 \, t \, ha^{-1} \, y^{-1}$ will occur within several years immediately after a cultivated field is abandoning. According to Uematsu et al. (2010), recent land-use changes involving abandonment have rapidly reduced the biodiversity of agricultural landscapes and decreased habitats of endangered and rare species as well.

One of the countermeasures for alleviating land abandonment launched by a municipal government in Yamagata prefecture, Japan in cooperation with the truck association is to cultivate rapeseed in abandoned fields as an energy crop.

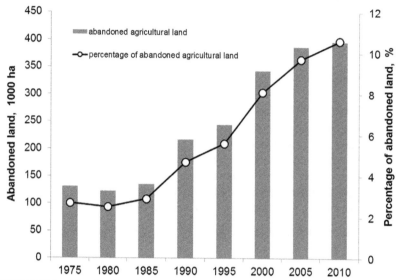

FIGURE 14.6   Abandoned agricultural land in Japan. *(Source: Census of agriculture and forestry, MAFF, Japan.)*

Rapeseed oil is used as cooking oil free of charge at public school lunch centers and public elderly care facilities. The used cooking oil is collected and processed into biodiesel that will be sold to truck association members. In this manner, biomass production may help to preserve farmland.

14.5. MONITORING SYSTEM FOR FOREST MANAGEMENT

Takeshi Matsumoto

14.5.1. Monitoring Methods

Investigating the forest enables us to grasp the current state of a forest's biomass resources. A direct and exact investigation is difficult because forests cover large areas and have huge quantities of resources. Sampling methods have been adopted in many cases for investigating forests. Furthermore, broader investigations at the municipal, prefectural, and national levels using remote sensing methods and other advanced technologies have been developed in recent years.

a. Ground Survey Method

(i) Complete Enumeration Method

This method measures the diameter at breast height (DBH) of all trees and heights for some selected trees in a target forest. The results will be used to develop a height–diameter curve, which can be used for estimating the height of all trees in the forest. The timber volume and stand volume are estimated using a two-way DBH and tree height volume table for each species in a certain region (Forestry Agency, 1970). Although the method is costly and time-consuming, it is an exact method that collects information on individual trees. Hence, this method is suitable for fine and expensive wood forests.

(ii) Sample Plot Method (by Non-Random Selection)

This method estimates the stand density and growth stock of a target forest based on field data collected on DBH, tree height, and number of stand trees in a selected sample plot, which is a typical representative point of the forest. In a uniform forest, the plot area tends to be larger but fewer plots are measured. In contrast, when the plot area is smaller more plots are measured, as in streaky forests. The following are examples of plot scale and accuracy:

- Circle (radius 4 m: using a fishing rod) $= 50\,\text{m}^2 \rightarrow 200$ times is 1 ha, a tree stand represents 200 trees in 1 ha.
- A 10-m square $= 100\,\text{m}^2 \rightarrow 100$ times is 1 ha, a tree stand represents 100 trees in 1 ha.
- A 20-m square $= 400\,\text{m}^2 \rightarrow 25$ times is 1 ha, a tree stand represents 25 trees in 1 ha.
- A 30-m square $= 900\,\text{m}^2 \rightarrow 11$ times is 1 ha, a tree stand represents 11 trees in 1 ha.

Sometimes, a rectangular plot is also used.

(iii) Sample Plot Method (by Random Selection)

This method differs from the above-mentioned non-random selection method because it applies sampling survey methods based on statistics that allows the calculation of standard errors and confidence intervals.

(iv) Plotless Method

Among the many available methods, the following three methods are described in this chapter.

Bitterlich method This method counts the number (N) of trees thicker than the collimation angle (after the investigator collimates trees around the sample point), and calculates the basal area of breast height [G ($\text{m}^2\,\text{ha}^{-1}$)] using the following formula (O'sumi, 1987):

$$G = k \times N, \tag{14.4}$$

where k is the basal area factor that is calculated using the following formula:

$$k = \frac{2500 \times d^2}{R^2}, \tag{14.5}$$

where d is the width of slit and R is the distance between the slit and the investigator's eye.

In most cases, mirror relascopes (Spiegel relascope) are used in the Bitterlich method; however, sometimes the surveyor's thumb is simply used. If the thumb is used, the investigator goes around on a sample point with an outstretched arm and raised thumb to count the number of trees thicker than the thumb's width. The basal area factor k is near 4; however, exact values should be calculated using the above-mentioned formulas with the slit width (d). The distance between the slit and the investigator's eye (R) is replaced by width of thumb (d), and the distance between the thumb and eye (R) respectively.

Stoffels method This method measures the distance of the jth nearest tree from sample point and estimates tree density D (number ha^{-1}) by using the following formula (O'sumi, 1987):

$$D = \frac{100^2}{\pi \times \bar{x}_j^2} \times \left(\frac{2j-1}{2}\right), \tag{14.6}$$

where $\bar{x}_j$ is the average distance to the jth nearest tree. The stand volume is calculated by averaging the single-stand volume and tree density.

Suzuki–Essed method The Suzuki–Essed method measures the distance of the third-nearest tree from the sample point and estimates tree density D (number ha^{-1}) by using Suzuki's (1965) formula:

$$D = \frac{8789}{\bar{x}_3^2}, \tag{14.7}$$

where $\bar{x}_3$ is the average distance to the third-nearest tree. The stand volume is calculated by averaging the single-stand volume and tree density.

(v) Parameters for Calculation of Forest Biomass

Various methods for measuring or estimating stem volume or stocks have been proposed. However, the forest biomass includes branches, leaves, and roots, which are often difficult to measure. The total forest biomass B (kg) is estimated from the stem volume V (m^3), expansion factor E, root to shoot ratio R, and wood density D (kg m^{-3}) by using the following formula (National Institute for Environmental Studies, 2012):

$$B = V \times D \times E \times (1 + R) \tag{14.8}$$

Table 14.5 shows the expansion factor E, root to shoot ratio R, and wood density D for each species.

TABLE 14.5 Expansion Factor, Root to Shoot Ratio, and Wood Density for Estimation of Amounts of Biomass

Category	Species	Expansion factor E		Root to shoot ratios R	Wood density, D (kg m^{-3})
		≤20 years	>20 years		
Conifer	Cryptmeria japonica	1.57	1.23	0.25	314
	Chamaecyparis obtusa	1.55	1.24	0.26	407
	Pinus densiflora	1.63	1.23	0.27	416
	Larix leptolepis	1.50	1.15	0.29	404
	Abies sachalinensis	1.88	1.38	0.21	319
	Picea glehnii	1.92	1.46	0.22	348
	Others	1.40	1.46	0.40	423
Broad leaf	Quercus acutissima	1.36	1.33	0.25	668
	Quercus serrata	1.40	1.26	0.25	619
	Others	1.40	1.26	0.25	619

Source: National Institute for Environmental Studies (2012).

b. Remote Sensing Method

(i) Aerial Photography

Aerial photographs are taken from airplanes, helicopter, balloons, and other airborne devices. Geographical Survey Institutes, the Forestry Agency, and other organizations, including private companies, take photographs to construct maps. Aerial photographs, which include important information such as forest type, tree species, geography, topography, soil, roads, and other features, must be viewed and interpreted using reading and stereoscopic techniques, particularly in the practical business of forestry, forest management, erosion control, and consultation, among many others. When GIS is used, aerial photographs need to be digitalized or undergo orthophoto conversion using an orthophoto engine that is sometimes included in the GIS software.

(ii) Satellite

Various earth observation satellites have been launched since the 1960s. Famous satellites include LANDSAT series (USA), SPOT series (France),

IKONOS (USA), and ALOS (DAICHI, Japan). These satellites are equipped with various sensors to measure various light or electromagnetic waves reflected by (or radiated from) objects on the Earth's surface. Each satellite is usually equipped with multiple sensors that detect various electromagnetic waves varying from ultraviolet (short wave) to microwave (long wave). A variety of indices for reading the vegetation information off satellite images have been developed. The Normalized Difference Vegetation Index (NVDI) is one such typical index. Plant leaves absorb most of the visible light and reflect a large portion of the near-infrared (NIR) light. The NVDI is calculated using the following formula (Kato, 2010), based on the NIR and Red (red band) readings:

$$ NVDI = \frac{(NIR - Red)}{(NIR + Red)}. \tag{14.9} $$

The NVDI ranges from -1.0 to 1.0.

(iii) Global Positioning System (GPS)

GPS is a global navigation satellite system (GNSS); the location of a user can be determined using GPS receivers that receive signals from at least four satellites at an altitude of 20 km. Two methods are used for positioning. Single-point positioning is performed using a single hand-held GPS receiver, which costs approximately 5000 to tens of thousand yen. However, its accuracy ranges from a few meters to 10 m. The differential method, which is classified as interferometric positioning or differential GPS (DGPS), uses two receivers resulting in higher accuracy than single-point measures. For the interferometric method, one receiver is set on a reference point, such as a triangulation station point, and another is set on an unknown point. This method requires expensive receivers and software, and complex calculations as well.

The DGPS method can use the correct information for GPS positioning, such as the medium-wave beacon for ships used by the Maritime Safety Agency and the MTSAT Satellite-based Augmentation System (MSAS). This method is capable of positioning with an accuracy of within a few meters.

14.5.2. Monitoring Methods Applicable for Forest Management

In Japan, the Forestry Agency using the standard Montreal Process (1995) has practiced "Forest resource monitoring surveys" to investigate the amount, quality, and dynamics of forest resources since 1999. This survey implements a systematic sampling method that differs from the traditional forest registration system (*Shinrinbo* system) (Yoshida, 2008). In this survey, the nation's land is covered by using 4-km distance grids to establish plots on the grid points. If a plot is forest land, further investigations will be carried out for the plot. There are approximately 15,700 points (plots) in Japan, and the investigations have been conducted on approximately 20% of the plots annually. This method is practiced by prefectural governments on privately owned forests, and by the Forestry Agency or regional forestry offices on national forests.

The survey also covers site condition, elevation, direction of slope, surface geology, soil erosion, and distance from road. Understory vegetation and degree of cover are investigated in a separate vegetation survey. Species, DBH, tree height, stump, and existence of withering are investigated, and a complete enumeration of stands is conducted. Birds and other forest creatures inhabiting forest plots, as well as damage from disease, harmful insects, and climate, are also included in the survey.

14.6. MONITORING SYSTEM FOR FARM MANAGEMENT

Seishu Tojo

14.6.1. Water Quality

Management of surface runoff is crucial to reduce pollutant loads from the watershed to surface water bodies and their surround wetland. Monitoring water quality of forest runoff is necessary for implementing good management practices to reduce pollutant levels in farm drainage water (Rice et al., 2002).

The rainfall is recorded with automatic rain gages, and water quantity is monitored with water level monitors at the sampling point. Typical water quality analyses include water temperature, hydrogen ion concentration (pH), electrical conductivity (EC), chemical oxygen demand (COD), suspended solids (SS), ammonium nitrogen, nitrate nitrogen, total nitrogen, and total phosphorus (Nakasone, 2003). According to Yoshinaga et al. (2007), the field results observed with a 1.5-ha paddy field in the foreshore of Biwa Lake located in central Japan show that the total nitrogen in a 4-month cropping period is 18.8 kg ha^{-1}, with 7.2 kg ha^{-1} from surface drainage and 11.6 kg ha^{-1} from percolation loss.

Pesticide and herbicide are indispensable chemicals even for biomass production on farm land. If applied excessively, they are easily carried by surface runoffs to pollute the receiving surface water bodies. These chemicals include tricyclazole (5-methyl-1,2,4-triazolo[3,4-b]benzothiazole), which is a systemic fungicide commonly used to control the rice blast in Asian countries (Phong et al., 2009). The study on tricyclazole spray in a basin reveals that high concentrations of tricyclazole are found in water of the paddy farms (Padovani et al., 2006). Analyses of pesticides and herbicides contained in the water sample are conducted in the laboratory using HPLC or GC after the chemicals are extracted by eluting the sample with an acetone–hexane mixture.

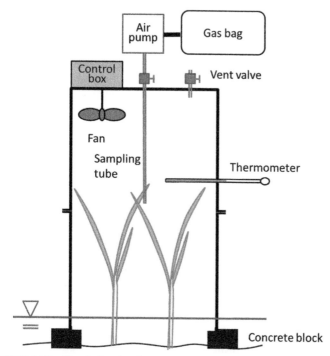

FIGURE 14.7 Closed chamber for measuring gas emissions from paddy field.

14.6.2. Gas Emission

Human activities in agriculture accelerate the production of environmental loading gas on a global scale. Increasing gas emissions from farmland may lead to adverse environmental effects, including global warming, acid rain, changes in biodiversity, and stratospheric ozone depletion.

The closed chamber schematically shown in Figure 14.7 is a device used for measurements of the gas flux emitted from paddy fields into the atmosphere. The CH_4 flux is determined by measuring the temporal increase of the CH_4 concentration of the air within the chamber that is placed on a paddy field to hold some growing rice plants inside. The chamber air sampling is usually collected within 10 min after the chamber has been set in place; the gas is stored in a gas sampling bag to be analyzed later in the laboratory by using a gas chromatograph equipped with a flame ionization detector (GC/FID) (Yagi and Minami, 1990). Methane flux rates from rice fields is likely overestimated by using the closed chamber technique because decreasing grain yield by chamber enclosure may result in more plant photosynthetic products released into soils to enhance CH_4 production (Yu et al., 2006).

Analysis of atmospheric air using open-path Fourier transform infrared (OP/FT-IR) spectrometry has been available for over two decades but has not been widely accepted because of the limitations of the software for commercial applications of the instrument. Most OP/FT-IR spectrometers use a photo-conductive mercury cadmium telluride (MCT) detector that is cooled down to about 80 K either with liquid nitrogen (LN_2) or by using a closed-cycle cooler (Griffiths et al., 2009). To simultaneously monitor concentrations of N_2O and CO_2 at a height of 3 m in the stable boundary layer, a concentration measurement precision of 1% for 3 min on average was achieved over a 97-m-long, open-air absorption path. The precision of N_2O concentration measurements with OP/FT-IR is an order of magnitude poorer than with GC/ECD. The associated errors are random and equal to 0.6% and 0.06% for the FTIR and GC/ECD systems respectively (Kelliher et al., 2002).

Emissions of ammonia (NH_3), which is the most abundant form of reduced reactive nitrogen in the atmosphere, have increased significantly as a result of intensive agricultural management and greater livestock production in many developed countries. Synthetic fertilizers and agricultural crops together contribute 9×10^{12} g NH_3-N y^{-1} or 12% of total NH_3 emissions (Schlesinger and Hartley, 1992). The absorption flasks sampler method is the typical method for capturing atmospheric ammonia. Two absorption flasks, each containing 30 mL of orthophosphoric acid or boric acid solution (0.1, 0.01, or 0.001 mol L^{-1}), are connected in series at the sampling point. Air is forced through the flasks through a 4-mm tube at 2 or 4 L min^{-1} to trap the ammonia contained in the air. Following the sampling step, the solution volumes are adjusted to 100 mL using deionized water; the ammonia trapped in the solution is determined using colorimetry or liquid chromatography (Misselbrook et al., 2005).

14.6.3. Field Monitoring Server

In order to facilitate field and environment monitoring of air and water qualities over long periods of time, a new method using a remote monitoring system as shown in Figure 14.8 has been developed. Field servers are one of the small monitoring sensor nodes that are equipped with a web server to be accessed via the internet using wireless LAN to provide a high-speed transmission network (Fukatsu and Hirafuji, 2005). The information technology (IT) field monitoring system consists of an intelligent sensor node web server that is equipped with in situ camera and sensor networks for agro-meteorological, soil, and plant growth monitoring. The actual field conditions are captured well by a combination of images, numerical results, and graphical data sets (Manzano et al., 2011).

Wireless sensor network (WSN) technology has the potential to reveal fine-grained, dynamic changes in monitored variables of an outdoor landscape. The design and implementation of a reactive, event-driven network for environmental monitoring of soil moisture has been proposed (Cardell-Oliver et al., 2005). The outdoor monitoring server system with a solar cell power

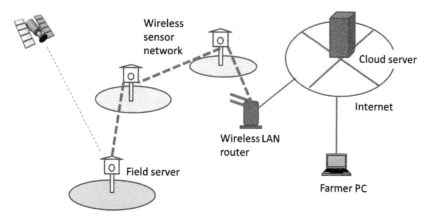

FIGURE 14.8   Field sensor network monitoring system.

supply collects environmental and soil information using a combination of WSN-based environmental and soil sensors, image information through CCTVs, and location information using GPS modules (Hwang et al., 2010).

Applying information technology to gather cumulative environmental and agricultural data can enhance high yields of biomass production while alleviating environmental burdens.

REFERENCES

Balat, M. (2011). Production of bioethanol from lignocellulosic materials via the biochemical pathway: A review. *Energy Conversion and Management, 52,* 858–875.

Binod, P., Sindhu, R., Singhania, R. R., Vikram, S., Devi, L., Nagalakshmi, S., Kurien, N., Sukumaran, R. K., & Pandey, A. (2010). Bioethanol production from rice straw: An overview. *Bioresource Technology, 101,* 4767–4774.

Buchholz, T., Luzadis, V. A., & Volk, T. A. (2009). Sustainability criteria for bioenergy systems: Results from an expert survey. *Journal of Cleaner Production, 17,* S86–S98.

Cardell-Oliver, R., Kranz, M., Smettem, K., & Mayer, K. (2005). A reactive soil moisture sensor network: Design and field evaluation. *International Journal of Distributed Sensor Networks, 1,* 149–162.

Corbière-Nicollier, T., Blanc, I., & Erkman, S. (2011). Towards a global criteria based framework for the sustainability assessment of bioethanol supply chains: Application to the Swiss dilemma: Is local produced bioethanol more sustainable than bioethanol imported from Brazil? *Ecological Indicators, 11,* 1447–1458.

Dale, V. H., Kline, K. L., Wiens, J., & Fargione, J. (2010). Biofuels: Implications for land use and biodiversity. Biofuels and sustainability reports. *Ecological Society of America.*

Forest Stewardship Council (FSC). <http://www.fsc.org/vision-mission.12.htm> Accessed August 2012

Forestry Agency. (1970). *Standing tree volume table (east Japan version and west Japan version).* Tokyo: J-FIC.

Fukatsu, T., & Hirafuji, M. (2005). Field monitoring using sensor-nodes with a web server. *Journal of Robotics and Mechatronics, 17*(2), 164–172.

Garland, C. D. (2008). *Growing and harvesting switchgrass for ethanol production in Tennessee.* Publication NO. SP701-A. Knoxville, TN: University of Tennessee Extension.

GBEP. (2011). *The global bioenergy partnership sustainability indicators for bioenergy.* Rome: Food and Agriculture Organization of the United Nations (FAO).

Gowdy, J. (1999). Hierarchies in human affairs: microfoundation and environmental sustainability. In J. Kohn (Ed.), *Sustainability in question: The search for a conceptual framework* (pp. 67–84). Northampton, MA: Edward Elgar.

Griffiths, P. R., Shao, L., & Leytem, A. B. (2009). Completely automated open-path FT-IR spectrometry. *Analytical and Bioanalytical Chemistry, 393,* 45–50.

Hwang, J., Shin, C., & Yoe, H. (2010). Study on an agricultural environment monitoring server system using wireless sensor networks. *Sensors, 10,* 11189–11211.

IPCC. (1995). *IPCC Second Assessment Report: Climate Change 1995 (SAR).*

IPCC. (2007). *IPCC Fourth Assessment Report: Climate Change 2007 (AR4).*

Itsubo, N., Tahara, K., Narita, A., Inaba, A., & Aoki, R. (2007). *Outline of LCA.* Japan Environmental Management Association for Industry.

Kato, M. (Ed.). (2010). *Forest remote sensing* (3rd ed). From base to application. Tokyo: J-FIC [in Japanese].

Kelliher, F. M., Reisinger, A. R., Martin, R. J., Harvey, M. J., Price, S. J., & Sherlock, R. R. (2002). Measuring nitrous oxide emission rate from grazed pasture using Fourier-transform infrared spectroscopy in the nocturnal boundary layer. *Agricultural and Forest Meteorology, 111,* 29–38.

Ko, J. K., Bak, J. S., Jung, M. W., Lee, H. J., Choi, I. G., Kim, T. H., & Kim, K. H. (2009). Ethanol production from rice straw using optimized aqueous-ammonia soaking pretreatment and simultaneous saccharification and fermentation processes. *Bioresource Technology, 100,* 4374–4380.

Koga, N., & Tajima, R. (2011). Assessing energy efficiencies and greenhouse gas emissions under bioethanol-oriented paddy rice production in northern Japan. *Journal of Environmental Management, 92,* 967–973.

Lendle, A., & Schaus, M. (2010). Sustainability criteria in the EU renewable energy directive: Consistent with WTO rules? ICTSD information note No. 2. *International Centre for Trade and Sustainable Development (ICTSD).*

Lewandowski, I., & Faaij, A. P. C. (2006). Steps towards the development of a certification system for sustainable bio-energy trade. *Biomass and Bioenergy, 30,* 83–104.

Manzano, V. J. P., Mizoguchi, M., Mitsuishi, S., & Ito, T. (2011). IT field monitoring in a Japanese system of rice intensification (J-SRI). *Paddy Water Environment, 9,* 249–255.

Mayer, A. L. (2008). Strengths and weaknesses of common sustainability indices for multidimensional systems. *Environment International, 34*(2), 277–291.

Merriam-webster, <http://www.merriam-webster.com/dictionary/sustained%20yield> Accessed August 2012.

Ministry of the Environment of Japan. (2012). *Calculation and report manual of greenhouse gas emissions, Ver. 3.2.*

Misselbrook, T. H., Nicholson, F. A., Chambers, B. J., & Johnson, R. A. (2005). Measuring ammonia emissions from land applied manure: An intercomparison of commonly used samplers and techniques. *Environmental Pollution, 135,* 389–397.

Nakasone, H. (2003). Runoff water quality characteristics in a small agriculture watershed. *Paddy Water Environment, 1,* 183–188.

Nansai, K., & Moriguchi, Y. (2009). Embodied energy and GHG emissions intensities based on the 2005 Japanese input-output tables (3EID). *National Institute for Environmental Studies.*

National Institute for Environmental Studies. (2012). *National Greenhouse Gas Inventory Report of Japan.* <http://www-gio.nies.go.jp/aboutghg/nir/2012/NIR-JPN-2012-v3.0E.pdf> Accessed 23.08.2012.

Nguyen, T. A. D., Kim, K. R., Han, S. J., Cho, H. Y., Kim, J. W., Park, S. M., Park, J. C., & Sim, S. J. (2010). Pretreatment of rice straw with ammonia and ionic liquid for lignocellulose conversion to fermentable sugars. *Bioresource Technology, 101,* 7432–7438.

OECD. (2001). *Executive summary of environmental indicators for agriculture.* In *Method and results,* Vol. 3. Paris: OECD.

O'sumi, S. (1987). *Lecture on forest measurement.* Tokyo: Yokendo [in Japanese].

Padovani, L., Capri, E., Padovani, C., Puglisi, E., & Trevisan, M. (2006). Monitoring tricyclazole residues in rice paddy watersheds. *Chemosphere, 62,* 303–314.

Park, J., Kanda, E., Fukushima, A., Motobayashi, K., Nagata, K., Kondo, M., Ohshita, Y., Morita, S., & Tokuyasu, K. (2011a). Contents of various sources of glucose and fructose in rice straw, a potential feedstock for ethanol production in Japan. *Biomass and Bioenergy, 35,* 3733–3735.

Park, J., Shiroma, R., & Tokuyasu, K. (2011b). Bioethanol production from rice straw by a sequential use of *Saccharomyces cerevisiae* and *Pichia stipitis* with heat inactivation of *Saccharomyces cerevisiae* cells prior to xylose fermentation. *Journal of Bioscience and Bioengineering, 111*(6), 682–686.

Phong, T. K., Nhung, D. T. T., Motobayashi, T., Thuyet, D. Q., & Watanabe, H. (2009). Fate and transport of nursery-box-applied tricyclazole and imidacloprid in paddy fields. *Water Air Soil Pollution, 202,* 3–12.

Program for the Endorsement of Forest Certification (PEFC). <http://www.pefc.org/about-pefc/overview> Accessed August 2012.

Rice, R. W., Izuno, F. T., & Garcia, R. M. (2002). Phosphorus load reductions under best management practices for sugarcane cropping systems in the Everglades Agricultural Area. *Agricultural Water Management, 56,* 17–39.

Roy, R., Tokuyasu, K., Orikasa, T., Nakamura, N., & Shiina, T. (2012a). A techno-economic and environmental evaluation of the life cycle of bioethanol produced from rice straw by RT-CaCCO process. *Biomass and Bioenergy, 37,* 188–195.

Roy, R., Tokuyasu, K., Orikasa, T., Nakamura, N., & Shiina, T. (2012b). Evaluation of the life cycle of bioethanol produced from rice straws. *Bioresource Technology, 110,* 239–244.

Saga, K., Yokoyama, S., & Imou, K. (2007). Net energy analysis of bioethanol production system from rice cropping. *Journal of Japan Society of Energy and Resources, 29*(1), 30–35.

Sarkar, N., Ghosh, S. K., Bannerjee, S., & Aikat, K. (2012). Bioethanol production from agricultural wastes: An overview. *Renewable Energy, 37,* 19–27.

Schlesinger, W. H., & Hartley, A. (1992). A global budget for atmospheric NH_3. *Biogeochemistry, 15,* 191–211.

SETAC. (1993). *Guidelines for life cycle assessment: A code of practice.*

SETAC. (1996). *Towards a methodology for life cycle impact assessment.*

Sheehan, J. (2009). Biofuels and the conundrum of sustainability. *Current Opinion in Biotechnology, 20,* 318–324.

Shinozaki, Y., & Kitamoto, H. K. (2011). Ethanol production from ensiled rice straw and whole-crop silage by the simultaneous enzymatic saccharification and fermentation process. *Journal of Bioscience and Bioengineering, 111*(3), 320–325.

Silva Lora, E. E., Escobar Palacio, J. C., Rocha, M. H., Grillo Renó, M. L., Venturini, O. J., & Almazán del Olmo, O. (2011). Issues to consider, existing tools and constraints in biofuels sustainability assessments. *Energy, 36*, 2097–2110.

Simpson, T. W., Sharpley, A. N., Howarth, R. W., Paerl, H. W., & Mankin, K. R. (2008). The new gold rush: Fueling ethanol production while protecting water quality. *Journal Environmental Quality, 37*, 318–324.

Sustainable Green Ecosystem Council (SGEC). <http://www.sgec-eco.org/index%28e%29.html> Accessed August 2012.

Suzuki, T. (1965). *A new Essed's distance method.* Bulletin of the Nagoya University Forest, 51–58 [in Japanese with English summary].

Takata, Y., Obara, H., Nakai, M., & Kohyama, K. (2011). Process of the decline in the cultivated soil area with land use changes in Japan. *Japanese Journal of Soil Science and Plant Nutrition, 82*(1), 15–24.

Uematsu, Y., Koga, T., Mitsuhashi, H., & Ushimaru, A. (2010). Abandonment and intensified use of agricultural land decrease habitats of rare herbs in semi-natural grasslands. *Agriculture, Ecosystems and Environment, 135*, 304–309.

United Nations. (1992). *The declaration of forest principle.* <http://www.un.org/documents/ga/conf151/aconf15126-3annex3.htm> Accessed August 2012.

USDA. (2010). *Programmatic environmental impact statement.* Biomass crop assistance program, Farm Service Agency, US Department of Agriculture.

Yadav, K. S., Naseeruddin, S., Prashanthi, G. S., Sateesh, L., & Rao, L. V. (2011). Bioethanol fermentation of concentrated rice straw hydrolysate using co-culture of *Saccharomyces cerevisiae* and. *Pichia stipites. Bioresource Technology, 102*, 6473–6478.

Yagi, K., & Minami, K. (1990). Effect of organic matter application on methane emission from some Japanese paddy fields. *Soil Science and Plant Nutrition, 36*(4), 599–610.

Yang, C., & Sagisaka, M. (2009). Evaluation of bioethanol production system from rice straw. *Journal of Life Cycle Assessment, Japan, 5*(4), 501–509.

Yoshida, S. (2008). Comparison between the past and present system used for national forest inventory in Japan. *Journal of Japanese Forestry Society, 90*, 283–290 [in Japanese with English summary].

Yoshikawa, S., Yamamoto, H., Hanano, Y., & Ishihara, A. (2004). Hilly-land Soil Loss Equation (HSLE) for evaluation of soil erosion caused by the abandonment of agricultural practices. *Japan Agricultural Research Quarterly, 38*(1), 21–29.

Yoshinaga, I., Miura, A., Hitomi, T., Hamada, K., & Shiratani, E. (2007). Runoff nitrogen from a large sized paddy field during a crop period. *Agricultural Water Management, 87*, 217–222.

Yu, K. W., Chen, G. X., & Xu, H. (2006). Rice yield reduction by chamber enclosure: A possible effect on enhancing methane production. *Biology and Fertility of Soils, 43*, 257–261.

Chapter 15

Local Activity of Biomass Use in Japan

Hiroshi Yoshida, Toshio Nomiyama, Nobuhide Aihara,
Ryoichi Yamazaki, Sachiho Arai and Hiroyuki Enomoto

Chapter Outline

Research Approaches to Sustainable Biomass Systems. http://dx.doi.org/10.1016/B978-0-12-404609-2.00015-5

15.1. OVERVIEW OF THE PERFORMANCE OF BIOMASS TOWNS

Hiroshi Yoshida

15.1.1. Establishment of Biomass Towns

Over the last decade, the government of Japan has made efforts to promote biomass utilization in a local society under the Biomass Nippon Strategy (BNS) that was enacted in 2002 and revised in 2006. The BNS stipulates that biomass utilization should be promoted in the following five directions: (i) fostering public understanding; (ii) developing a system to use biomass comprehensively (e.g. in the form of biorefineries); (iii) coordinating stakeholders (national government, local municipalities, suppliers, and consumers) to share burden and responsibility; (iv) adjusting competitive conditions (carefully supporting biomass businesses at their inceptions); and (v) considering an international perspective (e.g. the use of clean development mechanisms or joint implementation).

Related to (iii) above, the role of local municipalities is emphasized as a principal coordinating agent to realize local energy produced for local consumption through the establishment of biomass towns. When the local biomass plan submitted by a municipality is approved by central government, the municipality is certified as a biomass town. In general, a biomass town contains various components for energy use (e.g. biofuel production, power or heat generation) and material use (e.g. organic fertilizer, livestock feed, construction material, bioplastics, charcoal, other chemical materials) while meeting a design requirement of utilizing more than 90% of waste biomass and more than 40% of unused biomass in terms of carbon equivalent. The biomass town is expected to create various biomass-based industries and jobs in the neighboring region to achieve comprehensive (or "cascade" or "circular") use of biomass.

15.1.2. Features of Biomass Town Plans

Major features of biomass town plans can be derived from the data set of 318 certificated biomass towns available online at the website of the Japan Organics Recycling Association (JORA, 2012). Note that all the data compiled by the JORA are based on biomass town plans; thus, the data illustrate the planned but not necessarily the implemented biomass utilization in each biomass town. The data set contains the general information summarized below.

First, biomass resources devoted to practical use are mostly waste and presently unused biomass. In particular, animal wastes are used in 297 towns (93.4%), followed by agricultural residues (296 towns; 93.0%), rice straw and husk (275 towns; 86.5%), forest wastes (263 towns; 82.7%), and sewage sludge (260 towns; 81.8%). These data reflect the design requirement for the target use rate of over 90% for waste biomass and over 40% for unused biomass. Second, biomass is more commonly used as raw material for products than for energy;

main end-use products are fertilizer/compost (318 towns; 100.0%) and animal feed (230 towns; 72.3%) as compared with biodiesel fuel (BDF; 257 towns; 80.8%), solid biofuels in the form of wood pellets (251 towns; 78.9%), and biogas (178 towns; 56.0%). Third, every biomass town combines use of multiple types of biomass resources (feedstocks) to produce various types of end-use products. The average number of feedstock types per town is 8.6, which is composed of 5.7 for waste biomass, 2.6 for unused biomass, and 0.2 for resource (dedicated energy) crops. The average number of end-use products per town is 5.6, which is composed of 2.3 for material use and 3.3 for energy use. Fourth, the pattern of biomass use is location specific, subject to biomass resource availability in type and volume. Based on the principle of "local use of locally available biomass resources", the supply chain that consists of the procurement of feedstocks as well as the production and distribution of end-use products is formed within a municipal boundary and adjunct areas. A biomass town in remote areas is more likely to use forest waste and agricultural residues whereas a biomass town located in or near urban areas is more likely to use industrial and household solid wastes. For example, using forest waste is concentrated in mesomountainous and mountainous areas, and recycled cooking oil is more frequently seen in urban areas. However, use of animal waste is noted to be widespread throughout the country regardless of whether a local biomass town is urban or rural.

One profound phenomenon of the location-specific characteristics, related to the characteristic of the combined use of multiple feedstocks for multiple products, is that biomass towns in or near urban areas are more diversified in both used feedstocks and end-use products than those in remote areas. Areas of higher population density have advantages in diversification of biomass utilization due to easier access to a variety of biomass feedstocks and to end users.

The JORA data also show that only 25.2% of the biomass town (80 towns) have attempted to utilize biomass to produce liquid fuel. Cereal crops and sugar cane are used for ethanol production on experimental trials; the feedstocks with negligible market values including broken rice, wheat of unacceptable quality in markets, and residues from sugar processing have been used. For example, broken rice is more than five times cheaper than standard-quality rice. The BDF produced from oil crops such as rapeseed or sunflower is more popular than ethanol production. However, BDF is not directly produced from raw rapeseed harvests that can be processed for high-quality cooking oil. The feedstock of BDF comes from recycled oil collected from food industries and households. BDF has been produced on a small scale in some integrated local biomass recycling systems from rapeseed or sunflower that is sown as an alternative crop in deserted or unused paddy fields as guided by the national policy of rice production adjustment. The local resource recycling system uses resources available locally at recycling centers as its core to involve various types of stakeholders, i.e. farmers, manufacturers, consumers, and local government staff. Moreover, in this localized system, various activities are undertaken that include using rapeseed/sunflower flower fields as tourist attractions, producing multiple processed products such as cooking oil, fertilizer, animal feed and BDF, and collecting recycled oil and other types of household wastes.

In summary, the most common patterns of biomass utilization for a local society in Japan are characterized by (i) production of fertilizer (compost) through conversion from animal waste, sewage sludge, and rice straw or husks; (ii) production of animal feed from leftover food; (iii) production of BDF from recycled cooking oil; and (iv) production of wood pellets from timber waste and forest residues. Most biomass towns have been designed based on the combination of the four patterns, i.e. combination of (ii) and (iii) in or near urban areas, and combination of (i) and (iv) in mesomountainous or mountainous areas. Presently, bioethanol production from resource crops is limited due to lower economic return to investment. Indeed, bioethanol production is still in the trial stage and requires a set of new technologies through a series of production stages from the development of high-yielding inedible biomass crops to be converted into ethanol.

15.1.3. Unsuccessful Performance of Biomass Town Plans

The BNS has led to the development of biomass town plans in more than 300 municipalities; however, their performance has been less than satisfactory. Due to the budget-tightening initiative and the relatively poor performance of biomass-related projects, the government put an end to applications for new biomass towns in April, 2012, when the number of biomass towns had reached 318. The Administrative Evaluation Bureau (AEB) evaluated biomass use projects in February 2011 (AEB, 2011), and the findings are summarized as follows. First, basic conditions for using biomass have been developed in general, as indicated by the increase in the number of biomass-related facilities. Second, however, the data that should have been recorded are limited by (i) the costs of projects, (ii) the outcomes of related businesses, (iii) the progress of biomass town plans, and (iv) the achieved reduction of GHG emissions. Third, the performance of biomass-related projects was much lower than expected. For example, 43% (92 of 214) of biomass-related projects surveyed in 136 biomass towns did not have proper accounting records, with only 16.4% (35 projects) having achieved some outcomes. Of 785 components included in the initial plans, only 277 components (35.3%) had been undertaken, whereas 221 components (28.2%) had already been abandoned or were unlikely to be implemented. Moreover, reductions of GHG emissions were calculated by only three facilities, and only 10.4% of facilities had reduced GHG emissions.

The available literature (e.g. Tomari, 2012) and other information from various sources have led to the following reasons for such unsatisfactory performance of biomass town plans. First, the lack of rural infrastructure such as roads made it difficult to collect biomass resources and transport them to centralized conversion facilities. For example, a biomass town located in a mesomountainous or mountainous area initially aimed at utilizing forestry waste in the form of wood pellets failed to meet the objective due to the inadequate forestry road network. Second, initial plans are too ambitious; they are characterized by a lack of feasibility as regards the technical, economic, and managerial aspects. An installation of sophisticated facilities based on advanced technology was designed, but they are not sustainable in operation without outside subsidies. Less attention was given to financial, managerial, marketing, and other considerations and conditions, as well as the capacity of the municipal government involved during the planning stage. Third, various projects involving biomass town plans have been undertaken rather independently without horizontal linkages, whereas system evaluations of the projects as a whole had not been prepared. Fourth, related to the third point, a coordination failure occurred at various stakeholder levels among various administrative offices, i.e. municipal government, and liaison between municipal government and the private sector or non-profit organizations involved. Fifth, the generation of new biomass-based industries has been hindered by the current legislative framework. For instance, energy-related regulations have been established, originally with little attention to the supply of gas or electricity by farmers or households so that their entry into energy markets cannot be initiated legally.

These reasons are associated with lack of profitability as well as inadequate provisions of the legislative and institutional framework that have been mentioned as the major bottlenecks to constructing effective biomass systems in Chapter 2. These two factors will be addressed in the next two sections by discussing the current situation and future challenges to implementing localized utilization of renewable energy, including bioenergy, through case studies for two biomass towns: Kuzumaki town and Higashiomi city. Both municipalities have been regarded as advanced in undertaking local environmental and energy programs in Japan; they had autonomously promoted producing biomass energy utilization using locally available resources before the BNS was enacted in 2002. With emphases on the legislative and institutional aspects, the study on Kuzumaki attempts to identify issues for improving local energy self-sufficiency. With the focus on the economic aspect, the study on Higashiomi examines the profitability of rapeseed used for BDF production via cooking oil within the city's resource circulation cycle.

15.2. CASE STUDY OF KUZUMAKI TOWN, IWATE

Toshio Nomiyama and Nobuhide Aihara

15.2.1. Overview of Kuzumaki

Kuzumaki town is located in a mountainous area in the northern part of Iwate prefecture that is located in the northern part of East Japan. Specifically,

Kuzumaki is located between the city of Morioka, the capital city of Iwate, and the Kitasanriku area with the 40th parallel running through the center of the town (Figures 15.1 and 15.2). Kuzumaki is a key trading area that connects the coastal areas of the Pacific and inland areas. The importance of this area as the route for material supplies connecting the Pacific coastal areas and inland areas was recognized during the East Japan Earthquake on March 11, 2011.

The population of Kuzumaki is 7417 with 2877 households as of April 1, 2011 in an area of 434.99 km². Forests account for 86% of the area, and 95% of the land is located over 400 m above sea level. The annual average temperature is about 8°C. Kuzumaki is characterized by a typical mesomountainous region or mountainous area.

The main industries are dairy farming and forestry. About 11,000 cattle are raised, including 10,000 dairy cattle and about 1000 beef cattle. Milk production is about 40,000 tons a year (about 110 tons per day). Kuzumaki is known in Japan as the largest dairy farming town in the Tohoku region. Forestry, another main industry, is also very active; laminated Japanese larch trees are shipped out as construction materials. Charcoal manufacture is also a traditional industry with skills to produce high-quality charcoal passed on through generations. Wine production using wild crimson glory vines has also started in recent years. The annual gross agricultural output from these main industries is estimated to be about 4.9 billion yen.

Kuzumaki is viewed as a unique town due to its municipal strategy for managing energy and the environment. While calling itself a town with milk,

FIGURE 15.1 **Location of the town of Kuzumaki.** (*Source: Website of the town of Kuzumaki.*)

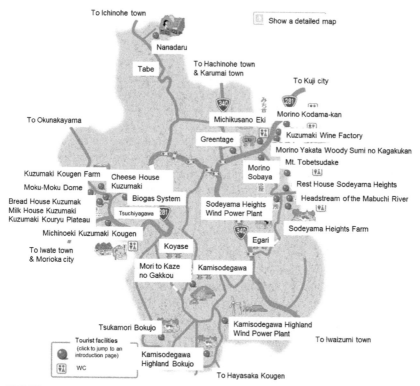

To Ichinohe town

Show a detailed map

Nanadaru

Tabe

To Hachinohe town
& Karumai town

To Kuji city

To Okunakayama

Michikusano Eki

Morino Kodama-kan

Greentage

Kuzumaki Wine Factory

Morino Yakata Woody Sumi no Kagakukan
Mt. Tobetsudake

Kuzumaki Kougen Farm   Cheese House
Moku-Moku Dome   Kuzumaki

Morino
Sobaya

Rest House Sodeyama Heights

Bread House Kuzumak
Milk House Kuzumaki
Kuzumaki Kouryu Plateau

Biogas System

Tsuchiyagawa

Sodeyama Heights
Wind Power Plant

Headstream of the Mabuchi River

Michinoeki Kuzumaki Kougen

Sodeyama Heights Farm

To Iwate town
& Morioka city

Koyase

Egari

Mori to Kaze
no Gakkou

Kamisodegawa

Tsukamori Bokujo

Kamisodegawa Highland
Wind Power Plant

To Iwaizumi town

Tourist facilities
(click to jump to an
introduction page)

Kamisodegawa
Highland Bokujo

WC

To Hayasaka Kougen

FIGURE 15.2 Detailed map of the town of Kuzumaki. (*Source: Website of the town of Kuzumaki.*)

wine and clean energy, Kuzumaki was also declared as a new energy town in 1999. Although the current municipal policy may be seen as conventional like those implemented in other small municipalities in Japan, Kuzumaki is distinct from other municipalities in that it has a longer history of dealing with local energy and environmental issues. "The policy to protect the environment" had been presented by Kuzumaki long before the national new energy policy was established in the 2000s. The town's unfavorable geographic and topographic conditions were considered as the motivation for the municipal government to tackle energy and environmental issues more seriously than governments of other municipalities with more favorable natural conditions. With a less favorable infrastructure such as electricity distribution systems, mitigating local energy and environmental problems is imperative to sustain the town's main industries and to ensure energy security for its residents.

In fact, Kuzumaki has been working on local energy production and consumption to deal with local energy problems. One such effort is to reduce the consumption of conventional electric energy and energies produced from fossil fuel such as oil and natural gas. Second, Kuzumaki has been promoting the use of new energies, such as wind power generation, photovoltaic power generation, and woody biomass by industries operating within the town boundaries. These activities underlie the fact that the residents are highly motivated to improve their environment, which seems to be working as an incentive for the town hall, the administrative body of the town, the town council, and the legislative body to make their best efforts to develop the town's policies to reflect the feelings of the residents.

The remainder of this section explores the current situation and challenges for localizing energy utilization based on interviewing the town hall of Kuzumaki and the manufacturers of woody biomass or other renewable energy located in Kuzumaki, supplemented by information collected from various sources.

15.2.2. Electric Energy

A system for local production and consumption of electric energy based on photovoltaic and wind power generation facilities was developed after fiscal year 2003. However, the Electricity Business Act stipulates that public power supply companies have an exclusive right to run the business for transmitting electricity to private households. Therefore, electric energy cannot be considered as strictly locally produced and manufactured in terms of actual values. As shown in Table 15.1, however, values of local energy production and consumption can be calculated theoretically. The breakdown shows that while the nominal self-sufficiency rate was less than 100% in the fiscal year 2003, its rates have remained greater than 100% since the fiscal year 2004. The interview-based investigations reveal that these rates remained over 100% in the fiscal years 2009 and 2010 as well.

However, some people in the town think that they have not been experiencing the effects of the established local production and consumption of electric energy. One reason is that the current Electricity Business Act stipulates that the supply of electricity is the exclusive preserve of public power supply companies. In addition, the Act also stipulates that the nominal self-sufficiency rate must not be reflected in the price of purchasing electricity in a town because the price of electricity is regulated by the national government. Thus, the residents of Kuzumaki, the town council, and the town hall are starting to form a common recognition that the necessity to separate power generation and power transmission is the most important issue to be considered for developing future electric energy policy.

An event that further solidified this recognition was the 4-day blackout of the entire town of Kuzumaki after the onset of the East Japan Earthquake on March 11, 2011. The initial understanding among the people was that no blackout would occur, or the power supply would be quickly restored even if a blackout occurred because the power generation facilities located in the

TABLE 15.1 Amount of Energy Consumption and Production in the Town of Kuzumaki (Electricity)

	MWh (per fiscal year)					
	2003	2004	2005	2006	2007	2008
Amount of consumption						
Amount of electricity supplied from outside the town	31,850	32,223	33,188	34,438	35,240	34,161
Amount of electricity used for generating power	998	2315	1587	1923	1865	1853
Total amount of consumption	32,848	34,538	34,775	36,361	37,105	36,014
Amount of production						
Amount of photovoltaic power generation (production of electricity)	52	44	49	45	30	37
Amount of wind power generation (production of electricity)	28,600	61,149	36,092	46,800	45,967	45,221
Total amount of production	28,652	61,193	36,141	46,845	45,997	45,258
Amount supplied to outside the town						
Amount of photovoltaic power generation supplied to outside the town	16	13	16	13	8	8
Amount of wind power generation (to outside the town)	20,688	59,105	31,363	45,164	44,178	43,872
Total amount of electricity sold (outside the town)	20,704	59,118	31,379	45,177	44,186	43,880
Nominal amount of locally produced and consumed electricity	−4196	26,655	1,366	10,484	8892	9244
Nominal self-sufficiency rate (%)	87	177	104	129	124	126

From Nomiyama and Aihara.

town were still generating electricity. However, in reality, they experienced a blackout that lasted 4 days. After such an unpleasant experience, the people in Kuzumaki have further strengthened their awareness of the necessity to separate power generation and power transmission, although the idea had been advocated long before the earthquake.

15.2.3. Gas and Solid Fuel Energy

The local production and consumption of gas and solid fuel energy in Kuzumaki are based on the use of woody biomass. Specific products from woody biomass include fuel woods, charcoal, chips, and pellets. Unlike electric energy, the amount of woody biomass energy produced and consumed locally is negative, as shown in Table 15.2. That is, the town is receiving a negative amount from outside communities. The self-sufficiency rate, however, is slightly over 100% because some fuel woods, chips, and pellets produced are consumed within the town boundaries.

However, a large portion of the total amount of charcoal produced from woody biomass in the town is shipped out of the town. This is because the people of Kuzumaki profit more economically by selling charcoal to other communities than by using the charcoal as an energy source in their own town.

Different researchers have evaluated this situation in a variety of ways based on their perspectives. The perspective of regional development, as suggested by the name "Regional development and use of biomass", states that the role of woody biomass is extremely effective in this case. On the other hand, from the perspective of local production and consumption of energy, the role of woody biomass in this case is extremely small. Resolving the differences between these two perspectives becomes a difficult problem when interpreting the numerous problems that various regions around Japan are facing.

When an area is located in either a mesomountainous or a mountainous region with only forestry-based industries, as is the case for Kuzumaki, the thoughtless introduction of new technology may result in unnecessary loss of the forestry or forestry-related industry in the area. If this were to occur, it would mean the loss of the town. Therefore, the current practice to use woody biomass in Kuzumaki should be considered very effective in terms of regional development.

Given the energy outlook for the future, a concept based on the perspective of local consumption of energy produced locally is necessary to achieve this goal. However, thoughtless introduction of new technologies may result in the loss of the town, as discussed in previous paragraphs. Therefore, we need to prioritize newly created strategies to promote beneficial and cost-effective uses of woody biomass to take advantages of the natural resources of Kuzumaki while avoiding the problems caused by practicing ideas without careful consideration and evaluation beforehand.

TABLE 15.2 Amount of Energy Consumption and Production in the Town of Kuzumaki (Gas and Fuel Energy)

	MJ (per fiscal year)									
	2001	2002	2003	2004	2005	2006	2007	2008	2009	2010
Amount of consumption										
Gas	2,270,815	2,203,908	2,240,484	2,162,872	2,160,195	2,244,944	2,203,908	2,333,707	2,234,239	2,137,001
Other fuels (such as kerosene)	1,642,395	1,378,439	1,404,974	1,365,869	1,948,249	2,311,364	2,059,977	2,284,829	1,555,806	1,791,831
Total amount supplied from outside town	3,913,210	3,582,347	3,645,458	3,528,741	4,108,445	4,556,308	4,263,885	4,618,535	3,790,045	3,928,832
Amount of production										
Wood biomass										
Firewood (consumed in the town)	631,617	682,146	1,099,013	322,124	713,727	2,198,026	1,206,388	1,452,718	764,256	966,373
Charcoal	4,089,696	4,050,024	4,084,998	4,425,603	3,479,217	2,640,537	2,278,385	3,020,930	2,676,613	2,318,057
Chips	0	0	0	0	593	603	47	0	0	1110
Pellets	7856	7741	17,812	14,892	17,353	24,602	24,602	26,964	22,749	27,817

(Continued)

TABLE 15.2 Amount of Energy Consumption and Production in the Town of Kuzumaki (Gas and Fuel Energy)—cont'd

	MJ (per fiscal year)									
	2001	2002	2003	2004	2005	2006	2007	2008	2009	2010
(Consumed in the town)	377	1000	4838	4297	4150	5019	4276	4986	5429	5445
Total amount of production	4,729,169	4,739,911	5,201,823	4,762,620	4,210,889	4,863,767	3,509,422	4,500,612	3,763,618	3,313,357
Amount supplied to the outside the town										
Woody biomass										
Charcoal	4,089,696	4,050,024	4,084,998	4,425,603	3,479,217	2,640,537	2,278,385	3,020,930	2,676,613	2,318,057
Pellets	7479	6741	12,973	10,595	13,203	19,583	20,026	21,978	17,320	22,371
Total amount supplied outside the town	4,097,175	4,056,765	4,097,971	4,436,198	3,492,420	2,660,120	2,298,411	3,042,908	2,693,933	2,340,428
Nominal amount of locally produced and consumed electricity	−3,281,216	−2,899,200	−2,541,606	−3,202,319	−3,389,975	−2,352,661	−3,052,873	−3,160,831	−3,020,360	−2,955,903
Self-sufficiency rate (%)	16	19	30	9	17	48	28	32	20	25

From Nomiyama and Aihara.

15.2.4. Issues and Outlook

One problem faced by Kuzumaki in undertaking energy-related programs is the diverse opinions expressed by local residents. As mentioned earlier, the residents of Kuzumaki are highly motivated to improve the local environment, which is an incentive for the town's administrative offices to make efforts to include the wishes of the residents in the town's policies. However, their views are sometimes contradictory. For example, concerning the program to install photovoltaic power generation systems in private households, some residents will install the system, whereas other residents are either unwilling or financially incapable of installing the system. Therefore, the town as a whole loses the opportunity to benefit the environment despite a strong motivation to do so. By subsidizing residents to install photovoltaic power generation systems, the town is facing financial problems.

There is a positive outlook despite these difficulties. The residents, the town hall, and the town council are actively sharing the promotion of public awareness among residents and the necessary measures to protect the environment; they are making efforts to include environmental improvement in the town's policy. The current goal is to realize local production and consumption of electric energy. Dissatisfaction and requests from residents stem from the fact that they cannot feel the benefits and advantages of producing electric energy in their town for local consumption. Their concern has been exacerbated by the town's 4-day blackout after the onset of the East Japan Earthquake last year. However, information about the town's advanced efforts to produce electricity has spread outside Kuzumaki, making the general public and government agencies believe that Kuzumaki was unaffected by the earthquake. This misunderstanding impeded timely relief operations urgently needed by Kuzumaki.

Separation of power generation and power transmission will boost benefits and advantages, and assure residents that similar misunderstandings can be avoided in the future. There are many obstacles to implementing the separation of power generation and transmission. The first is the current legislative framework for electricity business that may be removed by revising the Electricity Business Act but can only be initiated presently by the national government. This means that legislative actions in the National Diet are required. The second obstacle is weak municipal financial capacity. Even if the Electricity Business Act were revised, it would be extremely difficult for a town with limited financial capabilities to manage separate power generation and transmission systems by running the power transmission business as an independent enterprise of the town. This financial obstacle could be solved by the following strategies: organizing a union with neighboring municipalities under the Local Autonomy Act and running the power transmission business, to use private capital to run the power transmission business as the third sector, or to implement measures to encourage participation of private companies (electric telecommunication carriers, cable television providers, etc.) to run their own power transmission businesses. However, the selection of any one of the strategies leads to another problem, i.e. the development of human resources to maintain, manage, and operate the facilities that will become obsolete or deteriorated in the long run. Presently, manufacturers of power-generating facilities manage the facilities and power companies manage the power transmission facilities. However, once the power generation system is separated from the transmission system, how to secure qualified personnel to run the power generation and transmission systems will become a serious problem.

Furthermore, results of analyzing the example in Kuzumaki suggest that promoting the use of woody biomass and renewable energies requires holistic studies based on theoretical and technological considerations along with practical policies and adequate strategies for capitalizing theories and technologies to benefit the society. Specifically, studies should shed light on the development of human resources and costs required in order to promote the use of woody biomass and renewable energies. Without addressing these considerations, we would not be able to find ideal solutions.

15.3. CASE STUDY OF HIGASHIOMI CITY, SHIGA

Ryoichi Yamazaki and Sachiho Arai

15.3.1. *Nanohana* Project

Higashiomi city, which is certificated as a biomass town, is located in Shiga prefecture in West Japan. Higashiomi is considered to have the most advanced energy system in Japan; detailed descriptions of the town's energy system can be found in the publications by Yaguchi (2007, pp. 169–188) and Hirano (2008, pp. 212–221). The resource circulation cycle implemented by the city's *Nanohana* (rape blossom) Project covers cultivating rapeseed and extracting oil, consuming rapeseed oil in the region, and collecting used oil, and processing the recovered oil into BDF for fuels in a so-called cascading utilization of rapeseed oil. In addition, the activities of Higashiomi are related to the two main purposes of the revised BNS: "local energy production for local consumption" through "establishment of biomass towns", and "realization of biomass energy industry."

However, based on detailed analyses on the city's resource circulation cycle, Hirano (2008) determined that this circulation system was unrealistic at the time when the analyses were carried out:

1. The rate of rapeseed oil self-sufficiency in Higashiomi is only 4.3% (estimate). Moreover, 80% of locally produced rapeseed oil is consumed outside the city.
2. Higashiomi collects approximately 25% of the wasted rapeseed oil discharged from kitchens. This figure is quite high among the data compiled

for other municipalities that collect used cooking oil. However, the amount of locally produced rapeseed oil contained in the collected used oil is insignificant for the above reason.

The problem lies in the fact that the amount of locally produced rapeseed oil is insufficient to support the intended program. Additionally, the locally produced rapeseed oil is not distributed in the city because the city has established a system for collection of used cooking oil to produce BDF for fuel even before starting rapeseed oil cultivation. This problem seems to be solvable through efforts to enhance local consumption. Furthermore, it is considered not impossible to solve the problem that the amount of locally produced rapeseed oil is small. Hirano (2008) stated the following:

To achieve 100% rapeseed oil self-sufficiency on the premise that rapeseed oil produced in Higashiomi will be consumed in the city, approx. 225 hectares of rapeseed planting area is required. Though it may seem that the required area is significantly large, it is equivalent to only 2% to 3.5% of the city's current cultivated area. Given the fact that the area used for rapeseed cultivation accounted for 4.1% of the cultivated area in Japan during the peak period, the required area is not so large. If the labor required for rapeseed cultivation is similar to that for the cultivation of wheat (competing crop), the profitability similar to that of wheat is secured (the Subsidy for Production Area Development is granted) and there is a distribution route similar to that of wheat, it seems that securing 300 hectares of rapeseed planting area is not a difficult task.

15.3.2. Rapeseed Cultivation in the Aito Region

The Aito Region in Higashiomi city started testing the growth of rapeseed oil on 0.05 hectares of paddy field that was used for crop rotation in 1998. As the region aimed to be a major production area of rapeseed by receiving subsidies from the government, the planted area increased to 15 ha in 2007 and decreased to 12 ha in 2011. In 2007, rapeseed was cultivated by community farming groups (three in 2009), large-scale individual farmers, and a roadside station farmer. In addition, the nationwide trend of the farmland area used for rapeseed cultivation shows that the average area from 2004 to 2007 was 839 ha (Report on ProductionResults of Local Special Agricultural Products), and it increased to 1690 ha (Statistics on Crop) in 2010 (note: two different statistics are referenced). The smallest planted area during the period 2004–2007 was 774 ha (2005) whereas the largest planted area was 989 ha (2007), with the total harvested area during the period being 665 ha, accounting for only 79% of the planted area.

The economic logic behind the trend of the rapeseed planting area in Higashiomi should be examined. Before the town implemented the rape seed energy program, the rapeseed had been planted in the city by a large-scale individual farmer (hereafter called Farmer O) and community farming groups using different systems. According to Hirano (2008), the cultivated area of Farmer O was 14 ha when the survey was conducted in 2007. The breakdown by crop was 6 ha for paddy rice, 2.5 ha for wheat, 3.6 ha for rapeseed, and 7.55 ha for soybeans. Farmer O employed a 3-year block rotation system to rotate four crops every 3 years in the order of "rice–rice–(wheat–soybean)" or "rice–rice–(rapeseed–soybean)". The problem is whether it is possible to incorporate rapeseed in this rice-crop rotation farming system. On the other hand, community farming groups do not cultivate soybean as a second crop after cultivating rapeseed and wheat (Hirano, 2008; Himi, 2012).

Table 15.3 shows that the subsidy for rapeseed cultivation under the financial assistance program changes periodically. Based on the double cropping of rapeseed and soybeans, and an expected yield of 100 kg per 10a, the total amount of subsidies granted under various financial assistance programs was 81,700 yen per 10a during the peak period from 2004 to 2005, and decreased considerably to 50,000 yen per 10a in 2009. However, the total subsidy increased to 71,175 yen per 10a in 2011 after rapeseed cultivation became eligible financially to receive subsidies under "Income Compensation Payment for Crop Farmers", and the "Income Compensation System for Individual Farmers". Changes in the subsidy amount seem to show a similar trend of producing other types of biomass under "New Agricultural Basic Law". Nonaka (2009) expects the rapeseed oil biomass business to have a ripple effect on agricultural areas during recessions, and offer employment opportunities to the disabled. Nonetheless, it is considered that assistance from the central government and local governments to the cultivation of rapeseed as a typical resource crop clearly reflects the current trend of biodiesel from rapeseed in Japan.

15.3.3. Comparison of Profitability Between Rapeseed and Wheat Cultivation

Comparison of profitability between rapeseed and wheat cultivations was carried out for 2004 when the largest subsidy was offered for rapeseed cultivation under the financial assistance programs, and for 2010 when the subsidy was the smallest. The results are listed in Table 15.4. As shown in the "References" section in the table, the data used in the table were collected from various sources. Hence, the data must be checked for whether they match the above estimates and can account for the actual trend of the area used for rapeseed cultivation. The estimates were made on the premise that the planting system does not affect the yield and expenses. Two key points, i.e. whether the profit is positive or whether it is higher than that for wheat, need to be considered when evaluating the profitability for rapeseed cultivation.

Based on double cropping of rapeseed and soybean, the total income from rapeseed cultivation in 2004 was approximately 92,000 yen per 10a that included 10,100 yen per 10a from sales and 81,817 yen per 10a from subsidies. The production cost (including labor cost) was 40,000 yen/10 ha using the data estimated by Hirano (2008) because no other statistical data on the cost of

TABLE 15.3 Subsidies to Rapeseed Cultivation under the Financial Assistance Programs and the Trend of their Amounts in Higashiomi City

	2004	2005	2006	2007	2008	2009	2010	2011
				(yen per 10a)				
(National) Measures for promotion of contracted cultivation of rapeseed (117 yen kg^{-1})	11,700	11,700						
(National) measures for development of production areas of high-quality rapeseed (103 yen kg^{-1})			10,300	10,300	10,300			
(National) subsidy for production area development (rapeseed–soybean)	40,000	40,000	40,000	35,000	35,000	35,000		
(National) project for self-sufficiency improvement in field applications (rapeseed–soybean)							35,000	
(National) income compensation payment for crop farmers								21,175
(National) income compensation payment for farmers utilizing rice fields (rapeseed–soybean)								35,000
(Prefectural) incentive for improvement of rapeseed production	30,000	30,000						
(City) subsidy to rapeseed cultivation			20,000	16,670	15,000	15,000	15,000	15,000
Total	81,700	81,700	70,300	61,970	60,300	50,000	50,000	71,175

Estimated yield = 1000 kg ha^{-1}. Blank = N/A.
Source: The table was created by adding data since 2010 to the data given by Hirano (2008, p. 217). However, the estimated yield is different from that used by Hirano (2008).

TABLE 15.4 Comparison of the Profitability Between Oilseed Rape Cultivation and Wheat Cultivation

	2004	2010
	(yen)	
Oilseed rape		
Yield (kg per 10a)	101	94
Sales price (yen kg^{-1})	100	160
Sales revenue (yen per 10a)	10,100	15,040
Subsidy I (yen per 10a; oilseed rape + soybean)	81,817	50,000
Subsidy II (yen per 10a; single crop of oilseed rape)	61,817	35,000
Total revenue I (yen per 10a; oilseed rape + soybean)	91,917	65,040
Total revenue II (yen per 10a; single crop of oilseed rape)	71,917	50,040
Labor cost + property expenses (yen per 10a)	40,000	37,013
Profit I (yen per 10a; oilseed rape + soybean)	51,917	28,027
Profit II (yen per 10a; single crop of oilseed rape)	31,917	13,027
Wheat		
Yield (kg per 10a)	269	227
Sales price (yen kg^{-1})	148	73
Sales revenue (yen per 10a)	39,812	16,571
Subsidy I (yen per 10a; wheat + soybean)	40,000	50,000
Subsidy II (yen per 10a; single crop of wheat)	35,000	35,000
Total revenue I (yen per 10a; wheat + soybean)	79,812	66,571
Total revenue II (yen per 10a; single crop of wheat)	74,812	51,571
Labor cost + property expenses (yen per 10a)	46,332	36,104

TABLE 15.4 Comparison of the Profitability Between Oilseed Rape Cultivation and Wheat Cultivation—Cont'd

	2004	2010
	(yen)	
Profit I (yen per 10a; wheat + soybean)	33,480	30,467
Profit II (yen per 10a; single crop of wheat)	28,480	15,467

Yields and costs are assumed to be constant regardless of planting system.
Profit I is the profit of double cropping and Profit II is that of single cropping. The difference between them comes from the difference of the total amount of subsidies.
Rapeseed
(i) Yield (amount of crops/planted area): Yield data in 2004 is the data of Shiga obtained from the "Report on Production Results of Local Special Agricultural Products". Yield data in 2010 are the data of Higashiomi City obtained from the "Statistics on Crops".
(ii) Sales price: Interview with Nanohana-kan in Higashiomi City.
(iii) Labor cost + property expenses: The figure of these costs in 2004 is the estimated figure of such costs incurred by Farmer O in Higashiomi City in 2005 shown by Hirano (2008). The figure in 2010 is obtained from the national data in the section "Production Cost of Rapeseed Oil" in the survey "Production Cost of Rapeseed and Buckwheat". However, we recalculated labor costs in 2010 using the hourly rate of 1000 yen h^{-1} used by Hirano (2008).
Wheat
(i) Yield: Data of Higashiomi City obtained from the "Statistics on Crops".
(ii) Sales price: Data of the national average obtained from the "Statistical Survey on Prices in Agriculture".
(iii) Labor cost + property expenses: Data of Shiga obtained from the "Production Cost of Rice and Wheat". However, we recalculated labor costs in 2004 and 2010 using the hourly rate of 1000 yen h^{-1} Hirano (2008) used to estimate labor cost.

rapeseed production in 2004 were available. Fujii and Kawamura (2005) and Fujiki and Awaji (2006) showed production costs of 35,485 and 34,348 yen per 10a respectively. The production costs reported by Fujii and Kawamura did not include the depreciation costs for farm machines owned by farmers such as drying machines and tractors, and hence their results underestimated the true total cost. Therefore, Hirano set the production cost at 40,000 yen per 10a that is slightly higher than the production costs shown above. In this way, a profit of approximately 52,000 yen per 10a, shown as Profit I, is generated from rapeseed cultivation. However, as mentioned above, a comparison of profitability between rapeseed cultivation and wheat cultivation can be made between the cropping seasons for both crops that are complementary to each other. When the sum of the sales revenue for wheat cultivation (39,812 yen) and the Subsidy for Production Area Development (40,000 yen) is approximately 80,000 yen per 10a, the production cost is approximately 46,000 yen per 10a in 2004, and Profit I is approximately 33,500 yen per 10a. So, rapeseed cultivation has a higher profit of approximately 18,500 yen per 10a than wheat cultivation.

If the community farming groups do not cultivate soybean as a second crop after cultivating rapeseed, a comparison should be made between single cropping of rapeseed and wheat. In this case, the amounts of the Subsidy for Production Area Development for rapeseed cultivation and wheat cultivation decreased to 20,000 and 35,000 yen per 10a respectively, hence Profit II for rapeseed cultivation and wheat cultivation also decreases to 32,000 yen per 10a and 28,500 yen per 10a respectively. However, rapeseed cultivation has positive profit that is higher than wheat cultivation for both single cropping and double cropping.

15.3.4. Further Analyses of Profitability: The Case of Change in Labor Cost or Subsidy

If the rapeseed production cost shown by Hirano is further examined in detail, the labor cost of 7950 yen (7.95 hours) with an hourly rate of 1000 yen per hour needs to be checked to find whether it is reasonable. According to the "Wage Survey of Outdoor Workers by Occupation", the average wage of light laborers (men and women) in Shiga in 2000 was 12,220 yen/7.5 hours that represents an hourly wage of 1629 yen. The labor cost for rapeseed production is estimated to be 12,951 yen per 10a (1,629 yen × 7.95 hours); it is higher than that estimated by Hirano by 5000 yen per 10a when the calculation is based on the above figures. The labor cost was estimated for unskilled labor, which is the realistic labor cost if it is calculated using the wages for complex labor by taking into account the fact that cities in this region are characterized by an abundance of off-farm employment opportunities ("Kinki-type Regional Labor Market" (Yamazaki, 1996)). Though the wages for complex labor vary depending on gender, age, occupation, and size of business, the hourly wage is estimated to be 2778 yen per hour based on the total wages for complex labor of 5 million yen per year divided by the average annual working hours in Japan (1800 hours). The labor cost for rapeseed oil production is thus estimated to be 22,085 yen per 10a (2778 yen × 7.95 hours); it is higher than Hirano's estimation by 14,000 yen per 10a. If the wages that serve as the basis for the calculation of labor cost are changed accordingly, the resulting labor cost should be higher. As shown in Table 15.5, the profit naturally decreases when the labor cost increases, but it was still positive in 2004.

The area used for rapeseed cultivation expanded in response to the relatively high profit for rapeseed cultivation during the period when solid financial subsidies were offered to rapeseed cultivation by central and local governments.

Table 15.4 shows that the amount of sales revenue for rapeseed cultivation in 2010 was approximately 15,000 yen per 10a. Because there were no subsidies granted to single cropping of rapeseed from central government, the total revenue is estimated to be 65,000 yen when calculated by adding the subsidy under the Project for Self-sufficiency Improvement in Field Applications (35,000 yen per 10a) and the Subsidy for Rapeseed Cultivation from the city

TABLE 15.5 Changes in Profit of Rapeseed Cultivation in Response to Changes in Labor Cost

	2004	2010
	(yen per 10a)	
Profit I (rapeseed + soybean) Survey on production cost	51,917	28,027
Unskilled labor	46,916	23,146
Complex labor	37,782	14,230
Profit II (single crop of rapeseed) Survey on production cost	31,917	13,027
Unskilled labor	26,916	8,146
Complex labor	17,782	−770

Yields and costs are assumed to be constant regardless of planting system.
Profit I is the profit of double cropping and Profit II is that of single cropping.
The difference between them comes from the difference in the total amount of subsidies.
For details on the survey on production cost, see under "Rapeseed" (iii) in Table 15.4. The hourly rate of labor cost is 1000 yen h^{-1} in both 2004 and 2010.
"Unskilled labor" profit is the profit calculated by estimating labor cost using the average wage of light laborers in Shiga shown in the "2001 Wage Survey of Outdoor Workers by Occupation" and the hourly rate of 1629 yen h^{-1}.
"Complex labor" profit is the profit calculated by estimating labor cost at an hourly rate of 2778 yen h^{-1} based on the premise of annual wage of 5 million yen y^{-1} and annual working hours of 1800 h y^{-1}.
Working hours per 10a were 7.95 hours in 2004 and 7.76 hours in 2010.
Sources:
Hirano (2008), obtained by the interview with Nanohana-kan, Higashiomi City.
"Statistics on Crops". "Production Cost of Rapeseed Oil" in the "Production Cost of Rapeseed and Buckwheat".
"Wage Survey of Outdoor Workers by Occupation", "Report on Production Results of Local Special Agricultural Products".

(15,000 yen per 10a) to the sales revenue based on double cropping of rapeseed and soybeans. Profit I is estimated to be 28,000 yen per 10a when calculated by subtracting the cost of 37,000 yen per 10a based on the section "Production Cost of Rapeseed" in the survey known as "Production Cost of Rapeseed and Buckwheat" from the total revenue. Profit I is lower than the level of 2004 by approximately 24,000 yen per 10a but is still positive. On the other hand, Profit I for wheat cultivation is estimated to be 30,500 yen per 10a when calculated by subtracting the cost of 36,100 yen per 10a from approximately 66,600 yen per 10a, which is the sum of the sales revenue of wheat cultivation (16,600 yen) and the subsidy under the Project for Self-sufficiency Improvement in Field Applications (50,000 yen). Thus, the profit for wheat cultivation was slightly higher than that for rapeseed cultivation by approximately 2400 yen per 10a in

2010. This means the relationship between the profit for wheat cultivation and that for rapeseed cultivation becomes almost balanced, or the trend was reversed as compared with the profits in 2004.

When single cropping of rapeseed and single cropping of wheat are compared, the subsidies under the Project for Self-sufficiency Improvement in Field Applications for rapeseed cultivation and wheat cultivation decrease to 20,000 and 35,000 yen per 10a respectively. Then, Profit II for rapeseed cultivation and for wheat cultivation is estimated to be 13,000 and 15,500 yen per 10a respectively. In this case, the profit for rapeseed cultivation remains positive and the profits for rapeseed cultivation and for wheat cultivation are almost balanced or reversed when compared with the profits in 2004.

If labor costs are estimated differently from those for 2004 (see Table 15.5), Profit I based on double cropping of rapeseed and wheat is positive whereas Profit II based on single cropping of rapeseed remains positive at 8000 yen per 10a if the labor cost is calculated using the wages for unskilled labor; the profit becomes −770 yen per 10a when the labor cost is calculated using the wages for complex labor.

Himi (2012) introduced the status of rapeseed cultivation by one of the community farming groups in Higashiomi city (hereafter called Agricultural Cooperative S). Agricultural Cooperative S is a specific agricultural organization with 17 hectares of cultivated area and 31 participating farmers. Its predecessor was a community farming cooperative established in 1980 with the objective of promoting group utilization of farm machines, group crop rotation, and improvement of efficiency through cooperative work. Agricultural Cooperative S cultivated wheat on 10 hectares and rapeseed on 7 hectares in 2010. However, as for rapeseed cultivation, there was almost no difference between its expenditure and income that includes subsidies. Thus, the Agricultural Cooperative plans to stop cultivating rapeseed. The observations made by Himi seem to match the calculation results shown in Table 15.5. As mentioned above, the total amount of subsidies increased for the first time in 2011. Attention will be paid to how this will affect the trend of the rapeseed planting area in the future.

15.4. TOWARD THE CREATION OF AN EFFECTIVE BIOMASS SYSTEM: LESSONS FROM GERMANY

Hiroyuki Enomoto

The case study of Kuzumaki demonstrates that the current legislative framework for electric utility is the bottleneck to realizing local electricity self-sufficiency. Even though the town has a sufficient electricity supply capacity to meet its total consumption, the residents cannot use locally produced electricity due to the current Electricity Business Act. Moreover, the fact that the shipment of high-quality charcoals out of the town has reduced the opportunity for the residents to consume woody biomass produced locally. This observation suggests that the design of the local biomass system needs to consider local-specific social and economic situations. On the other hand, the case study of Higashiomi revels that relying on subsidies to ensure the profitability of rapeseed cultivation casts great doubts on the economic viability of rapeseed cultivation and the future sustainability of long-term BDF production. This implies that the production of resource crops will rely on an agricultural policy that determines the magnitude of the subsidy. Under the New Agricultural Basic Law, the current Japanese agricultural policy is characterized as responsive to changes in economic situations; therefore, the production of resource crops is subject to budgetary instability. This would be less likely to provide steady opportunities for farmers to produce resource crops in the years to come.

Lack of profitability without subsidies or limited economic viability is recognized as the major challenge to be solved to expand biofuel production. With current technologies, almost all developed countries rely on subsidies or tax exemptions for the development of biomass energy utilization; Japan is no exception in this regard. A key to improving profitability is technical innovation, specifically the development of second-generation biomass technology that would lead to more efficient production processes using inedible parts of crops (e.g. rice straw and husks) or waste biomass (e.g. forestry residues).

In contrast, an inadequate legislative framework is not necessarily a serious issue in other countries, although it has been identified as one of the many bottlenecks in the development of renewable energy in Japan. One notable example is the case of Germany. The German legislative framework, which is based on a constant assessment of the current situation and the implementation of prompt responses, can be regarded as behind the successful renewable energy policy. Because the German experience can provide profound implications for Japan to establish an effective legislative framework that would facilitate the adoption and development of renewable energy including biofuels, the basic features of the German legislative framework are worth considering in the following paragraphs.

Among major industrialized countries, Germany generates the highest share of total primary energy supply from renewable energy sources. According to 2011 statistics (IEA, Energy Balances of OECD Countries 2012 edition), the share of renewable energy is 12.6% for Germany as compared with 6.3% for the USA, 4.3% for the UK, 7.8% for France, and 3.7% for Japan. The legislative framework established in Germany is a major reason for its success in promoting the use of renewable energy while implementing economic reforms similar to those implemented in Japan. Specifically, the early adoption of a feed-in tariff (FIT), as well as its constant development, has been a source of this success. More than 50 countries including Japan have established a legislative framework adopting this practice. Also, the Renewable Energies Heat Act that was enacted in 2008 has attracted global attention in the design of heat supply programs.

The German FIT, which was enacted in 1991 and revised several times later, is regarded as the most successful method for promoting the supply of electricity produced from renewable energy sources. In addition to the simple step of introducing an FIT at an early stage, it has other notable features such as an accurate assessment of the facility initial cost for each source to supply renewable energy, the related management costs and the status of the adoption of renewable energy, and constant system improvement as necessary with frequent revisions of relevant laws every year. In fact, the law has been revised every year.

The first act with the aim to promote the use of renewable energy was the Electricity Feed-in Act (EFA) established in 1991. This act obliges electricity providers to buy renewable energy with the purchase price set as a percentage of the retail price of electricity; however, variations of this percentage may entail management risk that can be minimized by fixing the price through fees stipulated by the act. Hence, the EFA was abolished in favor of the new Renewable Energy Sources Act in order to promote the entry of new players into the market. In addition, lessons learned from the existing system lead to the introduction of ideas such as limiting the purchase obligation to 20 years. This arrangement ensures that efforts are made to increase production efficiency in order to maximize profit within this period. Because production efficiency, production cost, and the extent of adoption differ depending on the energy source, improvements have been made by setting different fees for each source, adopting quantitative limits for the purchase obligation and introducing reduction rates for the fees. Later, the fees and fee reduction rates will be repeatedly revised for each renewable energy source in consideration of the newly developed situation. Furthermore, because energy-intensive businesses are heavily burdened by the increasing cost of electricity following the implementation of FIT, they receive preferential treatment that can also be regarded as a suitable response to the current situation.

In Japan, the nuclear incident caused by the Great East Japan Earthquake on March 11, 2011 incited the Diet to finally pass an act on regulating the fixed-price feed-in tariff (Act on Special Measures Concerning the Procurement of Renewable Energy by Electric Utilities) in August 2011. It is interesting to evaluate Japan's new system with respect to the German experience. Article 10 of the Japanese law contains provisions for reviewing the system by stipulating that the Basic Energy Plan of the government will be revised and that the government will make necessary revisions every 3 years. This article also states that the law is to be substantially revised by March 31, 2021. According to the German experience, the basic interval for making revisions in Japan may be considered too long to legally respond to issues that emerge in practice. However, since the Japanese system can be improved through revisions implemented by the government before the end of this period, necessary measures and substantial revisions of the law would need to be implemented proactively regardless of the stipulated interval. It seems legitimate to summarize that the establishment of the legislative framework for the development of renewable energy utilization is still at an early stage in Japan.

REFERENCES

Administrative Evaluation Bureau (AEB). (2011). Ministry of Internal Affairs and Communications, Japan. *Baiomasu no rikatuyou ni kansuru seisakusho (Policy evaluation of biomass utilization)*. <http://www.soumu.go.jp/menu_news/s-news/39714.html#hyoukasyo> Accessed 02.08.2012.

Federal Ministry for the Environment, Nature Conservation and Nuclear Safety (Germany). <http://www.bmu.de/english>

Fujii, Y., & Kawamura, H. (2005). Current state of Colza cultivation in Shiga Prefecture and view in the future. *Journal of Agricultural Science, 60*(12), 22–24 [in Japanese].

Fujiki, N., & Awaji, K. (2006). Regional resources recycling and value added formation by utilizing rapeseed. *Japanese Journal of Farm Management, 44*(2), 66–69 [in Japanese].

Government of Japan. (2006). *Baiomasu nippon sogo senryaku (Biomass nippon comprehensive strategy)*. <http://www.maff.go.jp/j/biomass/pdf/h18senryaku.pdf> Accessed 29.06.2012.

Himi, O. (2012). *A study of actor and profit of multi purpose rice*. Masters Dissertation. Tokyo University of Agriculture and Technology [in Japanese].

Hirano, N. (2008). *Symbiotic agriculture in the consuming Central Region*. Nourin Toukei Kyoukai [in Japanese].

Japan Organics Recycling Association (JORA). (2012). *Baiomasu taun koso bunseki DB (Database for analysis of biomass town proposals)*. <http://www.jora.jp/biomasstown_DB/index.html> Accessed 25.07.2012.

Nonaka, A. (2009). Actual situation of research on rape seed biomass and target of agricultural economics. *Journal of Agricultural Science, 64*(12), 4–16 [in Japanese].

Tomari, M. (2012). *Baiomasu: Honto no hanashi (Biomass: Real stories)*. Tokyo: Tsukijishokan [in Japanese].

Yaguchi, Y. (2007). Nourin Toukei Kyoukai [in Japanese]. In (Series Ed.) & S. Hattori (Vol. Ed.). *Annual bulletin of Japanese agriculture: Vol. 54. Economic structure of the biomass production and utilization with the regional circulation system* (pp. 162–189).

Yamazaki, R. (1996). *Regional characteristics of the local labor market and the agricultural structure*. Nourin Toukei Kyoukai [in Japanese].

Index

Printed and bound by CPI Group (UK) Ltd, Croydon, CR0 4YY

03/10/2024

01040425-0013